Communications
in Computer and Information Science   2324

Series Editors

Gang Li , *School of Information Technology, Deakin University, Burwood, VIC,
Australia*
Joaquim Filipe, *Polytechnic Institute of Setúbal, Setúbal, Portugal*
Zhiwei Xu, *Chinese Academy of Sciences, Beijing, China*

Rationale

The CCIS series is devoted to the publication of proceedings of computer science conferences. Its aim is to efficiently disseminate original research results in informatics in printed and electronic form. While the focus is on publication of peer-reviewed full papers presenting mature work, inclusion of reviewed short papers reporting on work in progress is welcome, too. Besides globally relevant meetings with internationally representative program committees guaranteeing a strict peer-reviewing and paper selection process, conferences run by societies or of high regional or national relevance are also considered for publication.

Topics

The topical scope of CCIS spans the entire spectrum of informatics ranging from foundational topics in the theory of computing to information and communications science and technology and a broad variety of interdisciplinary application fields.

Information for Volume Editors and Authors

Publication in CCIS is free of charge. No royalties are paid, however, we offer registered conference participants temporary free access to the online version of the conference proceedings on SpringerLink (http://link.springer.com) by means of an http referrer from the conference website and/or a number of complimentary printed copies, as specified in the official acceptance email of the event.

CCIS proceedings can be published in time for distribution at conferences or as postproceedings, and delivered in the form of printed books and/or electronically as USBs and/or e-content licenses for accessing proceedings at SpringerLink. Furthermore, CCIS proceedings are included in the CCIS electronic book series hosted in the SpringerLink digital library at http://link.springer.com/bookseries/7899. Conferences publishing in CCIS are allowed to use Online Conference Service (OCS) for managing the whole proceedings lifecycle (from submission and reviewing to preparing for publication) free of charge.

Publication process

The language of publication is exclusively English. Authors publishing in CCIS have to sign the Springer CCIS copyright transfer form, however, they are free to use their material published in CCIS for substantially changed, more elaborate subsequent publications elsewhere. For the preparation of the camera-ready papers/files, authors have to strictly adhere to the Springer CCIS Authors' Instructions and are strongly encouraged to use the CCIS LaTeX style files or templates.

Abstracting/Indexing

CCIS is abstracted/indexed in DBLP, Google Scholar, EI-Compendex, Mathematical Reviews, SCImago, Scopus. CCIS volumes are also submitted for the inclusion in ISI Proceedings.

How to start

To start the evaluation of your proposal for inclusion in the CCIS series, please send an e-mail to ccis@springer.com.

Anabela Marto · Rui Prada · Patrícia Gouveia ·
Ruth Contreras- Espinosa ·
Alexandrino Gonçalves · Eduarda Abrantes ·
Roberto Ribeiro
Editors

Videogame Sciences and Arts

14th International Conference, VJ 2024
Leiria, Portugal, December 5–6, 2024
Proceedings

 Springer

Editors

Anabela Marto [ID]
CIIC, ESTG, Polytechnic University of Leiria
Leiria, Portugal

Patrícia Gouveia [ID]
ITI/LARSyS Instituto Superior Técnico
and Faculdade de Belas-Artes
Lisbon, Portugal

Alexandrino Gonçalves [ID]
CIIC, ESTG, Polytechnic University of Leiria
Leiria, Portugal

Roberto Ribeiro [ID]
CIIC, ESTG, Polytechnic University of Leiria
Leiria, Portugal

Rui Prada [ID]
INESC-ID and Instituto Superior Técnico,
Universidade de Lisboa
Lisbon, Portugal

Ruth Contreras- Espinosa [ID]
Universidad de Vic - Universidad Central de
Cataluna
Barcelona, Spain

Eduarda Abrantes [ID]
CIIC, ESTG, University of Leiria
Leiria, Portugal

ISSN 1865-0929   ISSN 1865-0937 (electronic)
Communications in Computer and Information Science
ISBN 978-3-031-81712-0   ISBN 978-3-031-81713-7 (eBook)
https://doi.org/10.1007/978-3-031-81713-7

This Springer imprint is published by the registered company Springer Nature Switzerland AG
The registered company address is: Gewerbestrasse 11, 6330 Cham, Switzerland

If disposing of this product, please recycle the paper.

Foreword

Gaming Entanglements: Human-Algorithmic Interaction for Diversity and Inclusion

Contemporary media ecologies involve a convergence between gaming, interactive media, online networks, and older forms of storytelling and aesthetic fruition available in books, television, cinema, visual arts, and design. Currently, transmedia digital literacies take advantage of various new ways of creating knowledge and are connected to communication challenges and creative expression. As a communicative medium, according to Yasmin B. Kafai and Quinn Burke, games allow us not only to "decompose a system, understand its structure and parts, but then also reassemble a model of that system using algorithms (code). It also means to express an idea intentionally for others, to create for them a first-person experience that one believes is worth having. It is to share a viewpoint from the inside" [8]. Gaming magic resides in emotional communication with participants and audiences and in making games people create world views and speculations. The previously quoted authors warn us of the danger of not involving the younger generations in critically examining the complex systems around them. In this context, we can think of political, health, education systems, free markets, neural networks, and emerging automation. Drawing from practice-based community experiences, a game-based pedagogy and assessment suggest that future students using a connected gaming culture are rewarded not for getting a convenient answer but for workable solutions. Gaming promotes design literacies of multiple dimensions in an inherently social activity [8], a systematic way for the code to orchestrate players' experiences [11].

As Shira Chess and Mia Consalvo stated, "critical theory in video game studies is at the forefront of understanding and explaining shifts in a variety of media content and media platforms", and "video games have had as much (if not more) impact on our current politics and culture wars than many care to acknowledge" [4]. According to these authors, another reason why it is important to highlight video game studies can be synthesized in a hashtag: #GamerGate. Ten years after the birth of "a reactionary movement that sought to police the identity of "true" gamers by harassing and threatening women and queer folks in the community" [11], authoritarian ideas must be fought and persecuted. As Anne Applebaum suggests, we may already be living through the twilight of democracy, where civilization seeks anarchy and tyranny through authoritarian ideas and populist worldviews [1]. At a time of global crisis, democracy must be reinvented [10] to avoid that day when "in the wake of the 2016 US presidential election, gamers were considered the core demographic (perhaps even the origin) of what the media dubbed the 'alt-right', that cesspool of hatred on the internet" [13]. Is this narrative repeating itself?

Shira Chess [3] highlights the interaction of oppressive cultural systems by referring us to inequalities of gender, ethnicity, social class, sexuality, and disability. Kishonna L. Gray uses transmedia studies to focus on intersectional technology to examine systemic exclusions legitimized by nation-states [7]. As Amanda C. Cote considers, the recurring representation of men as players, creators, and producers of games in the news, marketing, and other media excludes women, making connection with this identity harder to imagine. For inclusion to exist, it is necessary to do more than believe that it exists, and it will be crucial in the future to investigate the specific contexts in which people are consigned to the margins [5]. In September 2022, Jamal Michel [12] wrote on the Media Diversity Institute's online platform about the toxic remnants of Gamergate and how they still target women in digital spaces today. According to Vittoria Elliott, Gamergate spread all over the internet and it is everywhere now [6].

If we consider that games and democracy are inherently social and participatory and are both designed to emphasize decision-making as their core activity, we can then assume that these ludic experiential structures created by designers [9] are open for players' tactics under the proposed design strategy. Games can become a map of algorithmic agency that helps us avoid oppression algorithms and stimulate the creation of resistance algorithms to fight data colonialism and instrumental platform power. In this sense, we might avoid algorithmic discrimination that takes intersectional forms, affecting, for example, young single women and ethnic minority families belonging to the working class [2]. The relationship between humans and algorithms can take advantage of gaming systems to exploit the platform's affordances according to human needs and not the opposite. The symbiotic nature and entanglements of human-algorithmic interaction must be interrogated by critical play to insert players in renowned articulations that question technical artifacts and their political qualities. Hegemonic gaming could be instrumental in fighting against moral democracies [2], avoiding populist views, and promoting long-term strategic ones. When democracy is unappealing and deviates from common-sense expectations, it attracts unrepresentative groups - political outcasts, extremists, and other junkies [9]. Suggesting that democracy should be redesigned, and that game design can transform democracy, Josh Lerner advocates for the usage of gaming to instigate participation and generate change [9].

It is with hope in mind that gaming will evolve into an epistemology of more ethical and consistent emergent behavior that this year's 14th International Conference on Videogame Sciences and Arts engages in a dialogue between the fifteen accepted full papers and the three short papers.

In the accepted full papers group, we can find a paper about a virtual reality game for perspective-taking in young adults' virology education. Another contribution discusses the method for generating player-driven narratives in games. We will understand better how game-based activities transform learning and alleviate stress in institutionalized elderly and read about how to explore asymmetric competitive gaming for mixed-visual-ability pairs. Two other papers will teach us about the use of real-time image processing to enhance privacy protection in LBARGs and explore the stakeholders clash game, from board game design to online collaborative platforms for urban planning. Another paper deals with an NFT-based game ecosystem and, in another one, we will be confronted with a systematic review of designing, developing, and implementing effective digital game interventions for mental disorders. We can also find a paper about exploring information asymmetries in cooperative games and another about a Gomoku game testbed for Monte-Carlo tree search algorithms. One paper explores a deep learning anti-cheat system based on player behavior for Minecraft, while another studies types and patterns of betrayal in multiplayer games. Finally, in the full papers category, we can also read about art and animation and procedural content generation for Sprite sheet creation, a Black Mirror-based digital breakout that promotes creative thinking, enjoyment, and absorption of teachers in training, and, lastly, leveraging tabletop games as education scenarios for enhancing media literacy skills.

In the accepted short papers group, you will find a paper about an evaluation study of the camera perspective in one game, how to enhance player experience through generative artificial intelligence and custom interaction in game design, and a game proposal to support language therapy for preschool children.

We hope you enjoy reading about these works!

Patrícia Gouveia

Rui Prada

Ruth S. Contreras-Espinosa

References

[1] Applebaum, A. 2020. *Twilight of Democracy, The Failure of Politics and the Parting of Friends*, Allen Lane, Penguin Books, UK.

[2] Bonini, T. and Treré, E. 2024. *Algorithms of Resistance, The Everyday Fight Against Platform Power.* Cambridge, MA: The MIT Press.

[3] Chess, S. 2020. *Play Like a Feminist.* Cambridge, MA: The MIT Press.

[4] Chess, S., & Consalvo, M. 2022. The future of media studies is game studies. Critical Studies in Media Communication, 39(3), 159–164.

[5] Cote, A. C. 2020. *Gaming Sexism, Gender and Identity in the Era of Casual Video Games.* New York University Press.

[6] Elliott, V. 2024. Gamergate's Aggrieved Men Still Haunt the Internet, Ten years ago, much of the frustrations gamers were expressing came from anger over no longer being the target audience. Now those feelings are everywhere, from fandom to politics. In Wired Magazine. Retrieved October 2024: https://www.wired.com/story/gamergates-aggrieved-men-still-haunt-the-internet/

[7] Gray, K. L. 2020. *Intersectional Tech, Black Users in Digital Gaming,* Louisiana State University Press, Baton Rouge.

[8] Kafai Y. B., and Burke, Q. 2016. *Connected Gaming, What Making Video Games Can Teach Us About Learning and Literacy.* Cambridge, MA: The MIT Press, (p. xii and p. 46).

[9] Lerner, J. 2014. *Making Democracy Fun. How Game Design Can Empower Citizens and Transform Politics.* Cambridge, MA: The MIT Press, p. 207.

[10] Mbembe, A., 2022, January 11. How to Develop a Planetary Consciousness. Interview [Noema Magazine]. Retrieved October 2024 from: https://www.noemamag.com/how-to-develop-a-planetary-consciousness/

[11] McDonald, P. 2024. *Run and Jump, The Meaning of the 2D Platformer.* Cambridge, MA: The MIT Press (p. 86 and p. 10).

[12] Michel, J. 2022. GamerGate's Toxic Remnants Still Target Women in Digital Spaces. Retrieved October 2024 from: https://www.media-diversity.org/gamergates-toxic-remnants-still-target-women-in-digital-spaces/

[13] Phillips, A. 2021. *Gamer Trouble, Feminist Confrontations in Digital Culture.* New York University Press, p. 2.

Preface

The 14th International Conference on Videogame Sciences and Arts – VJ 2024 – took place in Leiria, Portugal, from December 5–6, 2024, organised by the Polytechnic Institute of Leiria in partnership with the Portuguese Society of Video Games Sciences (SPCV). The event was hosted at the School of Technology and Management, with support from the Department of Informatics and the Computer Science and Communications Research Centre, all part of the Polytechnic Institute of Leiria.

The annual SPCV conferences are international events aimed at fostering scientific exchange among researchers and professionals in the broad field of videogames. This multidisciplinary gathering brings together research from areas such as New Media, Culture, Technology, Education, Psychology, and the Arts.

In this year's edition, a total of 30 submissions were received, with 15 long papers and 3 short papers accepted for publication. For each study three reviewers were invited to review each manuscript in a double-blind process.

The integration of these scientific areas has enhanced both academic expertise and acceptance, while highlighting key sub-fields such as game experience, game learning, game development, and game culture. Accordingly, this year's conference aligned with these broader trends, offering sections dedicated to each of these sub-areas.

The following section includes a Foreword from the Scientific Chairs titled "Gaming entanglements: human-algorithmic interaction for diversity and inclusion". The manuscript then proceeds with five main sections: Player Experience and Accessibility, Serious Games, Edutainment, Game Design and Development, and Games and Artificial Intelligence.

We extend special thanks to the scientific board for their invaluable contributions in ensuring the highest scientific quality and relevance of this publication. We also thank the organising team for their dedication in preparing and coordinating the 2024 Conference, along with the Association of Portuguese Video Game Producers (APVP) for their support.

Anabela Marto
Rui Prada
Patrícia Gouveia
Ruth Contreras Espinosa
Alexandrino Gonçalves
Eduarda Abrantes
Roberto Ribeiro

Organization

Conference Chair

Anabela Marto — Polytechnic University of Leiria, PT

Scientific Chairs

Rui Prada — INESC-ID and Instituto Superior Técnico, Universidade de Lisboa, PT

Patrícia Gouveia — ITI/LARSyS Instituto Superior Técnico and Universidade de Lisboa, PT

Ruth S. Contreras Espinosa — Universitat de Vic - Universitat Central de Catalunya, ES

Organizing Committee

Alexandrino Gonçalves — Polytechnic University of Leiria, PT
Catarina Cardoso — Polytechnic University of Leiria, PT
Eduarda Abrantes — Polytechnic University of Leiria, PT
Roberto Ribeiro — Polytechnic University of Leiria, PT

Scientific Committee

Alan Carvalho — FATEC, São Caetano do Sul, Brazil
Ana Veloso — Universidade de Aveiro, Portugal
Anna Unterholzner — ITI/LARSyS, Portugal
Anabela Marto — Instituto Politécnico de Leiria, Portugal
André Pereira — KTH, Sweden
Andreas Theodorou — Umeå University, Sweden
Alexander Dockhorn — Gottfried Wilhelm Leibniz University Hannover, Germany
Bárbara Barroso — Instituto Politécnico de Bragança e Vale do Cávado, Portugal
Carla Sousa — Universidade Lusófona, Portugal
Eduarda Abrantes — Instituto Politécnico de Leiria, Portugal

Contents

Game Design and Development

Games and Artificial Intelligence

Player Experience and Accessibility

Played like a Damn Fiddle: Types and Patterns of Betrayal in Multiplayer Games

Pedro Valério[1]([⊠]) and Pedro Cardoso[2] [iD]

[1] University of Aveiro, Aveiro, Portugal
pedrovalerio@ua.pt
[2] University of Aveiro/DigiMedia, Aveiro, Portugal
pedroccardoso@ua.pt

Abstract. In multiplayer games, players will naturally need to interact with one another. Their relationships are not static and evolve over time, because players are people who make their own decisions and have their own perceptions of the game and other players. This means relationships are uncertain, because players can't know who is trustworthy and who will deceive them. As such, fear of betrayal often greatly affects the course of a game. This study analyses how betrayal manifests in different games and aims at defining a conceptual model capable of describing betrayal in games and analysing it formally. Additionally, this study also presents a set of patterns, designed from the created model. The conceptual model and these patterns were also subjected to tests with participants.

Keywords: Betrayal · Conceptual Model · Game Design · Video Games · Analogue Games · Multiplayer Games · Design Patterns

1 Introduction

Betrayal is a social interaction between two parties connected by a set of expectations towards each other, in which one party deliberately breaks these expectations. As such, it requires a situation in which a person A trusts person B to cooperate for A's benefit and not take advantage of them if the opportunity arises (Ben-Ner & Halldorsson, 2010). When person B consciously chooses to break the expectations placed on them, they betray. Additionally, because the victim trusted their traitor, it always comes as a shock to them (O'Neil, 2017). The act of trusting necessarily leaves one vulnerable to the other person's decisions. As such, the very act of trusting makes betrayal possible (Baier, 1986). This type of behaviour is normally perceived as negative and anti-social and may lead to severe consequences for the victim and even the traitor (Rachman, 2010). Multiplayer games, being played with many people, also feature moments of betrayal. Some argue that, even if games happen in a fictional context, betrayal is still an inexcusable action (Consalvo, 2009, Brooks, 2017). On the other hand, many games still make use of betrayal to provide entertainment and, sometimes, their core gameplay. This is mainly because betrayal is always, by nature, an unexpected interaction, often drastically changing the state of the game. This provides memorable moments of gameplay, derived from the very

© The Author(s), under exclusive license to Springer Nature Switzerland AG 2025
A. Marto et al. (Eds.): VJ 2024, CCIS 2324, pp. 3–20, 2025.
https://doi.org/10.1007/978-3-031-81713-7_1

uncertainty, risks involved in knowing who to trust and ultimately the consequences of mistrusting someone (Allison et al.; 2015, Carter, 2022).

In multiplayer games, players play not only with the game's inherent elements, like score, enemies, obstacles or objectives, but also with the other players themselves. Depending on the game, players will often need to form relationships of trust with other players to achieve a certain goal or protect themselves from dangers. This means they can also betray, which may manifest in different ways. Typically, players play towards an objective laid out by the rules of the game and, as such, all their actions should, in theory, serve to help them achieve that goal (Suits, 1978). On the other hand, players may be motivated to play in a certain way for their own *player-defined goals*, in which they define their own way of enjoyment, not necessarily following what is imposed by the game's objectives (Consalvo, 2007). Betrayal can be used to improve a player's state in the game, to achieve their objectives. But it can also be done only for their own personal fun, such as roleplaying as a traitorous character, without in-game benefit for them (Yee, 2006; Carter, 2022).

Furthermore, betrayal can be integrated in the game, tied to explicit mechanics or explicit roles of betrayal. In this case, the actions used are mechanics which are perceived by the game's systems and variables, interacting with the *game state* (Sicart, 2008). This is typical of most Social Deduction games, such as *Among* Us (2018) or *Secret* Hitler (2016), in which players are directed, by the rules of the game, to betray. However, it may also purely rely on social interactions and be completely external to the game's elements. This happens through so-called *soft mechanics* (Robinson, 2016), which depend only on the context in which actions happen and are given their meaning by the players and not any systems or mechanics of the game. This often manifests in communication with promises, lies and *ad-hoc* relationships between players in games like Diplomacy (1959), *EVE* Online (2003) or DayZ (2013).

This paper is divided into 4 main sections. In Sect. 2, the model of dichotomies is presented and explained. In Sect. 3, the model is further explained, exploring all possible pairs of dichotomies and how they relate to one another and how they manifest in game examples. In Sect. 4, from the model and combinations, a set of patterns was designed to identify recurring situations in games. In Sect. 5, the results of the practical sessions are presented and analysed.

2 Model of Dichotomies

Betrayal is a special interaction in multiplayer games because it results from a combination of typical game elements, like goals, obstacles, risks and rewards and social elements, which exist only between players, independent of rules or systems of the game. Most notably, the fact that each player has their own individual perception of the state of the game can be the cause for suspicions and deception to arise, which can lead to betrayal. With this into consideration, this study is guided by four main questions: 1) What characterises the experience of betrayal in multiplayer games? 2) How does betrayal manifest itself in multiplayer games? What are its causes, motivations, rules, etc.? 3) What kinds of betrayal are there in games? 4) How can we design betrayal in multiplayer games?

To tackle these questions, this work was divided into three main steps: **Step 1**: Creation of a conceptual model. After literature review about betrayal in games and state of the art of games with betrayal, this knowledge was used to conceive a model of five dimensions, each one capable of describing an individual aspect of betrayal and, ranging between two opposite concepts, thus, referred to as dichotomies. Because each dichotomy is independent from the other, they were also analysed in combinations of two, to understand how their variations affect a situation of betrayal. **Step 2**: Patterns of Betrayal. These dichotomies were, then, applied to a group of multiplayer games to further improve the model, identifying flaws, outliers and recurring situations. These games were analysed by 1) studying the game's rules and mechanics; 2) reading and interpreting reports from other players and 3) playing the game extensively when possible. The result was a set of patterns of betrayal (see Sect. 4), each one related to a combination of dichotomies (see Sect. 3). **Step 3**: Assessment. Finally, the model of dichotomies and resulting patterns were tested with participants in workshop sessions to evaluate them (see Sect. 5).

The main result was a conceptual model created to formally analyse betrayal and aid the design of betrayal mechanics and dynamics. The proposed model describes betrayal in five dimensions. Each one is a spectrum, ranging between two opposing concepts, thus, referred to as *dichotomies of betrayal*: Motivation (Intrinsic/ Extrinsic), Choice (Game role/ Player role), Mechanics (Direct/ Indirect), Relationship (Planned/Failed), and Focus (Victim-centred/ Traitor-centred) (Fig. 1).[1]

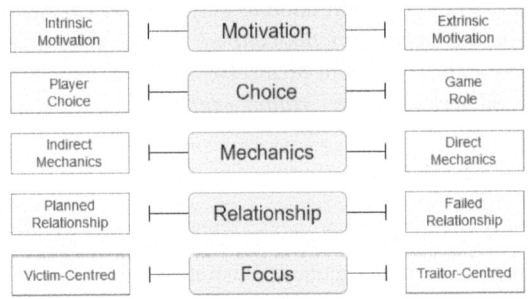

Fig. 1. Conceptual Model of Dichotomies of Betrayal

- **Motivation:** The decision to betray is conscious and intentional, which means it is always done to achieve something. This goal may relate more to external incentives, provided by the game itself, like obstacles, rewards and objectives – *extrinsic motivation* – or for more personal reasons, defined by the player, which are not necessarily related to any game elements – *intrinsic motivation*.
- **Choice**: Depending on the game, a player may be afforded more or less freedom in how they can play and achieve their objectives. This also applies to the relationships between players and how they are managed. In some games, players are fully in charge

---

[1] This model was first developed in a master's dissertation (Valério, 2024) and an expanded version of this section can be consulted elsewhere (Valério & Cardoso, 2024).

of their alliances and rivalries. In these, when a player betrays, they do so because they decide it themselves – *player choice*. However, some games restrict who players may cooperate with and may impose betrayal as a necessary action to win – *game role*.

- **Mechanics**: The mechanics of betrayal describe how betrayal affects the game and how much it depends on explicit rules, systems or other game elements. On one hand, betrayal may be fully independent of the game and occur only in the form of social interactions between players, through communication. In these cases, betrayal does not directly affect the game state and is only perceived as such because of the context in which it happens – *indirect mechanics*. On the other hand, it may be tied to specific game actions, which directly result in betrayal and produce a perceivable change in the game state – *direct mechanics*.

- **Relationship**: The necessary condition for betrayal to manifest is a mutual understanding between players. However, this relationship may be, from the start, planned to end in betrayal by one party. In this case, the relationship's purpose is to betray the victim, contrary to their expectations – *planned relationship*. However, betrayal may also manifest at some point after the relationship started. In these relationships, both parties started with equal views on each other and their relationship degraded into betrayal for external reasons – *failed relationship*.

- **Focus**: Betrayal always happens between two parties: victim and traitor. Both are affected in some way, usually positively for the traitor and negatively for the victim. However, the reason to betray may be more focused on either of these effects. Betrayal may be seen purely to directly hurt another player, in which case the effects on the traitor are their main concern – *victim-centred*. On the other hand, it may also be just a means for the traitor to benefit themselves, without any particular victim, who becomes simply collateral damage – *traitor-centred*.

3   Combinations of Dichotomies

This section explores combinations of dichotomies and how they relate to each other, providing real examples in various games, illustrating how different combinations of dichotomies influence how betrayal manifests. Figure 2 presents these combinations, each one illustrated in a matrix of two dichotomies.

3.1   Motivation and Choice

Requirement. When a traitor is assigned a *Game Role* and betrays for *Extrinsic Motivation*, they can be said to be playing as intended. This betrayal is carried out as it was meant to by the objectives and dynamics of the game. This is typical with *social deduction* games, such as *Among* Us (2018), Project Winter (2019) or *Secret* Hitler (2016), where a group of secret traitors is tasked with betraying the remaining players.

Tool. However, when betrayal comes from *Player Choice* and *Extrinsic Motivation*, it is not imposed on players and does not directly constitute their win condition. Instead, it is one of their possible strategies to get ahead in the game, but is entirely optional. Indeed, it is exactly because it is optional that it is a viable strategy. Since no player

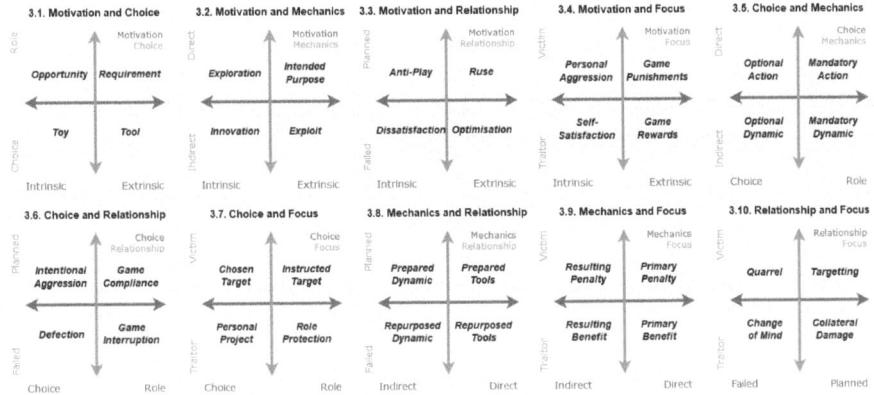

Fig. 2. Combinations of Dichotomies

can ever be certain of whether another will try to betray them, they also have no direct reason to immediately suspect anyone, which leaves them more vulnerable. This sort of management of relationships is common in games where every player starts neutral to each other, such as strategy games, like *Civilization VI* (2016), Age of Empires IV (2021) or Diplomacy (1959) or survival games like DayZ (2013), Rust (2018) or *EVE Online* (2003).

Opportunity. *Intrinsic Motivation* means that the goals of the traitor are independent of the objectives laid out by the game. When combined with *Game Role*, this manifests when players actually want to be assigned the role traitor. This motivation comes from outside a specific game session and comes, for example, from playing multiple sessions of the same game and wanting to experiment a different role, not necessarily to win, but to experiment the different mechanics, actions and systems that come with the role. For example, Impostors in *Among* Us (2018) can sabotage the ship and fast travel, which Crewmates are not. Being given the traitor role, they take that opportunity to interact with other aspects of the game.

Toy. When traitors are not required to betray by rules and betray for their own personal motivations, it results in a combination of *Intrinsic Motivation* and *Player Choice*. This type of betrayal can, in theory, happen in any multiplayer game. As long as a player can form relationships with another, they can choose to betray them. However, because they are motivated intrinsically, their reasons to do so cannot be predicted as they are fully personal. An example happens in games like DayZ (2013) or *EVE* Online (2003), where players may choose to betray only to roleplay as a traitorous character and behave accordingly to that persona.

3.2 Motivation and Mechanics

Intended Purpose. This combination can be said to be the most straightforward manifestation of betrayal. In the case of *Extrinsic Motivation* and *Direct Mechanics*, betrayal is fully incorporated into the gameplay, as an interaction tied to concrete game mechanics, meant to produce effects that aid the traitor in achieving their objectives. This is the most common type of betrayal in social deduction games, like *Among* Us (2018), Project Winter (2019) or Town of Salem (2014), where hidden traitors have additional abilities meant to secretly take advantage of non-traitors.

Exploit. However, even if meant, specifically, to further a player's progress in the game, where players betray out of *Extrinsic Motivation*, betrayal does not need to be fully integrated in the game as concrete mechanics. Particularly, betrayal can depend solely on social interactions, such as communication as a strategy in the game, using *Indirect Mechanics*. This happens more prominently in board games like *Secret* Hitler (2016), Werewolf (1986) or Diplomacy (1959), but also in digital games, with a greater amount of freedom, like DayZ (2013), Rust (2018) or *EVE* Online (2003).

Exploration. When betrayal depends on *Intrinsic Motivation*, it becomes more context-dependent and can manifest in very different ways for very different reasons. Even if motivated by elements external to the game, betrayal can still happen with *Direct Mechanics*. A common example is in hidden traitor games, like *Among* Us (2018), which offer different abilities to different roles. A player who played as the non-traitor role may be motivated to play as the traitor role in subsequent games, simply to explore a different playstyle, experimenting abilities, even if that does not aid their progress.

Innovation. If, on the other hand, there are no explicit mechanics of betrayal, relying instead on *Indirect Mechanics* and players do not play towards the game's objectives, instead with *Intrinsic Motivation*, betrayal occurs as a completely different type of gameplay. In these cases, betrayal is used as a way for the traitor to achieve their own personal goals and contribute to their own definition of fun. This is the most characteristic feature of survival games such as DayZ (2013), Rust (2018) and *EVE* Online (2003), where players manage their own relationships, but have no concrete objectives. Instead, they are free to discover their own preferred playstyle, resulting in emergent gameplay.

3.3 Motivation and Relationship

Ruse. *Planned Relationships* happen when the traitor knows, from the start they want to get close to a victim, only to take advantage of them. When done with *Extrinsic Motivation*, relationships are used as tools to further a player's goals in a game. This is the normal case with social deduction games, where gaining a victim's trust is the direct path to victory for hidden traitors. Also, in strategy games, relationships are often seen as another type of resource.

Anti-play. When, instead, betrayal comes from *Intrinsic Motivation* and *Planned Relationship*, it is used to accomplish personal goals, which have no relation to the game's objectives. For example, in *Dark Souls III* (2016), players can leave messages for other players to read. A common prank is to leave a deceitful message, seemingly leading players to hidden treasure only to cause them to kill themselves. Pranksters do this knowing fully well they will cause others to die and derive enjoyment from knowing that, although it does not help them in any way.

Optimisation. On the other hand, when betrayals result from *Failed Relationships*, it means that neither party intended to betray the other at first. When in *Extrinsic Motivation*, this means that the traitor decides their current relationship is no longer beneficial for them and decides to betray. In *Friday the 13th: The Game* (2017) and, similarly, in Dead by Daylight (2016), players are tasked with surviving and escaping danger as a team. However, during the game, an opportunity may arise for a player to escape alone, by leaving another player behind as bait.

Dissatisfaction. In *Failed Relationships*, when there is *Intrinsic Motivation*, however, players end their relationships for reasons external to the game. For example, in survival games such as DayZ (2013), Rust (2018) or Minecraft (2011), it is common for players to group together, to have better chances of surviving. These groups are often led by one or more players. A common occurrence is for power struggles to arise, when one or more players grow discontent with the current leadership, often based solely on the personality of the leaders, making betrayal motivated by social reasons.

3.4 Motivation and Focus

Self-satisfaction. Betrayal which is *Traitor-Centred* is always oriented towards the traitor and what consequences happen to them, instead of the victim. When this comes from *Intrinsic Motivation*, a traitor betrays only to achieve their own personal goals, regardless of other players and even the game's rules, resulting in an especially individualistic type of betrayal. An example happens in cooperative games like *New Super Mario Bros. Wii* (2009), when players go against their team members. In this particular game, players can, at any moment, kill their teammates, which brings no benefit to them in-game, but is often considered fun.

Personal Aggression. When, however, the focus is *Victim-Centred*, with *Intrinsic Motivation*, it means that the traitor means to harm the victim, not as a player in the game context, but as a person in the social context. This often happens, for example, as revenge for past aggressions. In the previous example of killing someone for fun in cooperative games, the victim may decide to betray their traitor as payback, even though they are both, as per the game rules, still on the same team.

Game Rewards. When motivation is, instead, *Extrinsic Motivation*, the focus of betrayal describes what types of in-game consequences were the underlying motive of betrayal. When *Traitor-Centred*, betrayal happens to benefit the traitor, relative to their state in the game. For example, in strategy games like *Civilization VI* (2016), players commonly use friendships and deals to actually steal resources from other players behind their backs.

Game Punishments. When the focus is *Victim-Centred*, with *Extrinsic Motivation*, betrayal happens because of the in-game penalties and negative consequences for the victim. A prominent example happens in Diplomacy (1959), when players promise to support another's attack only to not do so in the end, possibly resulting in a massive defeat for their victim. This is a common strategy of the game, meant to weaken opponents without direct conflict.

3.5 Choice and Mechanics

Mandatory Action. Betrayal may be designed into games as a core aspect of gameplay, independently of player motivations. In this case, it relies on *Game Role* and *Direct Mechanics*, which means that betrayal is tied both to objectives and mechanics or systems of the game. This is the case with most social deduction games, particularly in digital format, such as *Among* Us (2018), Project Winter (2019) and Town of Salem (2014), where different roles are allowed only a specific set of mechanics and where even communication and actions are managed by the game.

Mandatory Dynamic. On the other hand, *Game Role* and *Indirect Mechanics* happen more commonly in social deduction board games, where communication plays a bigger role and is less restricted. In games such as *Secret* Hitler (2016) or Werewolf (1986), betrayal is also more dependent on social interactions and each players' interpretations.

Optional Action. Even if betrayal is tied to *Direct Mechanics* of a game, some games may not impose it on players and, instead, make it an optional interaction, being a *Player Choice.* This is the case with *Civilization VI* (2016), which offers multiple hard mechanics that directly cause betrayal, being tracked as such by the game's systems and variables. These include espionage, breaking deals, stealing land from friends, breaking promises and so on. However, these are all optional. In this case, although betrayal is integrated in the game, it is only one of many possible strategies for the player.

Optional Dynamic. When, instead, betrayal comes from *Player Choice* with *Indirect Mechanics*, it is completely independent of the game's rules and systems. In these cases, it manifests as emergent gameplay in unscripted moments of gameplay. This is the main appeal of games like *EVE* Online (2003) and DayZ (2013). Both games are notorious for having emergent conflicts between players, ad-hoc alliances, deception and power struggles. These dynamics are, however, initiated and managed by the players themselves and not the game's systems.

3.6 Choice and Relationship

Intentional Aggression. Games that allow more freedom to players allow them to manage more aspects of gameplay. When this includes how they can manage their in-game relationships, betrayal often comes from *Player Choice.* On the one hand, players may see relationships as another element of the game and use them to further their own goals, sometimes, at the expense of another player, resulting in *Planned Relationships.* This is often the case with strategy games like *Civilization VI* (2016), Age of Empires IV (2021) or Diplomacy (1959), where players often try to hide their true intentions and, sometimes, resort to deception and betrayal to take down an opponent.

Defection. On the other hand, when betrayal comes from *Player Choice* but results from a *Failed Relationship*, it means that players willingly decided to betray an existing relationship of trust with another player, after it began. This can happen for numerous reasons. For example, allies in a game may begin to fear one another, mutually suspecting each other of betrayal; a player may decide to betray an ally if the costs of a relationship begin to outweigh its gains or simply because they become bored and want to introduce conflict to generate stimulation.

Game Compliance. Games which assign players a formal role of *traitor*, tasked with betraying a team of *non-traitors*, fit the *Game Role* in the choice dichotomy. In most cases when this happens, particularly when the traitor player is extrinsically motivated, betrayal will naturally emerge from a *Planned Relationship*. In social deduction games, like Project Winter (2019), players know, from the start, they need to betray. If they mean to achieve the objectives imposed on them by the game rules, they also know that any relationships of trust they form with non-traitors will end in betrayal. Indeed, because their main motivation (winning the game) depends on making use of the fact that their role is secret, any relationships they begin are, by definition, asymmetric, from the start.

Game Interruption. When games make use of *Game Roles*, relationships are generally planned, because players are usually aware, from the beginning of the game, of their role and how to achieve their objectives. Betrayals with game roles that result from *Failed Relationships* require a much more specific scenario. Namely, betrayal is imposed on players *after* they begin relationships of trust. A notable example of this is *Battlestar Galactica: The Board Game* (2008). This game is mainly divided into two phases. In each phase, players draw a role card, indicating whether they are traitors or not, which is known only to themselves. This means that players who were working together with honest intentions may, when they reach the second phase, have their role changed by the rules of the game.

3.7   Choice and Focus

Role Protection. Mafia-like and social deduction games derive their core gameplay from the fact that traitors are secretly assigned and must do what they can to hide their *Game Role*. For this, lying and deception are used to help the traitor, resulting in *Traitor-Centred* betrayal. While in board games, this must be done mainly through communication, many digital games of this type provide additional abilities to hide the traitors. For example, *Among Us* (2018) allows impostors to fast travel through the map, so they can try to convince non-traitors they were somewhere else.

Instructed Target. Lying and deceiving to protect one's role is a necessary step to avoid the lose condition in social deduction games. While this is necessary to protect the *Game Role*, a more *Victim-Centred* approach is usually required to win the game. This happens naturally in games with roles, because players assigned with roles know from the start who their targets are. As such, directly killing players in secret is a mechanic allowed only to traitors in most of these games like *Among* Us (2018), Werewolf (1986) and Town of Salem (2014).

Personal Project. In games where players can manage their own relationships and enemies, betrayal is normally an optional tool, depending on *Player Choice*. When it is used simply to benefit the traitor, it is *Traitor-Centred*. This type of betrayal is common in many cases and can vary greatly according to the remaining dichotomies. For example, it is common in situations where players compete for resources, such as strategy games like *Civilization VI* (2016) or survival games like *EVE* Online (2003) and DayZ (2013). In these, the main motivation behind betrayal is usually to acquire more resources, regardless of who suffers in the process.

Chosen Target. When betrayal is still a *Player Choice*, but is *Victim-Centred*, it becomes less impersonal, as it is motivated by the adverse effects on a specific target. Also in strategy games, this can usually happen when a player becomes fearful of an ally who is becoming too powerful. Anticipating a future conflict, the weaker ally betrays their partner before they become too powerful to stop.

3.8 Mechanics and Relationship

Prepared Tools. Betrayal always depends on social dynamics. However, some games incorporate betrayal as a concrete element of gameplay, tying it to specific mechanics that produce specific, foreseeable results. This betrayal makes use of *Direct Mechanics*. When players start *Planned Relationships*, they know what mechanics to use and exactly what effects they will produce. In *Civilization VI* (2016), the use of espionage mechanics, for example, is used directly to take advantage of another player behind their back. A common strategy is to become allied with that player specifically to spy on them, stealing resources or sabotaging them.

Repurposed Tools. However, these betrayal mechanics are not always a mandatory part of gameplay. Using the previous example, players may play a game of *Civilization VI* (2016) without ever betraying their allies. However, even if they had not planned to do so from the start, some reasons may lead them to betray, resulting in betrayal with *Direct Mechanics*, but *Failed Relationship*. For example, players may back out of a trade deal with an ally abruptly, because they think they no longer benefit from the deal. While the game's systems treat this as an aggressive action, the traitor only considered using this mechanic after the relationship began.

Prepared Dynamic. If betrayal depends more heavily on social aspects, the mechanics used to betray may not even be related. Instead, betrayal depends only on the context in which the in-game actions were used and is much more situational, meaning it uses *Indirect Mechanics*. *Planned Relationships* that use these are a common occurrence in board games such as Diplomacy (1959) and Risk (1957). Because of their medium, gameplay depends much more on social interactions, such as verbally communicating deals and ad-hoc alliances between players, as well as breaking them.

Repurposed Dynamic. However, *Indirect Mechanics* are not exclusive to board games. Examples of *Failed Relationships* with these mechanics often happen in team games, when a player decides to act against their teammates. For example, Impostors in *Among Us* (2018) may accuse fellow impostors to avoid suspicion on themselves. The game does not offer any concrete mechanics to do this, nor does this dynamic affect the gameplay in a different way. However, to the players involved, it is perceived as betrayal.

3.9 Mechanics and Focus

Primary Benefit. Betrayal which relies on *Direct Mechanics* is incorporated as an explicit game element. However, it may differ, depending on its focus. On the one hand, it may be designed to benefit the traitor, being *Traitor-Centred*. One clear case is the espionage system in *Civilization VI* (2016). When spying on other players, traitors can select what their spies will do. Options include stealing resources, syphoning funds or stealing technologies. These options are typically chosen, based on what will benefit the traitor the most and not so much on who the victim is. The chosen victim will be the one who has better resources to steal.

Primary Penalty. On the other hand, *Direct Mechanics* of betrayal can also be used to directly harm a player, being *Victim-Centred*. The espionage system of *Civilization VI* (2016) can also be used in this way. Other options include sabotaging constructions of the target. Unlike theft espionage, this brings no direct benefit to the traitor and is meant simply to delay other players and make them waste resources.

Resulting Benefit. In other games, betrayal may be detached from the game's mechanics, being *Indirect Mechanics*. In these, the meaning of "betrayal" is not inherent to explicit actions in the game but, rather, the context in which they happened. Likewise, when these betrayals are *Traitor-Centred*, the benefits for the player are not a direct consequence of the actions themselves. An example happens in *New Super Mario Bros. Wii* (2009). Although a fully cooperative game, players still have to share power-ups and lives. A more individualistic player can choose to steal a better power-up even if another player needed it more. This action is recognised as betrayal, not by the game, but by the players.

Resulting Penalty. Likewise, when betrayal relies on *Indirect Mechanics*, but is *Victim-Centred*, the effects felt by the victim are not a direct consequence of the in-game actions of the traitor. Also in *New Super Mario Bros. Wii* (2009), quarrels often arise between players, sometimes because another player stole their item as described. This often results in players killing each other by jumping on them or throwing them at a pit, by surprise. These actions are not directly aggressive and can be used cooperatively in other situations. However, due to the social circumstances in which they happened, they are considered betrayal.

3.10 Relationship and Focus

Change of Mind. Betrayal always results from the degradation of a relationship between two parties, where, at some point, one decides to betray. This may be due to a relationship having lost its utility and having become detrimental instead. This is the case of *Failed Relationship* and *Traitor-Centred* betrayal. In this type, the traitor betrays because they decide the current relationship is no longer in their best interest and end it. This is a common occurrence in strategy games like *Civilization VI* (2016), Age of Empires IV (2021), Diplomacy (1959).

Collateral Damage. Scams in *EVE* Online (2003), on the other hand, are a form of *Planned Relationship*, which is *Traitor-Centred*. The scammer's intent to betray exists even before they and their victim interact. Furthermore, their only goal is to benefit themselves, by stealing their victim's resources. It is not important for them who the victim is, so long as they fall for the scam.

Quarrel. Conversely, when betrayal results from a *Failed Relationship* and is *Victim-Centred*, it means that at some point, one player decides to betray, only to harm the victim. This also happens in *EVE* Online (2003), most notably in one of the most famous heists in the game's history, the heist on the in-game corporation 'Co2'[2]. In this case, this heist was performed by one of the company's own high-ranking employees, whose main motivation was stated to be disliking the corporation's leader's personality.

Targeting. Similarly, *Planned Relationships* in which betrayal is *Victim-Centred* are also oriented towards its effects for the victim, only this time from the start. This is the most normal case of betrayal in social deduction games, where the hidden traitors are usually tasked with eliminating the non-traitors. This often requires them to get close enough to them to kill them, which means they must fake trustworthiness just long enough to strike.

4   Patterns of Betrayal

Each case of betrayal can be classified into each of the five dichotomies, ranging between the two extremes. However, some combinations are more common than others, resulting in similar situations across different games. In this study, a set of different multiplayer games was analysed with the proposed conceptual model to identify recurring instances of betrayal, as explained in Sect. 2.

To study situations of betrayal, the selected games followed two guidelines: 1) must be a multiplayer game, since the focus of the study is betrayal between different players; 2) must not be fully competitive, since betrayal requires an initial state of cooperation, which does not naturally happen in games of pure competition.

As was explained in Sect. 3, games are not limited to only one type of betrayal. Instead, depending on the context and situation, separate cases of betrayal can reflect different classifications under the model in the same game. For this reason, each game was studied as a set of individual situations. The result was a set of recurring patterns of betrayal. Each of these patterns is characterised by two more prominent dichotomies, while the remaining ones are more variable depending on each specific situation.

4.1   Social Deduction

This pattern happens when betrayal is fully integrated into the game, both in rules and mechanics. It is typical of Mafia-like or Hidden Traitor games, where players are divided into two roles at the start of the game, the non-traitors and the secret traitors. For this reason, this betrayal makes use of *Game Role* and *Direct Mechanics*.

---

[2] Consult https://www.pcgamer.com/inside-the-biggest-heist-in-eve-online-history/. Date of Access: 29/07/2024.

In these games, betrayal is explicitly enforced, mainly by the fact that players of the non-traitor team are unaware of the roles of the remaining players, while the traitor team knows most or everyone's roles, depending on the game. This forms the core gameplay of these games, meaning that one team must discover the identities of other players to survive, while the other team must use lies and deception to hide their roles and achieve their own objectives.

This type of betrayal is very common in many board games, such as *Secret* Hitler (2016), Werewolf (1986) and *Battlestar* Galactica (2008), but is also increasingly found in digital games, like *Among* Us (2018), Project Winter (2019) or Town of Salem (2014).

4.2 Optional Strategy

This pattern happens when betrayal is only one of the many possible actions available to a player, to achieve a certain objective in a game, but is not imposed on them. Instead, betrayal becomes a tool, which is effective precisely because players never know whether someone will play treacherously or not. As such, betrayal is a fully *Player Choice* and is used only to improve the traitor's state in the game, making it *Extrinsic Motivation*.

In these cases, players will often need to manage their own relationships and decide for themselves who to trust. As a result, a viable strategy is to pose as an ally, only to take advantage of trusting players to improve one's own state in the game. For this to happen, however, games must also ensure that trusting players and cooperating can be beneficial, otherwise, players would have no reason to trust in the first place. For this reason, this pattern typically happens in games with a large variety of objectives and optional ways to improve which do not directly lead to the win condition.

This is the typical case of most strategy games like *Civilization VI* (2016), Age of Empires IV (2021) or Diplomacy (1959), but also other games with great player freedom, like survival games such as DayZ (2013) or *EVE* Online (2003).

4.3 Roleplaying

This pattern happens when players betray only for their own personal enjoyment. This enjoyment is independent from the game's rules, instead being the player's own personal goals, making it *Intrinsic Motivation*. Additionally, in these cases, the main goal is to entertain the player themselves, making it *Traitor-Centred*.

In these cases, players betray only because they want to have fun with the act of betrayal itself. This can be, for example, wanting to experiment with actual mechanics of betrayal, not because it helps them, but because they want to try them. This is common in Social Deduction games, where players want to play as the traitor role only to experiment a different role. On the other hand, it may be that players may only want to roleplay as a treacherous character and derive enjoyment from staying true to that type of character.

This can happen in most games as it is only dependent on the player's own personal definition of fun. However, it is most common for games with much freedom, like sandbox games, such as *EVE* Online (2003), DayZ (2013) or Minecraft (2011), where players decide their own goals or MMORPGs, where one of the main attractions is character-building.

4.4 Preemptive Action

This is a more defensive type of betrayal, where deception and treachery are used, mainly to end a relationship which is no longer beneficial to the traitor. This relationship is, then, broken intentionally by one of the players, making it *Player Choice*. Additionally, the main characteristic of this pattern is that betrayal is used to end a relationship, instead of the relationship being used to betray, making it a *Failed Relationship*.

This betrayal happens in cases where two players are in an existing friendly relationship. At one point, one of the players decides that the current state of affairs is no longer in their best interest and decides to end it. This may be motivated by a variety of things, such as feeling that the other player is dragging them down or not contributing equally. In other cases, players may choose to betray out of fear of being betrayed first, meaning this pattern is typical of cases of mutual suspicion.

This pattern can occur in many games, where players can both benefit and suffer from relationships. More commonly, it happens in strategy games, where players can benefit greatly from forming alliances but leave themselves vulnerable at the same time, such as *Civilization VI* or *Diplomacy*. In other cases, it also happens in team settings, where one player chooses to protect themselves at the expense of their teammate, such as selling out fellow traitors in *Among* Us (2018) or *Secret* Hitler (2016), only to avoid suspicion or leaving other survivors behind in *Friday the 13th: The Game* (2017).

4.5 Team Rebellion

This betrayal happens in situations and games where betrayal is not, at all, expected or intended. Instead, it happens fully emergently and depends completely on the context in which it happens, making it rely fully on *Indirect Mechanics*. Additionally, because it is not expected, it is a *Player Choice*.

When betrayal happens emergently, its nature as 'betrayal' is given meaning only by the perceptions and interpretations of the players involved. This means that this betrayal depends only on context and each specific situation is different, because it is played in different circumstances, possibly by different people. The actions used in this betrayal are not meant to betray, as per the rules and objectives of the game, but, instead, are performed in a way that allows players to take advantage of their fellow players.

This betrayal happens commonly in team games, as these are often meant to be played as a team, where the win condition does not depend on individual goals, but a shared goal. However, players still decide to betray their teammates. This can happen both for intrinsic motivations, such as simply enjoying causing chaos in Portal 2 (2011) or *New Super Mario Bros. Wii* (2009) or wanting to compete in some way, such as getting a higher individual score in Super Mario 3D World (2013) or *Left 4 Dead 2* (2008).

4.6 Harassment

This pattern happens when betrayal is a means to directly harm another player, less as a player and more as a person. This betrayal depends much more on the social space of games, making it a completely *Intrinsic Motivation*. The objective of harming another player makes it *Victim-Centred*.

In these situations, the game's current state, objectives or penalties are not relevant for the traitor, whose only intent is to harm another player. It is most typical in cases of revenge, where acts of aggression are meant to be personal. For example, a player who was attacked by another may, later, plot to hurt them as payback. One way to do this is fake being friendly only to get close enough to harm them. Other cases may happen when players simply dislike another player's personality and attack them because of it.

Because motivation is fully intrinsic, this pattern can happen in any situation, depending on the player. It is a common consequence of cases of Team Rebellion, where one betrays their own traitor as payback. However, it also happens, for example, in power struggles in games like DayZ (2013), where players are unhappy with their current leaders, such as the heist on the leader of Co2 in *EVE* Online (2003), explained in Sect. 3.

5 Assessment

To assess the model's usability and utility as a tool of analysis and game design of betrayal in games, it was subjected to an exploratory study, with a series of workshop sessions. In these sessions, participants could interact and experiment with the model. Two types of workshops were performed: *game concept creation* and *game analysis.*

The sample of participants consisted of students and professionals in game design, with knowledge in gaming, having been exposed to a large variety of games across different genres enabling them to identify recurring types of dynamics and situations; and with 2 to 5 years of experience in game design, in a total of 10 participants, distributed across sessions.

5.1 Game Concept Creation Workshop

In the first workshop, participants were tasked with creating a short multiplayer game concept with elements of betrayal, in groups of two. To aid these, participants were provided with the model of dichotomies, its patterns and their respective explanations and examples of games. Their design process was observed to understand how they used the model and how it influenced their process. This workshop was separated into 2 sessions, one with 4 participants and the other with 6, resulting in 5 game concepts.

At the end, a focus group discussion was held to discuss the experiment and collect feedback about the model and patterns. In general terms, participants responded favourably to the model and found it useful to help structure betrayal as an element of gameplay. Some also said it helped them think of different ways of betrayal they had not considered before. However, some participants reported difficulty in using it, mainly due to the lack of a practical artefact, finding having only the theoretical explanations of the model insufficient. This was especially true with the Relationship dichotomy. The patterns, on the other hand, were mostly not used in the process, with participants finding the dichotomies themselves more useful and easier to work with.

5.2 Game Analysis Workshop

In the second workshop, participants were presented with a set of examples of betrayal from multiple games and were asked to discuss and classify each presented situation in the five dichotomies and choose an appropriate pattern for it. This helped analyse each individual dichotomy and pattern in more detail and how well they could describe betrayal formally. This workshop was performed last for this reason, as it required some familiarity with the model and patterns. Nevertheless, a brief recapitulation of the model and patterns was provided. Only one session was performed, with a total of 4 participants from the previous sessions. A total of 15 situations across 7 different games were presented and each one was accompanied with an explanation of the context of the betrayal, an illustrative image and, when possible, a link to a video with the actual situation.

Similarly to the previous workshop, participants were mostly able to correctly interpret the dichotomies, although their opinions and individual classifications varied, sometimes being completely different.[3] Furthermore, consistent with the previous workshop, more difficulties were felt with the Relationship dichotomy, generating the most disagreement among participants.

Another important observation was that some examples were very nuanced and required prior knowledge of the game to fully interpret it. Once more, the patterns were more negatively received. In this case, most participants could not find an appropriate pattern for many situations, stating that the patterns were, sometimes too specific, others too general to fit.

6 Conclusions and Future Work

The proposed model of dichotomies can be a viable tool to formally describe betrayal and aid designing betrayal in games. The current dichotomies can be further improved in future iterations. Furthermore, adding new dichotomies can help to better analyse betrayal as a phenomenon. However, it was concluded during the trials that the current patterns are not sufficiently well-defined to be a viable auxiliary tool.

Despite the promising results, this study was limited in some ways. First and foremost, betrayal can be a highly nuanced and complex interaction, often depending completely on the context of each specific situation in which it occurs, as well as the group of people involved. For this reason, it becomes necessary to fully study a game to understand how and why betrayal manifests in each specific game. Ideally, all studied games would have been played extensively. However, due to time constraints, this was not possible for all games. Instead, second hand reports from other players were used.

Nevertheless, the current version of the model provides a good starting point for future improvements and serves as the basis of practical artefacts and frameworks to be used as a practical tool. This next version should revise the dichotomies and patterns to tackle the main issues found. Most notably, it would be beneficial to refactor the

---

[3] For example, selling out fellow impostors in *Among* Us (2018) was seen by some as extrinsic, since it was to avoid dying and, by others, as intrinsic as it was directly playing against the intended rules.

current dichotomies into more specific and specialised versions, possibly adding more dichotomies.[4]

While the existing patterns do present recurring situations of betrayal, they should be restructured into a more intuitive version. A possible way to do this could be to reframe them as *Archetypes* of traitor players, as opposed to attempting to describe situations. A possible new set of these could be: **The Secret Traitor**, who is assigned the role of traitor by rules of the game and betrays to play as per the rules of the game. **The Strategist**, who is not necessarily aligned with anyone but will form and break relationships and alliances for their own convenience, choosing to improve their state in the game through betrayal. **The Self-Interested**, who, although in a team setting, will always play for their own interests first, even if it means betraying teammates. **The Troll**, who betrays purely to enjoy the act of betrayal itself, either by choice, in a team setting or even if assigned the role of traitor. **The Vengeful**, who uses betrayal as a form of payback for something that was done to them, motivated solely for reasons external to the game. **The Paranoid**, who betrays out of fear of being left vulnerable by being in a relationship of trust with another player, betraying before they themselves are betrayed, even if unjustified.

To better define these archetypes, future research should focus more on the dynamics between players and how they evolve over time. As such, more research could be done on common experiments and scenarios of Game Theory that were not explored here, such as more variations of the Prisoner's Dilemma, Snowdrift Game, Public Goods Experiment and Tragedy of the Commons.

Acknowledgements. This work is [partially] financially supported by national funds through FCT – Foundation for Science and Technology, I.P., under the project UIDB/05460/2020.

References

Allison, F., Carter, M., Gibbs, M.: Good frustrations: the paradoxical pleasure of fearing death in DayZ. In: Proceedings of the Annual Meeting of the Australian Special Interest Group for Computer Human Interaction, pp. 119–123 (2015). https://doi.org/10.1145/2838739.2838810

Baier, A.: Trust and antitrust [Publisher: The University of Chicago Press]. Ethics **96**(2), 231–260 (1986). https://doi.org/10.1086/292745

Ben-Ner, A., Halldorsson, F.: Trusting and trustworthiness: What are they, how to measure them, and what affects them. J. Econ. Psychol. **31**(1), 64–79 (2010). https://doi.org/10.1016/j.joep.2009.10.001

Brooks, I.: Is betrayal in EVE online unethical? J. Virtual Worlds Res. **10**(3), 1 (2017). https://doi.org/10.4101/jvwr.v10i3.7259

Carter, M.: Treacherous Play. The MIT Press (2022). https://doi.org/10.7551/mitpress/12023.001.0001

Consalvo, M.: Cheating: Gaining Advantage in Videogames. The MIT Press (2007). https://doi.org/10.7551/mitpress/1802.001.0001

Consalvo, M.: There is No Magic Circle [Publisher: SAGE Publications]. Games Cult. **4**(4), 408–417 (2009). https://doi.org/10.1177/1555412009343575

---

[4] A proposal of new dichotomies in a future version can be consulted elsewhere (Valério, 2024; Valério & Cardoso, 2024).

O'Neil, C.: Betraying Trust. In The Philosophy of Trust. Oxford University Press (2017). https://doi.org/10.1093/acprof:oso/9780198732549.003.0005

Rachman, S.: Betrayal: a psychological analysis. Behav. Res. Ther.. Res. Ther. **48**(4), 304–311 (2010). https://doi.org/10.1016/j.brat.2009.12.002

Robinson, W.: Ruse, trust, and the fiction of betrayal: how game designers can author player experiences. Analog Game Stud. **11**(1), (2016). https://analoggamestudies.org/2016/07/ruse-trust-and-the-fiction-of-betrayal-how-game-designers-can-author-player-experiences/. Retrieved September 20, 2023,

Sicart, M.: Defining game mechanics. Int. J. Comput. Game Res. **8**(2), (2024). https://gamestudies.org/0802/articles/sicart. Retrieved January 5, 2024

Suits, B.: he Grasshopper: Games, Life and Utopia. Broadview Press (1978)

Valério, P. Tricked, Backstabbed and Bamboozled: A Conceptual Model of Betrayal in Multiplayer Games [Master's Thesis, University of Aveiro]. Repositório Institucional da Universidade de Aveiro (2024).

Valério, P., Cardoso, P.: It's a trap! A conceptual model of betrayal in games. In: Manuscript Accepted for Publication at Digicom 2024: 8th International Conference on Design and Digital Communication (2024)

Yee, N.: Motivations for play in online games. CyberPsychol. Behav. Behav **9**(6), 772–775 (2006). https://doi.org/10.1089/cpb.2006.9.772

Ludography

Age of Empires IV. (2021). Relic Entertainment, World's Edge
Among Us. (2018). Innersloth
Battlestar Galactica: The Board Game. (2008). Fantasy Flight Games
Civilization VI. (2016). Firaxis Games
Dark Souls III. (2016). From Software
DayZ. (2013). Bohemia Interactive
Dead By Daylight. (2016). Behaviour Interactive
Diplomacy. (1959). Wizards of the Coast
EVE Online. (2003). CCP Games
Friday the 13th: The Game. (2017). IllFonic, LLC
Minecraft. (2011). Mojang Studios
Portal 2. (2011). Valve
Left 4 Dead. (2008). Valve
New Super Mario Bros. Wii. (2009). Nintendo
Project Winter. (2019). Other Ocean Interactive
Risk. (1957). Hasbro
Rust. (2018). Facepunch Studios, Double Eleven
Secret Hitler. (2016). Goat, Wolf, & Cabbage
Super Mario 3D World. (2013). Nintendo
Town of Salem. (2014). BlankMediaGames
Werewolf. (1986). Dimitry Davidoff

Exploring Asymmetric Competitive Gaming For Mixed-Visual-Ability Pairs

Pedro Trindade(✉) ⓘ, David Gonçalves ⓘ, Pedro Pais ⓘ, Jão Guerreiro ⓘ,
Tiago Guerreiro ⓘ, and André Rodrigues ⓘ

Faculdade de Ciências, LASIGE, Universidade de Lisboa, Lisbon, Portugal
{pgtrindade,dmgoncalves,pgpais,jpguerreiro,
tjguerreiro,afrodrigues}@ciencias.ulisboa.pt

Abstract. Competitive games assume stereotypical players with equal abilities face mostly symmetric gameplay. For mixed-ability groups, equal challenges limit the design space and can be unappealing. Conversely, introducing asymmetric play raises concerns about fairness and balance. This work first explores competitive mixed-visual-ability games, focusing on understanding players' perspectives of competition, fairness, transparency, and asymmetric play. Through a mixed-methods study involving six sighted and four blind participants, we examined player experiences across four combinations of a/symmetric play. Our results reveal how disability disclosure can affect the experience, how design choices of asymmetry affect the perceived fairness, that asymmetric competition can be engaging, and the nuances between the perspectives of sighted and blind players. We highlight that while asymmetric design presents an opportunity for fostering competitive mixed-ability environments, it necessitates thoughtful design where skill and disability are not conflated, ensuring that different in-game challenges are either equally demanding or that the overall game experience reflects a balance of what is demanded of players.

Keywords: Competitive gaming · Accessibility · Mixed-ability ·
Visual impairments · Asymmetry

1 Introduction

Digital games play a significant role in fostering socialization and connecting individuals across different contexts, backgrounds, and age groups [18,35,43,47]. Competitive gaming was shown before to have various social benefits, acting as a platform for people to come together, bond, and build communities [9,28].

However, most games are designed without anticipating the needs of individuals with disabilities, rendering them inaccessible to a big portion of the population [5,7,23,25,49]. Prior work [17] has shown how blind players (and content creators) manage to play a subset of mainstream games by using existing game mechanics to overcome accessibility barriers (e.g., utilizing weapons to feel their surroundings). These players are the exception, and despite their efforts, they often have to use external resources and face challenges not intended by design (e.g., persistent trial-and-error), which in multiplayer competitive games creates an extreme imbalance between players (e.g., in Call of Duty [31] not being able to determine and adjust crosshair height). For blind players, playing with others typically means mainly engaging with games specifically designed for them (i.e. audio games) in a segregated community based on visual ability with limited to non-existent play with sighted peers [19].

Asymmetric game design provides players with distinct gameplay experiences, differing in-game abilities, challenges, and information [14,29], among others. Past work [20] has shown how to leverage asymmetry to create inclusive mixed-visual-ability cooperative games, ensuring balanced player contributions with players equally engaged in a single cooperative experience. We believe there is an opportunity to leverage the same design principles to create competitive games. However, the issue shifts from perceived equal contribution and autonomy to the fairness of asymmetric competition. In this work, we explore how different a/symmetry types affect the perceived fairness and experience of sighted and blind players in competitive games.

To achieve this, we developed a game prototype that had players compete to be the fastest in creating four magical items. Players had to complete a/symmetric challenges to create these items. We conducted a mixed-methods study with six sighted and four blind participants. In each session, participants played the game and were told they were competing against another player online. This was followed by: 1) the Mini PXI [1] questionnaire to measure the overall player experience, 2) telling the players they were competing with someone with different visual ability than their own (i.e. sighted or blind) and explaining how they were playing and competing, 3) a fairness questionnaire regarding the a/symmetries experienced, and 4) a semi-structured interview, focused on understanding the impact of the design choices on player experience, perceptions of fairness, mixed-ability competition and disability disclosure. Our research questions were:

- How do players perceive competition and fairness in mixed-visual-ability competitive gaming?
- What are players' perceptions about different symmetric and asymmetric approaches to mixed-ability gaming?

Our findings demonstrate that asymmetry in competitive mixed-visual-ability contexts has the potential to create engaging experiences where differences in abilities are not limiting. Notably, asymmetric game design tended to be perceived as fair when tailored to each player's abilities. On the other hand, symmetric play restricts sighted users to audio-only gameplay (while fair, hurts the experience), representing the status quo of no cross-play between communities. We highlight how disability disclosure and expectations of play affected sighted and blind players differently. Lastly, we discuss how skill and disability should not be conflated but represent a new design challenge for the field.

2   Related Work

In this section, we cover the state-of-art of visual accessibility in games and current alternatives to vision-centric games (i.e. audio games), past research centered on asymmetric game design, and the role of fairness in competitive gaming.

2.1   Playing Blind

For blind players, visuals in digital games need to be replaced with audio feedback through the use of text-to-speech, audio cues (e.g., footsteps), descriptions and/or sonification (e.g., unique sounds for different objects) [13,20] as well as haptic feedback [3,10,21]. Efforts to make blind-accessible games often involve adapting existing games that did not consider players with disabilities during the design process, like AudioQuake [5] (Quake) and Blind Hero [48] (Guitar Hero). However, a *posteriori* attempts to adapt games whose core gameplay is based on visual engagement may harm the player's experience [19,49]. In contrast, some recent games [12,44] are launched with accessibility options that make them fully playable by blind players. While these are remarkable advances, they are the exception, as most titles still have no consideration for blind players' needs and depend on vision [17,40].

Prior work has explored design solutions to augment or propose alternative techniques for environmental awareness [36,37] and navigation [39]. Notably, the RAD [42] is an auditory interface proposed by research to make racing games accessible to blind players. The approach created an asymmetric interface for blind players (e.g., audio cues for an upcoming curve), which made it possible to compete without significantly different lap times or driving paths against sighted players. Since then, the technique has been announced to be leveraged in the recently released commercial game Forza Motorsport [45][1]. However, even in this successful instance, it is still a challenge to have blind players competing with other cars on the track, given the difficulty to accurately convey other vehicles' position—disabling vehicle collision is available as an accessibility feature in Forza Motorsport, but it cannot be enabled in multiplayer matches. While asymmetry of the interface can effectively align audio ability with sight, it is most likely impossible to match the throughput of visual perception with audio in complex games. This means players are either at a disadvantage against their sighted peers or limited to more simple/controlled environments for competition.

Despite advances achieved by the industry and research, most games are still inaccessible to many people (especially blind) [19,40]. Also, there is no understanding of how one can design inclusive multiplayer games where abilities are not limiting to the experience. In the few mainstream games accessible to blind players, challenges are usually made accessible through options that allow players to skip, diminish, or totally alter the challenges posed [17]. It is unclear how

---

[1] John Walker (12/09/23). Forza Motorsport's Blind Drive Assist Is A Breakthrough For Gaming Accessibility. Kotaku URL: https://kotaku.com/forza-motorsport-xbox-blind-accessibility-options-race-1850829331 (visited on 20/02/24).

multiplayer games could incorporate these options and changes while maintaining the shared experience engaging and balanced. This is a notable challenge in competitive games, where differences in the gameplay between players (e.g., aim assistance for some players) can lead to a sentiment of unfair play [11].

Audio Games. Audio games are games with audio-based gameplay. They provide information to the player through text-to-speech and unique sounds for each game element, allowing players to differentiate between various sound patterns. At AudioGames.net[2], a repository of sound-based games, a great variety of audio games genres and themes can be found, such as Manamon [46] (a Pokémon-like adventure game), and Crazy Party [41] (arcade-style minigames and card battler), among many others, of which some are competitive audio-only games. Audio games struggle to reach popularity among the mainstream gaming community because they are difficult and unappealing for sighted people due to the lack of visual information [4,19]. Furthermore, research examining the involvement of sighted players in audio games remains scarce [2,32].

Universally Accessible Games. Grammenos et al. [24] introduced the concept of Parallel Game Universes as a means to create universally accessible games. This approach involves tailoring the game experience to individual player abilities, allowing for more inclusive gameplay. While successful in simpler games (e.g. chess [22], tic-tac-toe [38] and space invaders [24]), scaling this methodology to more complex games may present challenges due to the potential need for significant simplification and the potential limitations in certain gaming contexts. Moreover, in a mixed-ability competitive gaming context, ensuring fairness presents a challenge due to the impracticability of uniformly applying rules [25] as not all challenges can be accessible without significant changes.

2.2 Asymmetry for Mixed-Ability Play

Asymmetric design creates different gameplay mechanics for different players [14,29]. Prior work has proposed a set possible *mechanical manipulations* that designers can leverage to create asymmetric player experiences [29], such as asymmetry of ability, challenge, interface, information, investment, and goal/responsibility. In the context of mixed-ability, asymmetries can be designed toward specific individual abilities. In prior work, asymmetry has been leveraged for mixed-ability cooperative games for blind and sighted people [19] and for wheelchair users and people without motor impairments [15]. The previous approaches have focused on cooperative play, designing for pairs of players with interdependent roles, and tailoring one role specifically for the target ability.

In Kinaptic [21], researchers design a competitive mixed-ability tag-like game for blind and sighted players. The study focused on providing different modalities for players (i.e. TV for visual feedback while blind players relied on a haptic

---

[2] AudioGames URL: https://www.audiogames.net/ (visited on 20/02/24).

device, wind, and 3D sound). The study evaluated how these alternative interfaces influenced players' ability to perceive and interact with the game environment, yielding mixed results. Despite the few works on mixed-ability gaming, our understanding of how players perceive and experience competitive mixed-ability games remains limited. Specifically, the intricate issues of fairness and transparency.

In asymmetric competitive games, round-based gameplay where teams/players switch roles (e.g. Predator: Hunting Grounds [30]), is often used to address any potential imbalances in the likelihood of one role having a higher chance of success (i.e. the game being imbalanced). This also opens up the design space as seen in previous works in VR asymmetric play, where roles are expected to be reversed, and thus imbalance between the two players is a core part of the game design. Gugenheimer et al. [26,27] found that variations in "power level", resulting from asymmetries in information, ability, and interface, could enhance player enjoyment. However, embracing this imbalance is only possible when roles are reversible. In mixed-ability scenarios, when roles are designed based on ability, some roles cannot be played by everyone, and thus purposeful imbalance is not desirable. However, as shown by work on cooperative games, asymmetric games can create engaging experiences for groups with mixed-abilities. The question is, how can we create competitive experiences that are perceived as fair by the players?

3   Competitive Mixed-Ability Testbed Game

We developed Cryptic Kitchen[3], a testbed game to explore players' perceptions of mixed-ability competitive gaming and different asymmetric design approaches. We aimed to provide a cohesive gaming experience where players were exposed to various game mechanics while their opponents could have the same, equivalent, or completely different ones. The goal was for players to compete and, after the gameplay session, be exposed to the other player's mechanics and informed of their different visual abilities to reflect on the variety of competition types created through asymmetric design. We carefully designed the game, aiming for a balance despite the asymmetries, as we sought to create an experience where players felt they were competing with their peers on equal terms.

While asymmetry is often used to create a purposefully different experience for both players, in this work, we aim for players to have the same in-game **abilities**, to be competing over the same **goal**, to have access to the same **information** and to require the same **time** investment [29]. We limited our exploration of asymmetry to **challenge** and **interface** as identified by Harris et al. [29], as we believed to be the most promising given the prior work on mixed-ability competition [21] and cooperation [15,19].

Below, we provide an overview of the gameplay, navigation, orders, and inventory system, followed by describing the four challenge combinations used (i.e. Symmetric Audio-Only, Symmetric, Asymmetric Interface, and Asymmetric

---

[3] Cryptic Kitchen. URL: https://techpeople.itch.io/cryptic-kitchen.

Challenge). We also provide a video demonstrating the challenges and navigation[4].

3.1   Design and Implementation

Testbed Gameplay. Inspired by the mechanics of the Overcooked [16] series,
Cryptic Kitchen (see Footnote 3) is a competition between two magical chefs.
Set within a 2D top-down environment, players **navigate**, gather ingredients,
use **stations** to imbue ingredients, which open a challenge (i.e. moments of
a/symmetric play described below), and fulfill **orders**. Blind players use the
keyboard and sighted also use the mouse.

Navigation. The map (Fig. 1) was designed to have an equal traveling distance and time for sighted and blind players. Sighted players freely navigate
the environment, while blind players navigate through a waypoint system (e.g.,
when pressing left the character automatically moves to the next waypoint on
the left), both using the keyboard WASD/arrow keys. The navigation system is,
therefore, only asymmetric in interface.

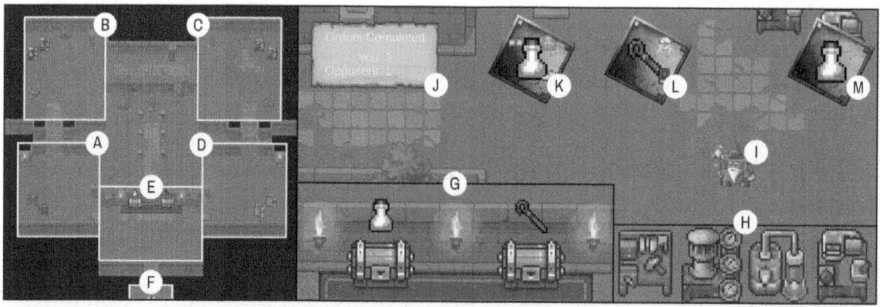

Fig. 1. Central Room (E); Delivery Room (F); Potency Room (D); Effect Room (C);
Element Room (B); Inscription Room (A); Ingredients chests (G); Cooking Stations
(H); Player Character (I); Scoreboard (J); First Order (K); Second Order (L); Item
In-Hand (M).

Inventory and Orders. We use Text-to-Speech (TTS) to grant access to inven-
tory/orders to blind players, meaning they hear one item description at a time.
Sighted players, on the other hand, rely on visual representations of their inventory and orders (Fig. 1), but can also view only one item at a time from the three
available (order one, order two, and inventory), which are hidden by default. The
inventory/order system is, therefore, only asymmetric in interface.

---

[4] Demonstrative video of the game prototype. URL: https://osf.io/zr43e.

3.2 A/Symmetric Cooking Challenges

Cooking Station. Interacting with a cooking station (e.g. Fig. 1-H) will open a challenge (Fig. 2). Upon completion, the item remains locked in the station for 30 s, incentivizing players to multi-task and start the next order in another station. Below, we detail the four challenges implemented, two symmetric and two asymmetric:

Symmetric Audio-Only (SAO) - Pairing Cards. Baseline condition equivalent to an audio game with no visuals. Players face the **same auditory challenge**, with the **same interface**. This game models a traditional memory card game. The player has to find 4 card pairs in a pool of 9 cards by selecting two cards with the same sound. Players used the numpad or number keys to select card positions.

Symmetric (S) - Isolate Pattern. Players face the **same challenge** and have the **same interface** (Fig. 2-C), but only sighted players can leverage the visual feedback. Players are tasked with adjusting three knobs, using the directional keys to isolate a sound (e.g., fire sound) and/or a visual pattern. The sound associated with each knob is only being played while the player is adjusting it, i.e. there are only two sounds simultaneously playing: the one required to be isolated and the knob sound.

Asymmetric Interface (AI) - Moving Goal. Players face **the same challenge** with **distinct interfaces** (i.e. visual-only or audio-only feedback). The player has to stay inside a moving score zone for a given amount of time using the directional keys (Fig. 2-D). Blind players have to follow the sound, using the spatial audio to determine the target position, and when on target, they hear a continuous scoring sound. Sighted players only have visual feedback (i.e. a moving target).

Asymmetry of Challenge (AC) Players have **distinct challenges** and, consequently, **distinct interfaces**. Blind players have the Find Sound challenge (Fig. 2-B), and sighted ones have the Find Silhouette (Fig. 2-A). **Find Sound.** The player is located in a random cell in a 5 by 5 grid and has to reach the goal which plays a unique sound. After every movement, a positive or negative cue indicates whether if they are closer or further from the goal. Moving against the grid limit plays a "bump" feedback sound. **Find Silhouettes.** The player is shown several different items floating across the screen and knows which ones they have to select. After a short delay, only generic silhouettes/shadows are displayed and the player has to select all the correct ones. Incorrect guesses temporarily block interactions. Pressing the spacebar shows the player the different items' colors/shapes for a short time but blocks selection.

Development and Playtesting. The game was developed in Unity[5] along with Photon Engine[6]. During the development process, we conducted playtest-

---

[5] Unity URL: https://unity.com/ (visited on 20/02/24).
[6] Photon Engine URL: https://www.photonengine.com/ (visited on 20/02/24).

ing with a group of selected individuals (sighted and one blind player), whose feedback was iterated to ensure usability.

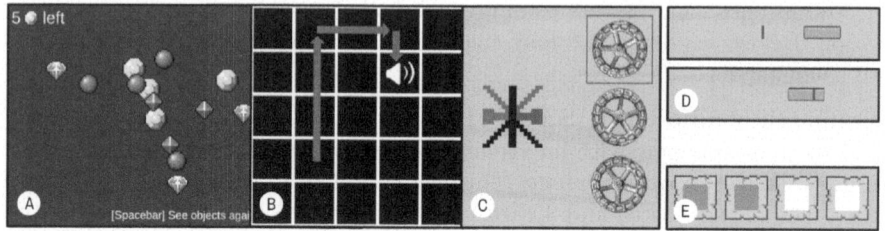

Fig. 2. Minigames developed: (A) Asymmetry of Challenge (Sighted) - Find Silhouettes: moving target and decoy items to track, before only the silhouettes are visible; (B) Asymmetry of Challenge (Blind) - Find Sound: grid layout with an auditory path leading to the target sound. (C) Symmetric – Isolate Pattern: three knobs (right) can adjust sound volume and the pattern image (left); (D) Asymmetric Interface – Moving Goal: the player marker (vertical bar) and the moving goal (horizontal bar); (E) Symmetric Audio-Only – Pairing Cards, the game's scoreboard indicating the number of card pairs found.

4   User Study

We conducted an in-person user study with sighted and blind participants. Our goal was to explore players' perspectives about different types of mixed-ability competition to understand how it can be designed to create an engaging and fair gaming experience. The study was approved by our school's Ethics Committee.

4.1   Participants

We recruited 10 participants, six sighted and four blind. Sighted participants were aged 22–29 ($M_S = 24.83$; $SD_S = 2.97$). Blind participants were aged 30–43 ($M_B = 37.5$; $SD_B = 5.32$). Participants filled in an online form with demographic information (i.e. age, gaming frequency, and competitiveness).

Our participant pool had sighted players identified as more frequent gamers than blind players. All blind participants engaged in gaming monthly, while the frequency of sighted participants was generally higher. Among sighted participants, only one individual, S1, identified as a casual player, while the remaining participants identified as either between casual and hardcore or hardcore players. Among the blind participants, B4 was between casual and hardcore, while the others identified as casual players. Participants S2, B3, and B4 identified themselves as very competitive individuals.

4.2 Procedure

Participants received a standalone game version before the in-person session containing only the four challenges corresponding to their visual abilities. Participants were encouraged to familiarize themselves with the different in-game challenges. During the in-person study play session, participants were asked to complete a questionnaire[7] about their play of each challenge and their perception of their difficulty level.

Participants were then guided through a step-by-step tutorial showcasing the controls and mechanics of the full game (i.e. navigation, orders, inventory, stations, and delivery). Participants were required to complete two orders during the tutorial. Next, participants were informed they would engage in an online competition with another participant, racing to be the first to complete four orders. Importantly, participants were not yet aware of their opponent's visual abilities, nor that both players would be engaging in different versions of the game based on their abilities.

Competition. In all sessions, the opposing player was a researcher playing remotely using the opposite game version (e.g., sighted player gameplay when their opponent was blind) following a script replicating playtesting sessions. This approach upheld a competitive experience while ensuring that each participant could progress through the step-by-step tutorial at their own pace, without any delays or interruptions, facilitating recruitment, scheduling and study execution, and streamlining the logistics of the study. During the gameplay session, the researcher offered minimal assistance to minimize their influence on the game's outcomes. Additionally, the game only ended once at least two deliveries were completed by the participant, even if they already lost (i.e. 10 min and 32 s for blind participants and 8 min and 26 s for sighted participants).

Upon completion of the game, participants were prompted to fill the Mini PXI [1] questionnaire to ensure that participants perceived the game as engaging, as a lack of engagement could potentially skew perceptions and impact the overall player experience. At this stage, participants were informed that the game was designed for mixed-visual-ability play. They were also made aware that their opponent possessed different visual abilities (i.e. sighted, or blind). We conducted semi-structured interviews[8] focusing on their views on mixed-ability competition, the perceived fairness of the main game mechanics, gameplay, and each a/symmetric challenge. During the interviews, participants were informed about the challenges (i.e., moments of asymmetric play) faced by their opponents. For blind players, the gameplay was described, while sighted players experienced their opponents' challenges firsthand. Participants' perceptions of fairness were additionally captured on a scale of 1, "In Disadvantage", to 5, "In Advantage", in comparison to their opponent.

---

[7] Minigames Questionnaire. URL: https://osf.io/zmch7.
[8] Semi-structured Interview Script. URL: https://osf.io/ed5kn.

4.3   Data Analysis

After transcribing interview audio recordings, we performed a mixed deductive-inductive codebook thematic analysis [8] over all open-ended questions of the interview. We familiarized ourselves with the data by iterative reading, then the first author developed codes based on research questions, data familiarity, and study observation. The team discussed interpretations and developed a preliminary codebook. The first author coded all interviews, adding new codes as needed. Themes were identified and named through iterative sessions, supported by quantitative engagement and fairness data.

4.4   Findings

The descriptive data presented below serves only to contextualize the play session, as our goal is to explore the perspective around the fairness of different a/symmetric design choices; it is not intended to be generalizable given the small sample and the disparity within it (e.g., game expertise).

The Mini PXI results show players overall classify their enjoyment as high ($M_O = 2.7$, $SD_O = 0.6$). Sighted players were, on average, faster in all challenges, with the largest discrepancy in the Symmetric ($M_S = 8.2$ s vs $M_B = 18.6$ s), and the smallest in the Asymmetry of Challenge ($M_S = 12.5$ s vs $M_B = 15.2$ s), Fig. 2-A and 2-B. It is worth highlighting that even in the Symmetric Audio-Only minigame, Fig. 2-E, sighted participants were on average 9 s faster ($M_S = 22.3$ s vs $M_B = 31,8$ s), which is likely to result from the demographic difference in the expertise and frequency of play.

We now present the themes that emerged from our qualitative analysis, highlighting participants' reflections on fairness, competition, and their experiences with the a/symmetric design choices.

Designing a Mixed-Visual-Ability Game. Blind players were required to rely heavily on their memory to keep track of crucial information, such as their current location, objectives, and challenges. This information is quickly perceived visually (e.g. entering the game with knobs and recalling the interaction method, Fig. 2-C), while audio requires memorization. The variety added by having multiple challenge types forced players to switch between gameplay interactions exacerbating the issue. In principle, while all challenges were designed to require the same proficiency and potentiate equal performance, it is only true when we consider them in isolation.

Similarly, while we made efforts to ensure fair navigation by meticulously ensuring that the same information is presented, at the same time, and equal path time travel, the persistent nature of visual feedback contrasts with the one-time audio cue announcements (e.g., room name when entering).

While designing for a balanced experience equal information and feedback can make the experience accessible and fair from the strict concept of equal information, it inadvertently creates an imbalance in the demands towards players.

Table 1. Minigames Fairness Questionnaire. Participants rated fairness on a scale of 1, "In Disadvantage", to 5, "In Advantage".

Participants	AC (SD)	AI (SD)	S (SD)	SAO (SD)
Sighted AVG	2,8 (1,1)	4,5 (0,5)	4,0 (0,8)	3,0 (0,0)
Blind AVG	2,8 (0,4)	3,0 (0,0)	2,5 (0,5)	3,5 (0,9)
Overall AVG	2,8 (0,9)	3,9 (0,8)	3,4 (1,0)	3,2 (0,6)

Fairness Perspectives. Regarding the game as a whole, some sighted participants perceived they had an advantage throughout the game, while others considered the game a fair experience. During the game, the scoreboard was hidden by default with only two sighted (S4 and S6) and one blind (B1) player using it. As such, some sighted participants believed they were falling behind.

"Overall, I think it was quite balanced. We were almost neck and neck. Actually, I thought the other person was ahead." (S1)

On the other hand, blind participants generally perceived the game as a balanced experience and attributed any differences in performance to a lack of training and familiarity with the game.

"I think it's very similar to the objective of both me and the person on the other side... adjusted to the needs of each individual." (B3)

However, participants had different perspectives when discussing the challenges (Table 1).

Symmetric Audio-Only Challenge is Fair but Visually Unappealing. The symmetric audio-only (i.e. Paring Cards), Fig. 2-E, with audio-only play was considered the fairest by all, except B4, who believed he had an advantage due to his familiarity with audio-based games. Importantly, not all sighted participants enjoyed the challenge due to a lack of visuals reiterating findings from past work [19].

Visual Feedback Perceived as an Advantage by Sighted Players in Symmetric Challenges. When the challenge was the same (i.e. Moving Goal, and Isolate Pattern), Fig. 2-D and 2-C, it did not matter how the audio feedback was designed, sighted players believed to always have an advantage due to visual feedback being faster to interpret than audio.

"Perhaps a blind person can play that game flawlessly. I couldn't, nor could I see how my progress would improve." (S5) when discussing the blind players' Isolate Pattern minigame, Fig. 2-C

"If he could discern the fire and wind like I visually see, [...] it would be exactly the same (in terms of fairness)." (S5)

Conversely, for blind players, asymmetric interface (i.e. Moving Goal), Fig. 2-D, was unanimously considered fair. We do not fully understand whether this discrepancy is an effect of sighted players being able to test the audio-base gameplay, or if it stems from sighted players' pre-assumptions of the performance of blind players (or themselves) with audio-only versus visual gameplay.

Information Availability. When discussing the Symmetric challenge (i.e. Isolate Pattern), Fig. 2-C, opinions were split among blind players.

> "I felt more disadvantaged. The game utilizes vision, whereas I rely solely on hearing." (B2)

Despite sighted players having access to more feedback and consequently more available information, two blind players commented that sighted individuals often disregard the auditory stimulus. Consequently, they considered that sighted players having both visual and audio feedback did not necessarily confer an advantage.

This points to design opportunities where the audio becomes an essential and sufficient part of the challenge but complemented by redundant and/or non-essential visuals (e.g., Isolate the Pattern could have the knobs visually turning to different sides, but without the transparency to isolate a pattern).

Asymmetry of Challenge was Perceived as Fair. On average, the asymmetry of challenge tended to be considered the fairest ($M_O = 2.8$, $SD_O = 0.9$) (Table 1). However, these results come from a disparity in the views of sighted participants, with three believing they had the disadvantage, two equal, and one the advantage. In contrast, only one blind participant believed he was at a disadvantage. Participants shared an overall perception that different challenges, as long as they are designed based on individual abilities, are fair, even if different. Depending on design choices, they may favor one player or the other, but by principle, it is the choice that should be made.

> "It's about **equivalence**, arriving through audio or through images. But ultimately, the goal is the same. Instead of silhouettes, we have sounds." (B1)

This further highlights the promise of an asymmetric-based design that caters to individual abilities rather than accommodating them via changes in interface and feedback.

Competitiveness and Transparency in Mixed-Ability Play. Some sighted participants disclosed they would have been less competitive and put less effort into winning if they knew their opponent's visual disability. This attitude contrasts with the perspective of blind participants, who saw themselves as equally capable competitors in mixed-ability gaming. This highlights the tension regarding the disclosure of disability in multiplayer games and how it can impact the experience of all involved players. We believe this type of mixed-ability game represents an opportunity to educate people to overcome their misconceptions about the abilities of blind players.

5   Discussion

Asymmetric competitive mixed-ability gaming can and did create an engaging experience for pairs. Players had a positive experience and expressed their appreciation for the opportunity to engage and interact with individuals with different visual abilities (RQ1). In line with prior work, our findings also reiterate: 1) the lack of games that can be enjoyed by both people with and without visual impairments [34,40,49]; 2) how audio games are not appealing to sighted play [4,19]; and 3) how simplifying games for equivalence and accessibility can reduce engagement [19,49].

5.1   How Do Players Perceive Fairness in Asymmetric Mixed-Ability Competitive Gaming (RQ1)

The discrepancy in perceptions of fairness raises questions about the role of knowledge and biases in shaping players' experiences. While in competitive gaming one would expect fairness to come from symmetric play, in mixed-ability gaming that is only consensual and achieved by creating the *de facto* audio games that sighted players will not engage with; the best alternative appears to be relying on the **asymmetry of challenge**. While it may be tempting to rely on **asymmetry of interface**, it will more likely inadvertently produce imbalances and will restrict the complexity of games capable of delivering equivalent challenges. Such constraints could result in simpler experiences that may not be equally engaging for all participants.

In particular, there are open challenges in designing asymmetric competitive games where players have game mechanics that directly interact with the other players' gameplay. Moreover, we only explored perceptions of pairs of players, but how can we scale mixed-ability gaming to mixed-ability groups composed of a varied number of players with and without disabilities? How can we design experiences that are balanced and perceived as fair by all?

Lastly, we believe the bias displayed by sighted players presents an opportunity to design thought-provoking games that purposely create imbalanced experiences (e.g., using asymmetric gaming or dynamic difficulty adjustment [6]). These designs can leverage sighted players' perceptions of their advantage to educate them about the fairness of play with people with disabilities.

5.2   Asymmetry for Competitive Mixed-Ability Play (RQ2)

Asymmetry in competitive mixed-ability games brings forth unique and dynamic gameplay experiences. However, ensuring a fair, balanced, and engaging gaming environment requires addressing the asymmetry between players with different abilities.

Players are not defined by their disabilities, and misconstruing disability with game skill or knowledge might lead to players with disabilities and their opponents being evenly matched when their skills are not. This may not be a problem if the goal is to maximize enjoyment and challenge [6], but in competitive games where skill is expected to be rewarded (e.g., leaderboards and ranking systems), this raises additional challenges. In traditional sports, we isolate competitors based on their disabilities (e.g., Paralympics) and although they are considered fair, they do not provide the ability for mixed-ability engagement. We believe that in gaming, we can do more to create fair, competitive experiences without isolating players by leveraging asymmetry. However, not conflating skill with disability remains an open challenge.

In competitive mixed-ability play, incorporating asymmetry allows both players to feel **included**, **challenged**, and **engaged**, providing **equity** in the gaming experience.

5.3   Tensions Between Transparency and Competitiveness

Online worlds can be a great place for disabilities to be invisible (by choice) [33,50]. Equally important is representation [33,50] without being a target of stereotypes that affect the experience or people's behaviors.

In mixed-ability asymmetric gaming, disability disclosure (i.e. making disabilities in/visible) can have consequences on the types of interaction between players that we can design. For example, how can we design games with matchmaking mechanisms for mixed-ability competitive games? Should we strive for transparency at all costs, or should we empower people with impairments to make the choice of disclosure? While disclosure is attributed to the user in cooperative and online worlds, should this remain a choice in competitive games where they are potentially competing in asymmetric challenges? Or should it be at least be perceivable in the ways in which each player is competing? There are tensions yet to be untangled between transparency and disclosure in mixed-ability competitive games that warrant further exploration.

6   Limitations

Despite the significant effort and time invested in developing the prototype, the gameplay experience may still differ from what participants are accustomed to in commercial games. The amount of training time participants had was short given the game's complexity, even though we attempted to mitigate this by sending the minigames in advance. The symmetric audio-only condition had a display of how many of the pairs of cards were completed for debugging and tracking purposes, and while it does not provide any feedback relevant to which or where the pair of cards were it can have influenced sighted participants' perceptions about the audio game. We also recognize that the design of each minigame cannot be disentangled from the results discussed. Additionally, sighted participants were, on average, 24 years old, compared to 37 for blind players, and played games more regularly, which may mean the differences in performance are due to skill and not an imbalance in the designed mechanics. The study's small sample size and diverse pool of participants should be considered when interpreting the results, as they are not meant to be generalizable, but a first exploration into the perceptions of mixed-ability competitive gaming. Our work contributes with insights into how players perceive fairness in symmetric/asymmetric competitive mixed-ability gaming, as well as the tensions of disability disclosure. We believe these findings pave the way for future research and development in inclusive play.

7   Conclusion

Gaming is an important part of many people's social lives, and we should strive to ensure gaming is an inclusive space for all. Competitive gaming poses additional challenges for mixed-ability gaming, particularly when past work's most promising approach for inclusive play relied on creating asymmetric experiences, and typically, competition expects symmetric play.

Our work is a first attempt at exploring competitive games for mixed-visual-ability groups. We focused on exploring the perceptions of fairness and engagement concerning asymmetric design choices. We conducted a mixed-method study with blind and sighted participants who played a competitive game and were exposed to four types of a/symmetries. Our findings have pointed to how symmetric challenge does not necessarily align with players' expectations of fairness and engagement, given the need to create asymmetric interfaces or restrict sighted players' access to a single modality. In line with past work on cooperative mixed-ability gaming but at odds with traditional notions of fairness, the most promising approach for competitive scenarios is to assume a high degree of asymmetry and have players compete in asymmetric challenges created based on their abilities. We highlight how asymmetric design is promising for supporting competitive mixed-ability but will require more careful design to guarantee that different in-game challenges are either equally demanding or that across the whole game experience, there is a balance of what we overall demand of players. We hope this work prompts new research in innovative game design approaches that embrace asymmetry to create more engaging and inclusive competitive experiences.

Acknowledgments. We thank the anonymous reviewers for their valuable feedback. This work was supported by FCT through project "Plug n' Play: Exploring Asymmetry and Modularity for Inclusive Game Design" ref. 2022.08895.PTDC (http://doi.org/10.54499/2022.08895.PTDC), scholarships ref. 2024.02901.BD, ref. UI/BD/151178/2021 (https://doi.org/10.54499/UI/BD/151178/2021) and ref. 2022.12448.BD, the Institutional CEEC, CEECINST/00032/2018/CP1523/CT0003 (https://doi.org/10.54499/CEECINST/00032/2018/CP1523/CT0003), and the LASIGE Research Unit, ref. UIDB/00408/2020 (https://doi.org/10.54499/UIDB/00408/2020)).

Disclosure of Interests. The authors have no competing interests to declare that are relevant to the content of this article.

References

1. Abeele, V.V., Spiel, K., Nacke, L., Johnson, D., Gerling, K.: Development and validation of the player experience inventory: a scale to measure player experiences at the level of functional and psychosocial consequences. Int. J. Hum Comput Stud. **135**, 102370 (2020)
2. Andrade, R., Rogerson, M.J., Waycott, J., Baker, S., Vetere, F.: Playing blind: revealing the world of gamers with visual impairment. In: Proceedings of the 2019 CHI Conference on Human Factors in Computing Systems, pp. 1–14 (2019)
3. Araújo, M.C.C., Façanha, A.R., Darin, T.G.R., Sánchez, J., Andrade, R.M.C., Viana, W.: Mobile audio games accessibility evaluation for users who are blind. In: Antona, M., Stephanidis, C. (eds.) UAHCI 2017. LNCS, vol. 10278, pp. 242–259. Springer, Cham (2017). https://doi.org/10.1007/978-3-319-58703-5_18
4. Archambault, D., Ossmann, R., Gaudy, T., Miesenberger, K.: Computer games and visually impaired people. Upgrade **8**(2), 43–53 (2007)
5. Atkinson, M.T., Gucukoglu, S., Machin, C.H.C., Lawrence, A.E.: Making the mainstream accessible: redefining the game. In: Proceedings of the 2006 ACM SIGGRAPH symposium on Videogames, Sandbox 2006, pp. 21–28. Association for Computing Machinery, New York (2006). https://doi.org/10.1145/1183316.1183321
6. Baldwin, A., Johnson, D., Wyeth, P., Sweetser, P.: A framework of dynamic difficulty adjustment in competitive multiplayer video games. In: 2013 IEEE International Games Innovation Conference (IGIC), pp. 16–19. IEEE (2013)
7. Bierre, K., Chetwynd, J., Ellis, B., Hinn, D.M., Ludi, S., Westin, T.: Game not over: accessibility issues in video games. In: Proceedings of the 3rd International Conference on Universal Access in Human-Computer Interaction, pp. 22–27 (2005)
8. Braun, V., Clarke, V.: One size fits all? What counts as quality practice in (reflexive) thematic analysis? Qual. Res. Psychol. **18**(3), 328–352 (2021). https://doi.org/10.1080/14780887.2020.1769238
9. Cheung, G., Huang, J.: Starcraft from the stands: understanding the game spectator. In: Proceedings of the SIGCHI Conference on Human Factors in Computing Systems, pp. 763–772 (2011)
10. Csapó, A., Wersényi, G., Nagy, H., Stockman, T.: A survey of assistive technologies and applications for blind users on mobile platforms: a review and foundation for research. J. Multimodal User Interfaces **9**(4), 275–286 (2015). https://doi.org/10.1007/s12193-015-0182-7
11. Depping, A.E., Mandryk, R.L., Li, C., Gutwin, C., Vicencio-Moreira, R.: How disclosing skill assistance affects play experience in a multiplayer first-person shooter game. In: Proceedings of the 2016 CHI Conference on Human Factors in Computing Systems, CHI 2016, pp. 3462–3472. Association for Computing Machinery, New York (2016). https://doi.org/10.1145/2858036.2858156
12. Dog, N.: The Last of Us Part II. Digital game [Playstation] (2020). https://www.playstation.com/en-us/games/the-last-of-us-part-ii/accessibility/
13. Friberg, J., Gärdenfors, D.: Audio games: new perspectives on game audio. In: Proceedings of the 2004 ACM SIGCHI International Conference on Advances in Computer Entertainment Technology, ACE 2004, pp. 148–154. Association for Computing Machinery, New York (2004). https://doi.org/10.1145/1067343.1067361
14. Fullerton, T.: Game Design Workshop: A Playcentric Approach to Creating Innovative Games. CRC Press (2014). https://doi.org/10.1201/b22309

15. Gerling, K., Buttrick, L.: Last tank rolling: exploring shared motion-based play to empower persons using wheelchairs. In: Proceedings of the First ACM SIGCHI Annual Symposium on Computer-Human Interaction in Play, CHI PLAY 2014, pp. 415–416. Association for Computing Machinery, New York (2014). https://doi.org/10.1145/2658537.2661303

16. Ghost Town Games Ltd., T.: Overcooked! 2. Digital game [Nintendo Switch] (2018). https://www.team17.com/games/overcooked-2/

17. Gonçalves, D., Piçarra, M., Pais, P., Guerreiro, J., Rodrigues, A.: "my zelda cane": strategies used by blind players to play visual-centric digital games. In: Proceedings of the 2023 CHI Conference on Human Factors in Computing Systems, pp. 1–15 (2023)

18. Gonçalves, D., Pais, P., Gerling, K., Guerreiro, T., Rodrigues, A.: Social gaming: a systematic review. Comput. Hum. Behav. **147**, 107851 (2023)

19. Gonçalves, D., Rodrigues, A., Guerreiro, T.: Playing with others: depicting multiplayer gaming experiences of people with visual impairments. In: The 22nd International ACM SIGACCESS Conference on Computers and Accessibility, ASSETS 2020, pp. 1–12. Association for Computing Machinery, New York (2020). https://doi.org/10.1145/3373625.3418304

20. Gonçalves, D., Rodrigues, A., Richardson, M.L., de Sousa, A.A., Proulx, M.J., Guerreiro, T.: Exploring asymmetric roles in mixed-ability gaming. In: Proceedings of the 2021 CHI Conference on Human Factors in Computing Systems, vol. 114, pp. 1–14. Association for Computing Machinery, New York (2021). https://doi.org/10.1145/3411764.3445494

21. Grabski, A., Toni, T., Zigrand, T., Weller, R., Zachmann, G.: Kinaptic - techniques and insights for creating competitive accessible 3D games for sighted and visually impaired users. In: 2016 IEEE Haptics Symposium (HAPTICS), pp. 325–331 (2016). https://doi.org/10.1109/HAPTICS.2016.7463198

22. Grammenos, D., Savidis, A., Stephanidis, C.: UA-chess: a universally accessible board game, p. 11

23. Grammenos, D.: Game over: learning by dying. In: Proceedings of the SIGCHI Conference on Human Factors in Computing Systems, CHI 2008, pp. 1443–1452. Association for Computing Machinery, New York (2008). https://doi.org/10.1145/1357054.1357281

24. Grammenos, D., Savidis, A., Georgalis, Y., Stephanidis, C.: Access invaders: developing a universally accessible action game. In: Miesenberger, K., Klaus, J., Zagler, W.L., Karshmer, A.I. (eds.) ICCHP 2006. LNCS, vol. 4061, pp. 388–395. Springer, Heidelberg (2006). https://doi.org/10.1007/11788713_58

25. Grammenos, D., Savidis, A., Stephanidis, C.: Designing universally accessible games. Comput. Entertainment **7**(1), 8:1–8:29 (2009). https://doi.org/10.1145/1486508.1486516

26. Gugenheimer, J., Stemasov, E., Frommel, J., Rukzio, E.: ShareVR: enabling co-located experiences for virtual reality between HMD and non-HMD users. In: Proceedings of the 2017 CHI Conference on Human Factors in Computing Systems, pp. 4021–4033 (2017)

27. Gugenheimer, J., Stemasov, E., Sareen, H., Rukzio, E.: Facedisplay: towards asymmetric multi-user interaction for nomadic virtual reality. In: Proceedings of the 2018 CHI Conference on Human Factors in Computing Systems, pp. 1–13 (2018)

28. Hamari, J., Sjöblom, M.: What is esports and why do people watch it? Internet Research (2017)

29. Harris, J., Hancock, M., Scott, S.D.: Leveraging asymmetries in multiplayer games: investigating design elements of interdependent play. In: Proceedings of the 2016 Annual Symposium on Computer-Human Interaction in Play, CHI PLAY 2016, pp. 350–361. Association for Computing Machinery, New York (2016). https://doi.org/10.1145/2967934.2968113

30. IllFonic: Predator: Hunting Grounds. Digital game [Microsoft Windows] (2020). https://predator.illfonic.com/

31. Infinity Ward, S.: Call of Duty: Modern Warfare 3. Digital game [Microsoft Windows] (2023). https://www.callofduty.com/en/modernwarfare3

32. Kirke, A.: When the soundtrack is the game: from audio-games to gaming the music. In: Emotion in Video Game Soundtracking, pp. 65–83 (2018)

33. Mack, K., Hsu, R.C.L., Monroy-Hernández, A., Smith, B.A., Liu, F.: Towards inclusive avatars: disability representation in avatar platforms. In: Proceedings of the 2023 CHI Conference on Human Factors in Computing Systems, CHI 2023. Association for Computing Machinery, New York (2023). https://doi.org/10.1145/3544548.3581481

34. Miesenberger, K., Ossmann, R., Archambault, D., Searle, G., Holzinger, A.: More than just a game: accessibility in computer games. In: Holzinger, A. (ed.) USAB 2008. LNCS, vol. 5298, pp. 247–260. Springer, Heidelberg (2008). https://doi.org/10.1007/978-3-540-89350-9_18

35. Musick, G., Freeman, G., McNeese, N.J.: Gaming as family time: digital game co-play in modern parent-child relationships. Proc. ACM Hum.-Comput. Interact. 5(CHI PLAY), 1–25 (2021)

36. Nair, V., et al.: NavStick: making video games blind-accessible via the ability to look around. In: The 34th Annual ACM Symposium on User Interface Software and Technology, pp. 538–551. ACM, Virtual Event USA (2021). https://doi.org/10.1145/3472749.3474768

37. Nair, V., et al.: Uncovering visually impaired gamers' preferences for spatial awareness tools within video games. In: Proceedings of the 24th International ACM SIGACCESS Conference on Computers and Accessibility, ASSETS 2022. Association for Computing Machinery, New York (2022). https://doi.org/10.1145/3517428.3544802

38. Ossmann, R., Miesenberger, K., Archambault, D.: A computer game designed for all. In: Miesenberger, K., Klaus, J., Zagler, W., Karshmer, A. (eds.) ICCHP 2008. LNCS, vol. 5105, pp. 585–592. Springer, Heidelberg (2008). https://doi.org/10.1007/978-3-540-70540-6_83

39. Piçarra, M., Rodrigues, A., Guerreiro, J.: Evaluating accessible navigation for blind people in virtual environments. In: Extended Abstracts of the 2023 CHI Conference on Human Factors in Computing Systems, CHI EA 2023. Association for Computing Machinery, New York (2023). https://doi.org/10.1145/3544549.3585813

40. Porter, J.R., Kientz, J.A.: An empirical study of issues and barriers to mainstream video game accessibility. In: Proceedings of the 15th International ACM SIGACCESS Conference on Computers and Accessibility, ASSETS 2013. Association for Computing Machinery, New York (2013). https://doi.org/10.1145/2513383.2513444

41. Pragma: Crazy Party. Digital game [Microsoft Windows] (2016). http://pragmapragma.free.fr/crazy-party/en/index.php

42. Smith, B.A., Nayar, S.K.: The RAD: making racing games equivalently accessible to people who are blind. In: Proceedings of the 2018 CHI Conference on Human Factors in Computing Systems, pp. 1–12. Association for Computing Machinery, New York (2018). https://doi.org/10.1145/3173574.3174090

43. Sobel, K., Bhattacharya, A., Hiniker, A., Lee, J.H., Kientz, J.A., Yip, J.C.: It wasn't really about the Pokémon: parents' perspectives on a location-based mobile game. In: Proceedings of the 2017 CHI Conference on Human Factors in Computing Systems, pp. 1483–1496 (2017)
44. Studio, S.M.: God of War Ragnarök. Digital game [Playstation] (2022). https://www.playstation.com/en-us/games/god-of-war-ragnarok/accessibility/
45. Turn 10 Studios, Playground Games, S.: Forza Motorsport. Digital game [Xbox] (2023). https://www.xbox.com/en-us/games/forza-motorsport
46. VGStorm: Manamon. Digital game [Microsoft Windows] (2016). http://www.vgstorm.com/manamon.php
47. Wen, J., Kow, Y.M., Chen, Y.: Online games and family ties: Influences of social networking game on family relationship. In: Human-Computer Interaction–INTERACT 2011: 13th IFIP TC 13 International Conference, Lisbon, Portugal, September 5-9, 2011, Proceedings, Part III 13. pp. 250–264. Springer (2011)
48. Yuan, B., Folmer, E.: Blind hero: enabling guitar hero for the visually impaired. In: Proceedings of the 10th International ACM SIGACCESS Conference on Computers and Accessibility, Assets 2008, pp. 169–176. Association for Computing Machinery, New York (2008). https://doi.org/10.1145/1414471.1414503
49. Yuan, B., Folmer, E., Harris, F.C.: Game accessibility: a survey. Univ. Access Inf. Soc. **10**(1), 81–100 (2011). https://doi.org/10.1007/s10209-010-0189-5https://doi.org/10.1007/s10209-010-0189-5
50. Zhang, K., Deldari, E., Lu, Z., Yao, Y., Zhao, Y.: "It's just part of me": understanding avatar diversity and self-presentation of people with disabilities in social virtual reality. In: Proceedings of the 24th International ACM SIGACCESS Conference on Computers and Accessibility, ASSETS 2022. Association for Computing Machinery, New York (2022). https://doi.org/10.1145/3517428.3544829

Behind Two Stories to Tell: An Evaluation Study of the Camera Perspective in the Game Mutation Madness

Liliana Vale Costa[(✉)] [iD], Ana Passos[iD], and Nelson Zagalo[iD]

Department of Communication and Art, DigiMedia Research Centre, University of Aveiro, Aveiro, Portugal
{lilianavale,apassos,nzagalo}@ua.pt

Abstract. A major cinematic component in game storytelling is the camera, offering the player a perspective towards the environment. Although the camera use in game storytelling has been undoubtedly important, there has been lack of understanding of its implications in narrative comprehension and relatedness to the game world. Forty-six young adults aged between 18 and 35 were selected to participate in an A/B evaluation of an in-game cinematic scene that was shown from a first- and third-person camera perspective, aimed to assess the effect of the Point of View (POV) camera on visual attention and story comprehension in the game Mutation Madness. Participants' self-reported evaluations of the scene were collected, together with eye-tracking recordings, suggesting that whereas the third-person camera perspective in game cinematics is suitable to show the omniscient knowledge of the story and evoke the sense of time and the agents of the story, the first-person camera perspective is used to guide the player to game events. The implications of camera use in the narrative-gameplay experience based on the POV analysis, visual attention and story comprehension are discussed.

Keywords: Game cinematics · Storytelling · Camera Perspective · Visual attention

1 Introduction

As games have been regarded as a medium of artistic expression [1], filmmaking and audiovisual language have been key to convey the game narrative and contribute to both positive spectatorship and game-playing experiences. Game cinematics, i.e., the use of cutscenes, in-game animations, or full-motion videos, are often used to introduce the game story and communicate its progress, establish a link between game levels, and set the gameplay rhythm.

Of particular importance in (re)telling the game narrative and reinforce the holistic spectator-player experience is the use of camera. For instance, fixed camera angles may constraint freedom and movements leading to an 'on-rails' experience, whilst the perspective can communicate the sense of agency, and the focus may draw the attention to important game artefacts, among others. Visual attention is also expected to develop

A. Marto et al. (Eds.): VJ 2024, CCIS 2324, pp. 40–51, 2025.
https://doi.org/10.1007/978-3-031-81713-7_3

differently while using first- and third- camera perspectives [2]. In other words, camera perspective may also affect the way players process and select information towards a certain environment [3].

Despite the increasing interest and body of literature on the use of cinematics in game-playing [4–6], there have been few empirical investigations of eye-tracking analysis in game cinematics [7]. Such knowledge would be extremely important for game designers and developers to enhance the emotional and immersive aspects of gameplay, linking them to classic filmmaking techniques such as shot sizes, angles, and movements.

Eye-tracking is a method that uses the eye-tracker device to measure eye gaze (fixation) and movement (saccades, regressions). Whilst fixations occur when an event or object grabs the players' attention holding their gaze for a longer time whether in interest or confusion [8], saccades are the rapid eye movements performed to avoid blur and these are used to understand the sequence of this motion. These latter may also include regressions to indicate processing difficulties. Pupil diameter is also a measure that can be used to assess either emotional arousal measures and/or mental workload.

Beyond the lack of these measurements in assessing the game-viewer experience in cinematics, little is known about the differences in cinematics production in the case of the first- and third-person camera perspectives. Measuring the impact of these on player focus and narrative engagement is important to assess the hypothesis suggesting that first-person camera perspective leads to increased focus on gameplay, while third-person camera perspective may allow for greater attention to dramatic elements, including the narrative and the game character.

The purpose of this paper is to assess the effect of the Point of View (POV) camera on visual attention and story comprehension in the case of the developed game entitled Mutation Madness. For that, this study will examine the contributions of cinematics to affect story comprehension, especially in young adults by analyzing the different strategies adopted in in-game cinematics, developing an A/B experiment of game cinematics production with a first- and third- person camera perspective, and assessing their visual attention and possible effects of the game-experience to narrative comprehension.

2 Related Work

In the light of the demand for game storytelling and the emergence of new formats (e.g., cinematic virtual reality [8] and alternate reality games), the viewer-player experience and dual role of gamer-spectators have led to an increasingly need to (re)think the articulation between filmmaking and gameplay strategies to provide a pleasant game experience. When designing and incorporating cinematics in the game design workflow, both filmmaking strategies and interactivity are essential to enable players' participation rather than mere passive spectatorship [9]. For that, staging, the dialogues, and backstories, editing, light setting, camera shooting, and post-processing are usually part of this process.

Camera is an important game component, in which its use may change the way players interact with the game scene and convey emotional experiences [4]. Indeed, camerawork may help creating closeness and empathic relationships with the characters whilst reinforcing the emotional intensity and enabling the player to engage with the game visuals [10].

Cozic [5] outlines the following types of camera systems for digital games:

- *Fixed Camera Views*: These game cameras are previously determined in a specific location - e.g., following the player's movement and reinforcing the sense of control. However, this type of camera view has constraints in terms of the purpose of transmitting emotions:
- *Fully Interactive Camera Systems* are relative to cameras able to rotate in different angles and dolly movements. Although this camera view gives a sense of freedom, lack of concentration and frustration with interactivity can also occur; and
- *Tracking Camera and Semi-Interactive Camera Systems* follow the player, leading him/her into a determined path.

It is worth mentioning that the player's experience and perspective towards the environment is also highly dependent on the use of different shot sizes (e.g., extreme long shot, mid-shot, close-up), angles (e.g., low angle, high angle, eye-level), movement (e.g., panning, zooming, tilting, dolly) and focus (e.g., deep, shallow, soft). Despite the camerawork is highly important to the game-playing experience, further experiments involving eye-tracking to detect the player's eye movement and gaze fixation [11] are needed.

In the context of games, both fixations and regressions (go-back fixations) have been associated to the game difficulty and UI navigation – i.e., longer fixation durations are often associated with confusion processing difficulty [2, 12]. Nevertheless, little has been explored about the differences in cinematics production in the case of the first-person or third-person camera perspectives.

Hence, the eye-tracking method was used to assess the player's visual attention to game events and narrative agents purposefully placed as areas of interest to understand the effect of the POV in the whole game experience.

## 3	The Experiment

To assess the effect of the POV camera on visual attention and story comprehension in the game Mutation Madness, a comparative analysis of game cinematics with first- and third person camera perspective was performed involving 46 young adults aged between 18 and 35 years old. This age bracket is very representative in the video gamer community [13], being this the target group of the game Mutation Madness,[1] developed under the research project PLAYMUTATION (https://playmutation.web.ua.pt/). The purpose of this project is to analyze the use of digital games to inform about the evolution of viral genomes, transmission, and mutations, as well as to promote self-care and infection prevention in young adults.

The Eye Tracking tests took place during July 2022 at the cinematics of the game Mutation Madness following the A/B evaluation design. In specific, two groups were randomly assigned to Experiment A – test the game cinematics using first-person camera perspective and then third-person camera perspective and Experiment B – test the game

---

[1] Mutation Madness, https://store.steampowered.com/app/2571650/Mutation_Madness/ (Last access: October 29th, 2024).

cinematics using third-person camera perspective and then first-person camera perspective. The participants' attention to pre-established areas of interest - game events and narrative events - was evaluated, using fixation time.

The difference between the two conditions was solely the use of the camera perspective and the areas of interest are presented on the screen for the same time in both conditions. There were, however, some difficulties in taking the same proportion of the screen size given the camera proximity depending on the perspective and these may affect the results obtained. However, eye-behaviors were complemented with participants' self-reports. It is worth mentioning that the areas of interest were determined by two researchers in the domain of digital games and established the elements essential to the gameplay and to understand the narrative. One of the researchers has been the designer and cinematic artist of the game Mutation Madness.

During the experiment, some questions were posed to participants, aiming to assess memorability relative to game and narrative events (e.g., What are the locations that Mike has to visit to fight the virus? What is the name of the Mike's cat?). After each experiment, participants answered a questionnaire about their preferences relative to their POV experience (namely the Game World, Characters, Emotion, and Looks and preferences). The eye-tracking planning structure is shown in Fig. 2.

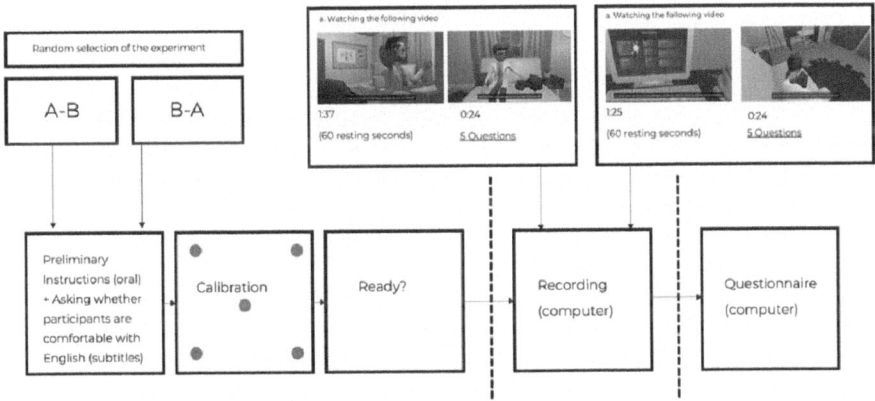

Fig. 2. Eye-tracking Testing Planning Structure

As shown in Fig. 2, the participants visualized the first- and the third-person camera perspective versions of the cinematics, and only after the experiment, they played the game Mutation Madness. No prior contact with the game was essential to not influence the results relative to memorability and visual attention. Considering this latter variable (visual attention) is an important aspect to understand the game information, the objective, and establish a sense of relatedness with the game world, eye-tracking data was used to assess it. Moreover, a sense of agency is enabled when used in the first-person camera perspective and, as such, this perspective is suggested to get the player more focused on the gameplay rather than on narrative elements.

The following section describes the procedures for the eye-tracking experiment.

3.1 The Eye-Tracking Experiment

The purpose of the eye tracking experiment was to identify eye movement and gaze fixation of the players in cinematics, using first- and third- person camera perspective. The eye-tracker is a useful tool of quantitative and qualitative data collection, giving further insight of the participants' experiences and cognitive processes [11, 14]. Additionally, it allows for the communication of product qualities that may draw or repel consumer attention [15, 16].

To understand how the POV affects the experience of game cinematics, areas of interest (AOIs) were set up. These AOIs can be defined as visually distinct image areas to assess the composition's elements' visual relevance [17]. The defined AOIs indicating the critical data for gameplay events and narrative agents are shown in Fig. 3.

As shown in Fig. 3, the AOIs were manually draw in specific video frames to assess the player's visual attention and information retention to gameplay events and narrative agents using first- and third- person camera perspective.

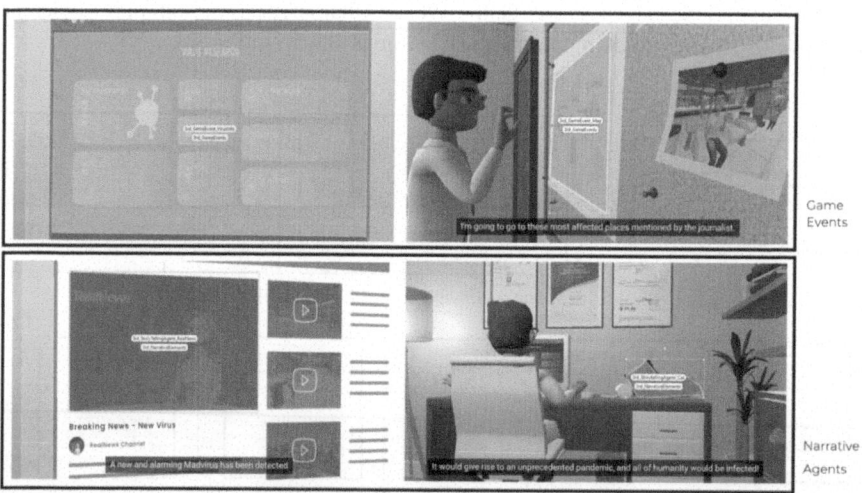

Fig. 3. Areas of Interest for Game Events and Narrative Agents

Whilst the term 'Gameplay events' is used to refer to the player's actions during gameplay (e.g., information to unlock about the virus, game map), the term 'narrative agents' is used to address the game characters and general aspects of the narrative.

Then, a post-experience questionnaire was administered to further collect the participants' insights on the game's cinematics.

Data collected during the study followed the procedures: (a) consent form; (b) voluntary participation; and (c) confidentiality and anonymity. When performing the eye-tracking tests, participants were informed about the procedures and purposes for data collection and analyses.

3.2 Participants

A convenience sample was used in the players' testing. The selected participants were based on proximity and had no restrictions in participating, other than age (aged over 18) given the target group of the game Mutation Madness and requiring normal or corrected-to-normal vision. Exclusion criteria included age that did not fit within the group of young adults (aged between 18 and 35), and colorblindness.

3.3 Apparatus

The model of the eye-tracker used was Tobii Pro X3- with a sample of 120 Hz and the software for data analysis was Tobii Studio - 3.4.8. The cinematics analysis was done using a monitor, a z230 workstation, hp 2311× display, with a screen resolution of 1920 × 1080 px and the speakers used for sound. Tracking was binocular with gaze accuracy of 0.6 at 30° gaze angle and 0.24 gaze precision. The eye-tracking technique was corneal reflection, and the stimulus size was presented in full screen.

3.4 Procedures

When viewing the cinematics, the players were approximately 500 mm (19.7 inches) away from the monitor. The calibration was done by following the movement of a red dot appearing on the screen. During the experiment, participants were asked to maintain a fixed, aligned posture, and no chinrest was used. Eye Tracking tests took between 20 and 30 min to complete. A 5-point calibration system was made.

3.5 Data Collection and Analysis

Participants' perception on the use different camera perspectives (first-person camera perspective and third-person camera perspective) was assessed in terms of the following dimensions: Game World, Characters, Emotion, and Looks and Preferences.

The questionnaire was created by the authors considering these important dimensions relative to the use of cameras in games and for these dimensions, participants were asked to assign either the first-person or third-person camera perspective:

- *Game World*: I felt like I was in the environment that was presented to me; I felt more in control of the environment that was presented to me; I felt greater relation with the presented and surrounding scenery.
- *Characters*: I felt closer to the character…; I identified more with the character…; It was easier to take on the personality of the character…; I felt more in control of what was unfolding…
- *Emotion:* I felt a greater sense of freedom…; I established a greater relationship with the plot and story…; I felt more disoriented…; and It was easier to follow the rhythm of the action…
- *Look and Preferences*: I prefer the appearance of…; The option that best suited this game was…

In terms of the game world, the sense of presence ("feel like being in the environment"), agency ("being in control") and awareness and relatedness with the environment were selected given that these seem to be interrelated with game immersion, namely with the perception of the surroundings, and control over actions [18, 19].

Relative to emotion, these variables were related to the sense of attachment to the character ("feel closer to the character", identification with the character, the sense of a parasocial relation), embodiment ("take on the personality", avatar) and sense of control [20, 21].

Finally, the category emotions allude to the camera navigability – i.e., free movement ("sense of freedom"), disorientation ("feel disorientation"), and rhythm of the action as well as the alignment with the plot and the story.

3.6   Self-reported Data on the POV Experience

To cross the participants' visual attention with information retention, first- and third-person camera perspective stimuli were presented, and the following questions were asked relative to essential elements to the gameplay (memorability) – e.g., *What are the names of the locations that Mike has to visit to fight the virus? What information does Mike have to collect about the virus?* and questions relative to narrative agents – i.e., *What is the name of Mike's cat? What is the name of the news channel that broadcasts the information about the virus? What is Mike's profession?* The goal of asking these questions were to assess the memorability to elements of gameplay in comparison with storytelling.

After the experiment, participants filled a questionnaire to assess the adequacy of first- and third- person camera perspective on establishing the game world, characters, emotions, looks and preferences. Table 1 illustrates the statements in which participants had to assign first-person or third-person camera perspectives that would best suit to their experience.

When selecting the terms 'First-person' or 'Third-person' for each statement according to which best suited their opinion, they were asked to compare each POV in terms of sense of connection and/or detachment and justify:

Select the terms 'First-person' or 'Third-person' according to which best suits your opinion: I felt a greater connection to the environment from the perspective...[First-person camera perspective] or [Third-person camera perspective] and more detached from the perspective [First-person camera perspective] or [Third-person camera perspective]. Justify.'

It is worth noting that participants could express their preferences for both or neither options.

Table 1. Statements to Assign the Self-reported POV Experience in a Post-Experience Questionnaire

POV Experience	First-Person	Third-Person
Game World	[]	[]
I felt like I was in the environment that was presented to me...	[]	[]
I felt more in control of the environment that was presented to me...	[]	[]
I felt greater proximity to the objects presented to me...	[]	[]
I felt greater relation with the presented and surrounding scenery...	[]	[]
Characters		
I felt closer to the character...	[]	[]
I identified more with the character...	[]	[]
It was easier to take on the personality of the character...	[]	[]
I felt more in control of what was unfolding...	[]	[]
Emotion		
I felt a greater sense of freedom...	[]	[]
I established a greater relationship with the plot and story...	[]	[]
I felt more disoriented...	[]	[]
It was easier to follow the rhythm of the action...	[]	[]
Looks and preferences		
I prefer the appearance of...	[]	[]
The option that best suited this game was...	[]	[]

4 Results

For each pre-established area of interest, fixation time was analyzed, being complemented with participants' self-reported data. Figures 4 and 5 shows the fixation duration of the participants of the Experiment A (testing the game cinematics using first-person camera perspective and then third-person camera perspective) and, then, the Experiment B (testing the game cinematics using the third-person camera perspective and then the first-person camera perspective).

Although, differences between durations about 200–300 ms may not be very significant, these slightly changes and self-reported data suggest that participants tend to rely more in narrative elements in a first-person camera perspective (e.g., cat, journalist) when compared to mise-en-scène associated to game progression (e.g., map).

Indeed, the POV confer different perceptions towards the game world, characters, emotions, and the looks and preferences of the game. Concerning the game world, the first-person camera perspective was mostly of the sensation of being there (82.61%, N = 38 participants), proximity (80.43%, N = 37), and relation (65.22%, N = 30).

First-person camera perspective also seems to be the best option when it comes to the possibility to take on the personality of the character (71.74%, N = 33).

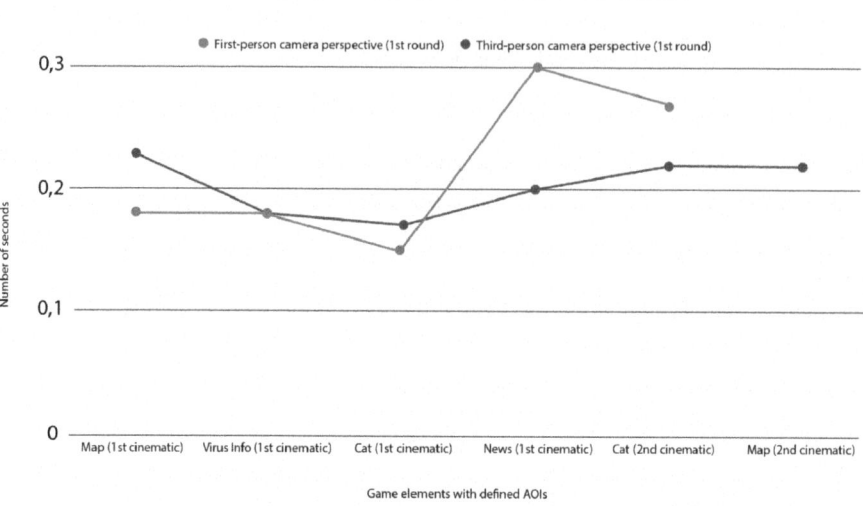

Fig. 4. Line Graph of Fixation Duration of the participants in Experiment A

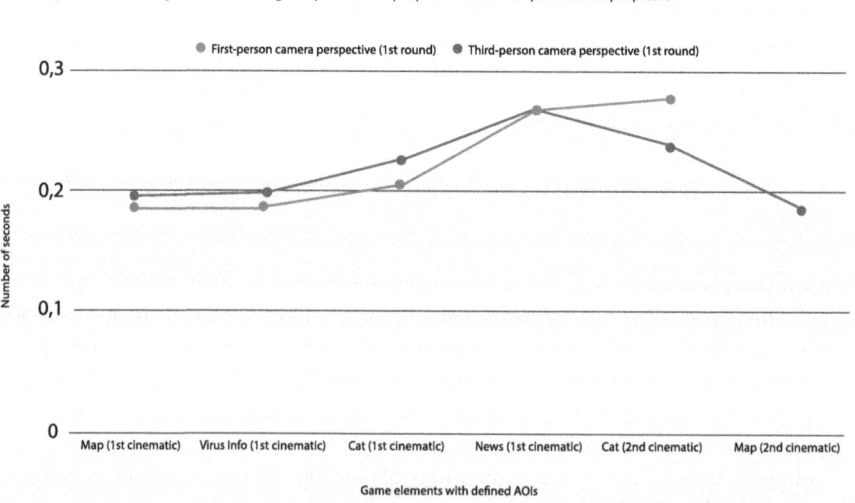

Fig. 5. Line Graph of Fixation Duration of the participants in Experiment B

The following are some of the participants' statements that illustrate these observations:

"Putting ourselves in the protagonist's perspective, we feel on the scene. In first person, we forget that we are a character in a video game and end up entering the universe of the video game."

"Being in first-person makes me feel like I am the main character in the game, making me want to help save people from the virus."

"I think that first-person environments give a realistic feel to the action."

"In first-person, I felt like it was me doing the actions, and in third-person, I felt like it was a 'movie', which I liked (...) maybe due to the fact that most of the games and narratives I watch are more in third-person and not so much in first-person."

In brief, players seem to feel a greater presence and proximity in the environment presented in first-person whilst strengthening the bond between the player and the character by assuming the character's identity. Nonetheless, disorientation also tends to occur in this perspective (73.91%, N = 34), probably owing to the non-representation of the whole game space.

When it comes to have a sense of control towards the game environment, it is mostly associated to the third perspective (65.22%, N = 30), sense of closeness to the character (54.35%, N = 25), sense of freedom (73.91%, N = 34), relationship with the plot and story (58.70%, N = 27), and game's rhythm (65.22%, N = 30).

The following are some of the participants' excerpts that illustrate these observations:

"In terms of the cutscenes, the third-person has the advantage of being more cinematic, it approaches cinema and acting, and in a cutscene, I prefer to be the spectator, in the game itself, prefer to be 'involved'."

"The third perspective allows me to have a broader sense of the context where the story is happening (...) personally, I think it makes the experience richer and more appealing."

"In the third-perspective, it is easier to be in control of objects and what surrounded the main character."

"In the third person, I had more freedom and more control of where to look (...) I had more freedom to pay attention to other aspects of the plot."

"I felt that in the third perspective I had more freedom to analyze everything that was in the scene so I could gain more information about what was being presented and connected to the storyline and world of the game.

Overall, findings reveal that the use of the first-person perspective in the game space is likely to (a) reinforce the player's presence int the game world; (b) ensure the proximity to the represented objects; (c) lack of awareness of the surroundings and, probably, a sense of disorientation/restrictions. By contrast, third-person perspective was mostly associated to an increase in the sense of freedom and control as well as detachment. These observations have important implications on the use of camera POV in different game scenarios and cinematics.

5   Conclusions and Future Work

This study set out to assess the effect of the Point of View (POV) camera on visual attention and story comprehension in the game Mutation Madness. For that, an A/B evaluation of game cinematics involving forty-six young adults aged between 18 and 35 revealed that participants tend to rely more in narrative elements in a first-person camera perspective, when compared to mise-en-scène associated to game progression.

From the participants' self-report, one may suppose that third-person camera perspective in game cinematics is suitable to show the omniscient knowledge of the story and evoke the sense of time and the agents of the story, whereas the first-person camera perspective directs more to game events.

There are, however, some limitations with this study and, as such, results should be interpreted with caution. There were some difficulties in taking the same proportion of the screen size when establishing the AOIs in the eye-tracking experiment given the camera proximity depending on the perspective and these may affect the results obtained. Self-reported data was also used, enabling us to substantiate the results and provide in-depth information.

Future work is being carried out to assess the player's visual attention to learning mechanics within the context of games.

Acknowledgments. This research was conducted in the scope of the project PLAYMUTATION Virus Epidemiologic-themed Digital Games and Youngsters' Attitudes to Viral Infections from the Digital Media and Interaction Research Center (DigiMedia), funded by Foundation for Science and Technology (FCT) in the scope of the project [UIDB/05460/2020].

References

1. Gee, J.P.: Why Game Studies Now? Video Games: A New Art Form. Games Cult. **1**(1), 1–4 (2006). https://doi.org/10.1177/1555412005281788
2. Almeida, S., Mealha, O., Veloso, A.: Video game scenery analysis with eye tracking. Entertain. Comput. **14**, 1–13 (2016)
3. Rothe, S., Buschek, D., Hußmann, H.: Guidance in cinematic virtual reality-taxonomy, research status and challenges. Multimodal Technol. Interact. **3**(1), 19 (2019)
4. Burelli, P.: Game cinematography: from camera control to player emotions. In: Emotion in Games: Theory and Praxis, pp. 181–195. Springer International Publishing, Cham (2016)
5. Cozic, L.: Automated Cinematography for Games (2007). http://eprints.mdx.ac.uk/4960/
6. Bojko, A.: Eye Tracking the User Experience: A Practical Guide to Research, 1st edn. Rosenfeld Media (2013a)
7. Dunlop, R.: Production Pipeline Fundamentals for Film and Games (2014). https://learning.oreilly.com/library/view/productionpipeline-fundamentals/9780415812290/
8. Ehmke, C., Wilson, S.: Identifying Web Usability Problems from Eye-Tracking Data, pp. 119–128. P119-Ehmke (2007)
9. Newman, R.: Cinematic Game Secrets for Creative Directors and Producers: Inspired Techniques from Industry Legends. Routledge (2009)
10. Bradley, A.: Creating God of War's Amazing Single-Shot Cinematography - Variety (2018). https://variety.com/2018/gaming/features/god-of-war-single-shot-camera-1202793441/

11. Salvucci, D.D., Goldberg, J.H.: Identifying fixations and saccades in eye-tracking protocols. In: Proceedings of the Eye Tracking Research and Applications Symposium 2000, October 2014, pp. 71–78 (2000). https://doi.org/10.1145/355017.355028

12. Winklbauer, A., Stiglbauer, B., Lankes, M., Sporn, M.: Telling eyes: linking eye-tracking indicators to affective variables. In Proceedings of the 18th International Conference on the Foundations of Digital Games, pp. 1–11 (2023)

13. VIDEOGAMESEUROPE (2023) All about Video games: Culture-Criativity-Technologies European Key Facts 2022. https://www.videogameseurope.eu/wp-content/uploads/2023/08/Video-Games-Europe_Key-Facts-2022_FINAL.pdf. Accessed 29 Sept. 2024

14. Hopper, L.M., Gulli, R.A., Howard, L.H., Kano, F., Krupenye, C., Ryan, A.M., Paukner, A.: The application of noninvasive, restraint-free eye-tracking methods for use with nonhuman primates. Behav. Res. Methods **53**(3), 1003–1030 (2020). https://doi.org/10.3758/s13428-020-01465-6

15. Granka, L.A., Joachims, T., Gay, G.: Eye-tracking analysis of user behavior in WWW search. In: Proceedings of Sheffield SIGIR - Twenty-Seventh Annual International ACM SIGIR Conference on Research and Development in Information Retrieval, pp. 478–479 (2004). https://doi.org/10.1145/1008992.1009079

16. Nielsen, J., Pernice, K.: How to Conduct Eyetracking Studies (2009). http://www.nngroup.com/reports/how-to-conduct-eyetracking-studies/

17. Bojko, A.: Eye Tracking the User Experience: A Practical Guide to Research, 1st edn. Rosenfeld Media (2013b)

18. Ryan, R.M., Rigby, C.S., Przybylski, A.: The motivational pull of video games: a self-determination theory approach. Motiv. Emot. **30**(4), 344–360 (2006)

19. Denisova, A., Cairns, P.: First person vs. third person perspective in digital games: do player preferences affect immersion? In: Proceedings of the 33rd Annual ACM Conference on Human Factors in Computing Systems, pp. 145–148 (2015)

20. Bowman, N.D., Oliver, M.B., Rogers, R., Sherrick, B., Woolley, J., Chung, M.Y.: In control or in their shoes? How character attachment differentially influences video game enjoyment and appreciation. J. Gaming Virtual Worlds **8**(1), 83–99 (2016)

21. Lankoski, P.: Embodiment in character-based videogames. In: Proceedings of the 20th International Academic Mindtrek Conference, pp. 358–365 (2016)

Serious Games

The Stakeholders Clash Game: From Board Game Design to Online Collaborative Platforms for Urban Planning

Micael Sousa[⊠] [iD]

CITTA—Research Centre for Territory, Transports and Environment, Department of Civil Engineering, University of Coimbra, Coimbra, Portugal
micaelssousa@gmail.com

Abstract. The Stakeholders Clash Game was a game design experiment that tested the implementation of modern board game mechanisms to deliver a digital, low-cost, and easy-to-use serious game approach for urban planners. The paper reports the game development process, including improvements and adaptations from playtesting it with users from the urban planning and game design field. It proposes a framework to develop other flexible, serious, game-based approaches.

Keywords: Board games · Game Design · Serious Games · Urban Planning

1 Introduction

The Covid-19 Pandemic affected almost all activities worldwide, including spatial and urban planning [1, 2]. During this period, social distance restrictions demanded new, fast-to-implement solutions. Using online video streaming and collaborative software tools was common during the pandemic social restrictions [3]. However, these tools were not designed to support collaborative spatial planning processes.

We propose a serious game design method where analogue game design elements inspired a digital game to simulate stakeholders' interactions during a hypothetical planning process. For this purpose, we create *The Stakeholders Clash* (TSC) game as a team-based game, where players assume decisions in a team of stakeholders during a generic urban development process. The game highlighted metaphors associated with each stakeholder group. We adopted a scoring system based on the urban grid size, the number of represented stakeholders, and interconnected relationships. Asymmetric scoring aims to simulate different claims and policies among citizens. It deliberatively provoked players to reflect on the roles and the overall game's relation to reality.

We used *Zoom* and *Google Slides* software in two sessions, first with planning experts and then with game design students. The TSC's serious game resulted from a codesign development process, allowing us to propose a method new for games.

A. Marto et al. (Eds.): VJ 2024, CCIS 2324, pp. 55–69, 2025.
https://doi.org/10.1007/978-3-031-81713-7_4

1.1 Online Serious Planning Games

Games can be ways to engage and attract citizens, stakeholders, and experts to spatial planning processes [4, 5]. However, game usage is not as widespread in spatial planning as it might be. Practitioners seem to have difficulties using game-based planning in practice [6–8]. Despite this, there is growing literature on using games for spatial planning [9]. Authors like Tan [10, 11] are among the most cited for collaborative urban design, now also resorting to digital apps [12], demanding considerable resources to implement the required physical and digital support tools. Developing the games can be complex and expensive [13, 14], partly explaining why entertainment games that simulate city development have been explored directly [15].

Another alternative is exploring analogue games, which are easier to adapt and develop [16–18]. Several examples appear in books that aim to disseminate game-based techniques for urban planning and democratic decision-making [10, 19–21]. These games result from using some game mechanics and planning over maps with coloured cubes, bits, and strings [22–24]. Although they might seem simple, the selection of game mechanics influences the output of a serious game and the users' experience [8, 25]. Using the same flexible approaches of analogue games in digital online tools seems promising. These design solutions provide inexpensive games while implementing codesign techniques like those recommended by Champlin et al. [26] that would engage users and adapt the game to efficient participation and results. Sousa [27, 28] tested similar online solutions for collaborative decision-making. However, they did not involve complex rules, win-and-lose conditions, or graphical urban models.

2 Game Development Methodology

The game TSC resulted from several design stages and playtest sessions defined in a sequence in Fig. 1, following Champlin et al. [26] recommendations (progressive codesign with experts aiming for serious game goals). The first game idea (preliminary game design) is the first stage of development (DS1), defined the first playable version through several interactions of trial and error until the game elements (mechanisms, pieces, and game economy/scoring) emerge (PS1). Then, the game was tested with urban/spatial planners (PS2), followed by a development correction stage (DS3). At this stage, if the game delivers experiences suited to approaching serious game goals. PS2 revealed the need to improve the decision-making experience by having more options and adding new constraints to the game setup (urban grid and pieces to move). The following steps focused on the playable dimensions (tested with game designers in PS3), helping to analyse the game and propose the final modifications for DS4.

Our proposal specifies the need to establish several stages of development (DS) after playtesting sessions (PS). We recommend at least one playtest focusing on the content and serious game purposes (in this case, urban planning) and another on playability, selecting users (players) and focusing on each serious game dimension. We present the methodology into two subsections, one regarding the game design process and the other the playtesting, as two core dimensions of serious game development [7, 26, 29].

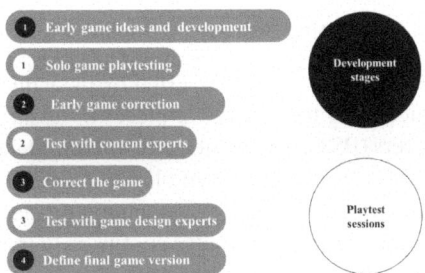

Fig. 1. Development process for TSC, Development Sessions (DS) and Playtest Sessions (PS)

2.1 Serious Game Development Process and Early Game Concepts

As in any standard serious game process, there is the need to balance the engagement/ entertainment dimension with the other, arguably, more serious objectives. The purpose of the TSC game was to deliver an experience where players assumed typified roles of stakeholders who would work together and make decisions to change an urban space. Each stakeholder had different options to affect the game state, which defined their scoring and the scoring of the other stakeholders. These conflictual scores were one of the game elements that simulated stakeholders' different claims and preferences. However, all stakeholders could collaborate to maximize their scoring since they benefitted from other stakeholders' actions.

From a mechanical point of view, we adopted a combination of game mechanisms from modern tabletop games associated with territory and transport building like:

- Tile placement: moving pieces that represent land uses or transport segments.
- Set collection/connections: combining transport pieces to connect land uses.
- Tableau building: available options for each player in each stakeholder team.

These previous mechanisms are framed and described by Englestein & Shalev [30].

During a fixed number of rounds, players activated the game mechanisms to move pieces over a conceptual map of a city (grid model of an urban zone) that defined the game board. Tile placement was implemented digitally by allowing players to move geometric elements (land uses and transport pieces). Tableau building helped to track stakeholders' options (representing those already used and those still available). Adjacency and connection of the geometric elements (land uses and transport) defined the scores through set collection mechanisms. Players could move the game pieces to overlay the map. Figure 4 represents how the blocks represent different land uses and transport segments, using the same dimension (d) of the blocks from the game board and specific colours for each land use (see also Fig. 4). The light brown squares inside the dashed grey lines (Fig. 5) represent empty or non-defined urban blocks. These squares are the places to overlay the game pieces representing land uses (green, yellow, blue squares and triangles), while the black frames with dashed white lines are the road system. As players overlay the land use pieces over the light brown squares, they can also overlay the red rectangles over the roads to define the transport network.

2.2 Learning from the Playtesting Sessions

The game's first version (from DS1) was playtested by the game developers (members of the research team), calculating the number of pieces necessary for a minimum and a maximum number of players (PS1). The number of different stakeholders (five) defined the minimum number of players. Adding more players required playing with teams of stakeholders that could be involved in real case planning process (see Sect. 2.3). More than four players per stakeholder team introduce confusion to the debate, increase waiting and downtime, and break the game flow with excessive downtime/pauses (more than 5 min per player turn). The game was turn-based (each player must wait for their turn to play). These requirements determined that 5 to 20 players could simultaneously participate in the same game. As stated before, each team of stakeholders had its scoring system (asymmetric mechanisms), resulting from the pieces over the map. Players could move different pieces into the collective game board (Map), improving their score for their stakeholder's team while also affecting the score of the other stakeholders. These positive and negative interconnections provided feedback loops that foster players to influence other players' moves and establish overall negotiation and collaboration dynamics.

The game was planned to last 60 min, considering a maximum of 20 players (maximum of 4 per team of stakeholders). Each player would take 2 min to move one of the pieces. Teaching the game took 10 min, and another 10 to discuss the dynamic (debriefing). A facilitator is needed to ensure the natural flow of supportive tasks. It's not a direct in-game persona. The facilitator explains the game, controls the time and helps players with their moves/decisions (could be removed if players master the game rules and mechanics). Finally, the facilitator conducts a debriefing to consolidate the outcomes of the play, which is required to achieve the serious game purposes.

2.3 The Stakeholders Clash Game Description

For the TSC case study, we defined a squared 4x4 grid with 16 blocks (Fig. 2). The block delimitations generated 40 road sections. Figure 2 also represents the type and shape of available pieces players can move to overlay and change the map, and the tile placement game mechanism (move and/or rotate).

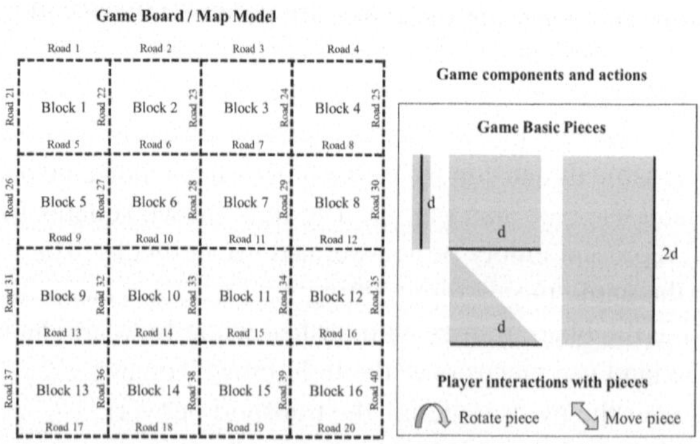

Fig. 2. Basic models to simulate an urban map, pieces for players to interact

Defining the game's pieces to move/rotate as squares and isosceles triangles with the same dimensions (d) of the blocks allows the simulation of single (square) or mixed (two triangles) land uses. This design and layout help players to know where to move the pieces on the game board (map). The rectangles represent constructions (land uses) that occupy more than one block. Players interact with the game by locating and moving/rotating a piece from the available pieces for their stakeholders' team to the game board. The overlaying of the pieces on the board changes the game state, meaning that a block has a new land use or the transport system is reaching new urban blocks. Rectangles are the only pieces that can overlay the roads. The modular dimension of the game board and game pieces also allows adaptations to more realistic models. It is possible to change the grids and the pieces according to the needs of the urban reality model. However, game designers/planners following our proposal must consider the players' ability to understand what game pieces to move and where, the geometric dimensional (lengths and areas) and topological (place to locate on the game board) relationships at stake, how it will affect the game state and scoring. Because we are dealing with orthogonal grids, distances are measured following Manhattan distance principles.

Next, it is necessary to set what the players can do in the game and their available options. At least there must be a piece per player to move. As team colleagues make choices, fewer options are available for the following turns as the game progresses, forcing the last players to influence their teammates' first choices. This delivers a sense of urgency and information about the game's progression toward the endgame state.

The generic interface layout appears in Fig. 5, with a column per type of stakeholder (team of players). The player interface can be modified for more players, demanding a minimum of one player per stakeholder (column) (Fig. 3). The game facilitator can distribute players evenly, writing each player's name per line in each column and forming teams that will act as a type of stakeholder (1). At the bottom of each column (stakeholder), players can see the game pieces (geometric shapes) to move/ rotate to place over the game board (3). The graphical interface and game board are set as a background in *Google Slides*. The chosen software allows us to add interactive (customizable) objects as game pieces (squares, triangles, and rectangles). Each team has different game pieces (3) to affect the game state and score (4).

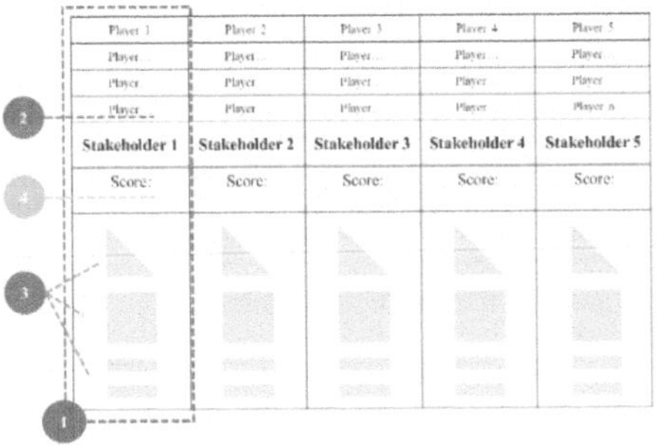

Fig. 3. Interface layout: 1 – Stakeholder Team (column); 2 – Line per player of the team; 3 – Available pieces per team; 4 – space to track the score of each team of stakeholders.

The scoring represents the game objectives for each stakeholder. Considering the number of available pieces and what stakeholders supposedly desire, represented in the game by how game pieces are connected by transport (set collection and connection mechanisms), we obtain the maximum score (four in the first version). The grid dimension (4 × 4) defines the scoring threshold, set initially as 4 (first game version).

Finding Meaningful Play to Build Meaning Interactions

The game must represent real stakeholders' roles (avoiding ludonarrative dissonance). Aiming for meaningful play demands a coherent game system, including all game elements, mechanisms, and intended metaphors. A geometric piece is recognized as a land use or a transport segment through metaphoric association. The words and iconography might not be enough to achieve meaningful simulation. Here, the game's theme was the development of a new urban zone (different land uses) near a heritage site and the respective sustainable public transport network. We defined the following stakeholders with a colour coding (Land uses: commerce was blue, housing was yellow, parks/heritage sites were green, and schools representing public facilities were grey; Transport system: transport segments were dark red) and scoring principles:

- Stakeholder 1 (St1): Shop Owners (blue) score according to pieces location.
- Stakeholders 2 (St2): Environmentalists (green) score according to pieces location.
- Stakeholders 3 (St3): Habitants (yellow) score according to pieces location.
- Stakeholders 4 (St4): Elected Politicians (no colour) score according to other stakeholders' scores.
- Stakeholders 5 (St5): Opposition Politicians (no colour) score according to other stakeholders' scores.

Figure 4 presents the shapes, colours, and iconographic symbols used to reinforce the meaning of each geometric piece. Transport pieces (dark red) placed on orthogonal connections over the roads represented the transport system that allows land uses to be connected (land uses connected to the transport network are considered adjacent). Adjacent/connect happens when land uses/transport pieces share edges).

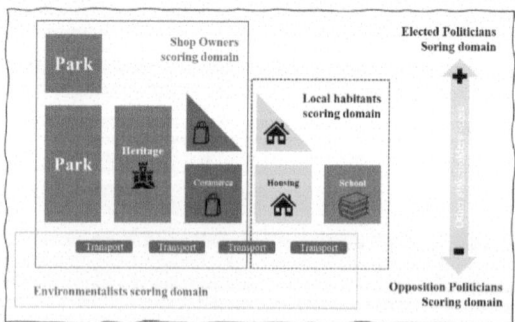

Fig. 4. Scheme of all Available pieces for players to move, colour codes and scoring domains per stakeholder. The quantity of pieces defines scoring possibilities and game economy balance (color figure online).

Scoring options aim for similar/balanced opportunities for all stakeholders to win the game while trying to represent what stakeholders might claim in a real-case situation. The scoring metric was set for a maximum of 4, related to the 4x4 grid dimension of the game board. Stakeholders scored in the following manner, according to the placement of pieces on the game board and the number of stakeholders in play:

- St1: One point per commerce connected/next to a park/heritage.
- St2: Two points per horizontal/vertical transport connection that crosses the map.
- St3: One point per housing connected.
- St4: Start with 5 points and get negative points per stakeholder below 4.
- St5: Gets 2 points per stakeholder below 4.

The scores of citizen stakeholders (St1 to St 3) defined the overall reference system. St1 and St3 score their points directly (1 point per connection). During the PS1, we realized they needed to double the scoring for St3. The available transport game pieces were insufficient to reach a maximum score of 4. This constraint made us double their score per continuous horizontal/vertical complete transport path. The politician stakeholders scored differently from the citizen stakeholders, although their scoring was interconnected. St4 started with 4 positive points (the maximum score any stakeholder could achieve) and lose points per other stakeholders below the maximum. St5 scores in an oppositive way to St4 (zero-sum game principles). The PS1 also revealed that the St5 must score the double to have equal chances to win the game.

From a thematic and metaphorical perspective, we adopted the following interpretations. Shop Owners (St1) desire to profit from the transport system passengers and access to parks and heritage sites. Environmentalists (St2) demand effective collective public transport to cross the city without constraints (horizontal and vertical connections from top to bottom and left to right of the game board). Local Habitants (St3) want schools near their homes. Elected Politicians (St4) start with public approval, wanting to please the stakeholders to continue their job. Alternatively, the Opposition Politicians (St5) might win the next election if stakeholders are not pleased. They are the only stakeholders that do not place new pieces on the map. Instead, they rely on political debates. They can call for voting to change a previously placed game piece. If they get the most the votes for stakeholders (1 to 3), St5 can change that piece.

Addressing the players' options and interactions for each role, the citizen stakeholders (St1, St2, and St3) can use/move pieces to increase their scoring and affect other stakeholders positively or negatively. St4 plays differently. St4's available pieces (options) help the citizen stakeholders to score. When they reach the maximum score, St4 does not lose points. The success of St4 affects the St5 score. St5 only proposes changing the previously placed pieces. Deliberately and despite the balancing requirements, the St4 role is easier/simpler to play than St5 because we decided to highlight the positive collaboration in political behaviour. These political dimensions emerge from the collaboration purposes of our serious game. The game intentionally promotes discussion of decision-making and its impact on the urban system. If not, the facilitator must refer to this during debriefing as part of the serious game experience.

Set Up the Game for the First Session with Players
After the first solo playtesting described previously, we completed the game board (grid

map) to set the challenge equally for each stakeholder (all could win the game as in modern board games). We added black roads and set up some existing land uses. This setup provided equal distances for the scoring of the shop owners and local habitants stakeholders (Fig. 5). We added additional text information and several crosses to mark the call for votes of the opposition stakeholders and marked with a dashed red line the game space player could interact (move the pieces from).

2.4 Game Playable Interactions and Results

The session with planning experts (SP2) occurred during an academic spatial planning academic conference in 2021. The session with game designers (3) happened during a graduation class in game design (PS3) in the same year. The two sessions occurred in very different academic backgrounds. PS2 players ($n_{ps2} = 9$) were spatial planning experts with no game design and/or serious game experience. PS3 background was played by graduation game design students ($n_{ps3} = 14$) (orthogames and serious games). PS2 was tested was intended to test more the simulation side (the content of the serious game) and PS3 the playability and game engagement (decisions, interactions, and challenges of games). Both sessions had a similar duration of one hour. The facilitator used *Zoom* software to communicate with the players. *Google Slides* was the game platform where players moved the game pieces to change the game state over a predefined background (grid board and interface with available game pieces).

Comments and Players' Perceptions and Game Adaptation
Besides the game outputs, the pieces over the urban map, and the scoring, the conversations and informal commentaries recorded during gameplay are relevant to examine the game experience in each game. Because the game comprised 20 moves, it was possible to classify collaborative behaviour per move. During both sessions' gameplay, more than 75% of the players asked other team members for suggestions before placing a piece. Simultaneously, more than 50% of other players tried to influence these moves, including players from the other teams. In PS3, this tendency for player influence and interactions was higher (60%).

During PS2 commentaries, some players argued the game was easier for Elected Politicians (St4) because they only needed other stakeholders to score well. As expected, the effective placement of the transport tiles benefited the first three stakeholders. Achieving goals was easy for St1, St2, and St3. During DS3, we introduced more pieces for each stakeholder, adding more options (higher player agency) and reducing the threshold for successful votes for Opposition Politicians (St5) (only one-third of approvals, not the majority). We added more pieces and park rectangles (occupying two continuous squares) that can block more road paths. The setup included a new school on the game board (initial setup). These previous changes obliged us to alter some of the scorings for other stakeholders. Elected politicians started with 6 points (St4), and Opposition Politicians (St5) scored 2 points per stakeholder below scores of 3 because playing this role was even more difficult than expected. We realized that players only understood the effects of the call for vote action (move previously placed pieces) in the final rounds of the game. This affected St5 score.

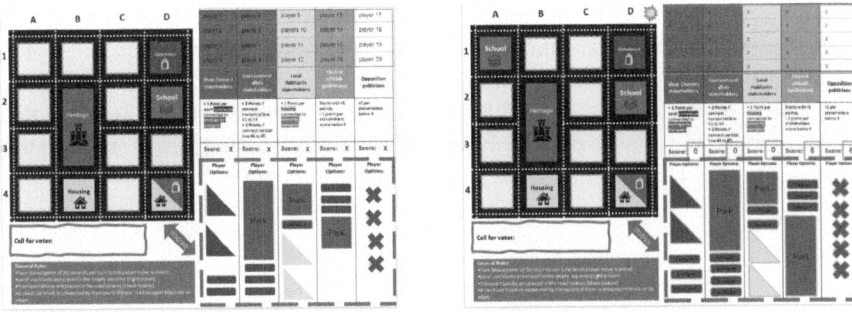

Fig. 5. Game set up for PS2 (left) and PS3 (right)

However, even during PS3, players stated that playing as Elected Politicians (St4) was still easier because it demanded straightforward decisions. Elected Politicians only needed to respond to the other stakeholders' claims without the same level of manipulation that Opposition Politicians (St5) were required to score well. During PS3, Opposition Politicians approved a voting call, which impacted Local Habitants' scoring. During PS2, the opposition vote call never passed. Regardless of this apparent failure, when the Opposition Politicians proposed voting for a change, the interactions changed (in both sessions). The discussions were much more vivid than in the turn of other players. The loss-aversion effects [31] were notorious for players' behaviour because approving the proposal of the Opposition Politicians could make some stakeholders lose points.

Game Results and Perceptions
Considering that the PS2 and PS3 games differed slightly, comparing their results brings additional information (Figs. 6 and 7).

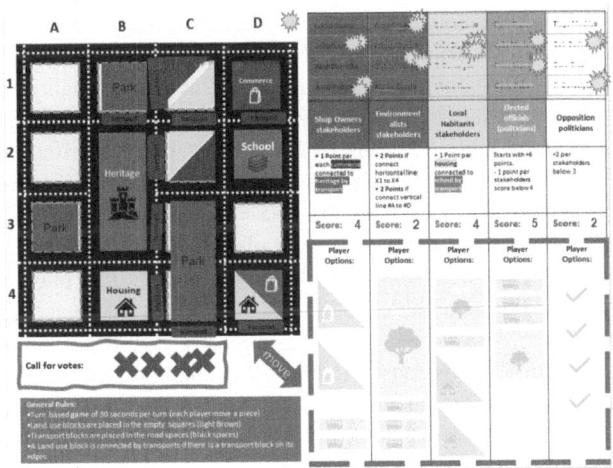

Fig. 6. Game results for PS 2.

As noticed in Figs. 6 and 7, there are explosion icons on the board. This improvisation signalled each player's turn and those that had already played. Internet connection problems affected gameplay. Some players lost their connections or could not use *Google Slides* with their devices. To deal with this problem, teammates played assumed their moves. Although the players previously signed up for each session, fewer attended than expected. This limitation forced some players to play more turns.

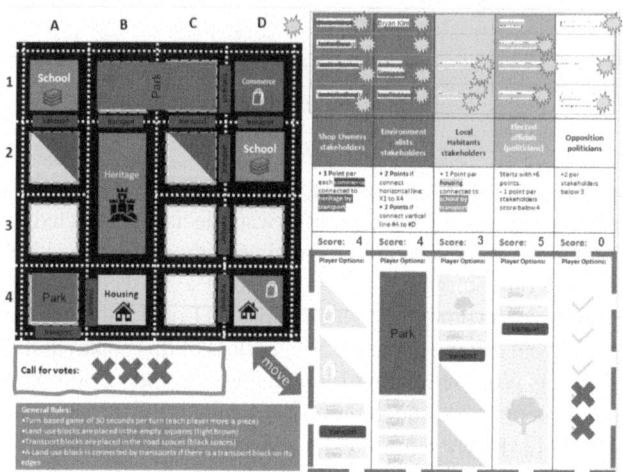

Fig. 7. Game results for PS 3

Table 1 gathers groups of commentaries players made during gameplay and the final discussion. We grouped the comments by issue typology, considering the commentary made by at least 50% of the participants, both orally and in the chat.

During gameplay, the game facilitator used the *Zoom* survey tool to get immediate feedback, according to a Likert scale from 1 to 7 (1 – low, 7 – high) about the following dimensions: "Classify the game fun/engagement"; "Classify the quality of the planning solution"; "Classify the simulation accuracy of the stakeholders' roles". We highlight the following data from Table 1 because it is coherent with the different participants' backgrounds in each session. S2 players considered the fun and engagement dimension higher ($\bar{x} = 5.88$; $\sigma = 1.27$) than S3 players ($\bar{x} = 4.67$; $\sigma = 0.47$). PS3 participants considered the game delivered a more coherent planning solution ($\bar{x} = 5.17$; $\sigma = 1.07$) than PS2 participants ($\bar{x} = 4.75$; $\sigma = 1.39$). Moreover, finally, PS3 players believed the game better simulated the stakeholder roles ($\bar{x} = 5.50$; $\sigma = 1.00$) than PS2 players ($\bar{x} = 4.67$; $\sigma = 1.25$).

Participants in both sessions considered the game engaging and able to represent the decision-making dynamics that can happen during a stakeholders' participation process. However, participants pointed out that using these methods can have some disadvantages, like oversimplification, improper solutions, or non-optimal decisions. From a design point of view and implementation, having the necessary resources and game design knowledge to balance the game can also be limiting.

Table 1. Final players' perceptions about advantages and disadvantages recorded during game-play sessions (# - number of comments; % - percentage).

PS/n	Advantages			Disadvantages		
	Typified Statements	#	%	Typified Statements	#	%
PS 2 n = 9	Collaboration and collective synergies	9	100	Prejudice and bias about game usage	8	89
	Interaction between citizens and/or practitioners	9	100	Difficult to implement in practice	8	89
	Learning about issues and claims	8	89	Improper solutions and decisions	6	67
	Data collection and opinion assessment	6	67	Oversimplification of reality and idealism	6	67
	Engagement that increases participation	6	67	Engage all participants simultaneously	5	55
PS 3 n = 14	Collaboration and collective synergies	14	100	Balancing simulation and gaming	10	71
	Engagement that increases participation	14	100	Oversimplification of reality and idealism	9	64
	Simulation of the situation and feelings	13	93	Prejudice and bias about game usage	8	57
	Explore different roles	10	71	Forcing to play when users do not like games	7	50
	Imagination and informality	10	71	Available resources	7	50

Some game advantages were perceived similarly in both sessions, like the establishment of collaborative and interactive synergies between participants and the engagement that leads to growing participation. Other positive dimensions were noticed more by one group than the other. PS2 players underlined the effect on learning issues and understanding conflictual stakeholders' claims, data collection, and opinion assessments. Alternatively, PS3 players focused on the simulation of the situation and the playable experience (personal feelings).

The identified disadvantages were more heterogeneous and varied. Despite this, both groups stated that prejudice about using games for serious purposes and oversimplifications of the game-based models could jeopardize the process. PS3 players were more concerned about the game design challenges, like balancing contents and playability (balancing simulation and gaming), and the effects that available resources might not be enough to build the desired serious game. PS2 players were concerned about the applicability and quality of the outcomes.

2.5 Game Results Discussion

As expected, each group perceived the play session based on their backgrounds, knowledge, and previous experiences. PS2 players were planners who considered serious game approaches surprising and with the potential to be used in practice but were concerned about the practical application and challenges of using games. PS2 players considered the game more engaging and less like a simulation. PS3 players were game design students, comfortable at playing, analysing, and designing games. Their perception of the playability potential was lower, even though they played a modified version that delivered a more playable approach (more options that could affect the game state and scores). However, the PS3 players considered that the game addressed the urban planning exercise (learning and decision-making) better than the PS2 players.

The experience results reinforce the challenges of developing a serious game concerning the balance between simulation and engagement (playability). As noticed, the perceptions depend on the users and their backgrounds. Sessions should include groups of users (players) focusing on the simulation ("serious" side) and others on playability (engagement and motivation). Using board game mechanisms in digital platforms like *Google Slides* and other software that allows users to move objects in collaborative digital environments can deliver flexible playable solutions. We argue that the proposed method can be replicated for spatial planning processes demanding low resources and development time. This option addresses the serious game challenges identified by Sousa et al. [9] for spatial planning if we acknowledge that games may demand high resources and a long time to develop.

We propose a method to replicate the TSC serious game for other spatial planning processes done online. This method is presented in Fig. 8, consisting of a five-step sequence for a game of moving pieces over a map. In step 1, planners must decide what platform(s) they wish to use. Step 2 is the first game idea and playable model. Then, the game should be tested multiple times with different users. For the first game model, it is necessary to define the map (game board), the rules and mechanisms of the game, what players can move (options), and the scoring to determine the results. The following steps identify the playtesting and the adaptations to fit the serious game goals and the playability to engage users (players).

Our proposal is bounded to board game mechanisms and based on moving objects like geometric pieces that represent different elements of an urban system. In step 2, game developers define the grid map to model the urban environment at stake. After having the game board (grid map), the time comes to set the general game rules and the game mechanisms that players will use to move objects and define scoring. The number of stakeholders is part of the rules, affecting the definition of mechanisms and objective (scoring thresholds). Finding adequate game mechanisms can be an iterative process of trial and error [9, 32]. However, game mechanism encyclopedias can help [30, 33]. Finishing step 2 generates the first game model for internal playtesting (step 3), desirably including game design experts and planning professionals. Step 4 defines the subsequent game testing, where planners must adapt the game to users, tweaking the game and making the necessary adaptations to fit players' backgrounds. Players may demand that the game developers change the complexity of the rules, the number of options, and the scoring thresholds. The available time and resources to implement the game might also

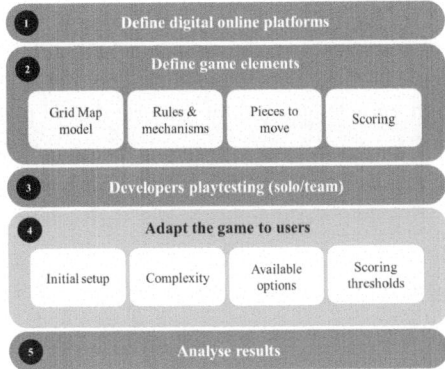

Fig. 8. Method to develop serious planning games (of moving pieces) in digital platforms.

affect the changes. Defining the initial setup has an impact, influencing all the previous game dimensions. Step 5 is when the final game analysis occurs. Planners can use several metrics to verify the serious game impact and complement the evaluation of the previous iterative development process [34]. We can follow Meyer et al. [29] framework as an example, evaluating the engagement elements and serious game outcomes (before, during, and after playing the game) as in any serious game.

It is also important to state that serious game developers' ideas and ideological biases might affect the game product, the playable experience, and the purposes of playing a game. In the case study, the stakeholders' scoring and, specifically, the relationship between elected and opposition politicians were not neutral representations. Playing the different roles provided different experiences to players. Game facilitation and debriefing should address these dimensions.

3   Conclusions

Serious games are growing in popularity among urban and spatial planners. However, there are several difficulties in implementing these methods in practice. Planners can use analogue or digital game solutions, both with advantages and disadvantages. Analogue ones can be cheaper and faster to implement, fostering unique collaborative and tangible experiences. Digital games might be expensive to develop, but they enable the participation of a larger quantity of participants who are distant from each other, which was very important during the COVID-19 pandemic.

We proposed a method based on *The Stakeholders Clash* (TSC) game that planners can use in practice, requiring low resources and allowing fast adaptation, even in real-time situations. The proposal replicates analogue game design to be played over digital platforms, opening new ways to profit from hybrid platforms. The principles of moving objects into maps and change planning solutions are simple enough yet able to simulate complex spatial realities, decisions, and interactions. Our experience revealed that users' backgrounds and experiences affect their perceptions of a serious game. The test groups have shown that experts in the simulation (planners) value more practical uses of serious games. On the other hand, game experts are more concerned with the playability of the product. This duality is the perfect metaphor that serious game projects must fulfill: a balance between these two dimensions. Our proposal is simple enough but still requires game design knowledge, and the generic digital platform can limit some game implementations.

References

1. Pánek, J., Falco, E., Lysek, J.: The COVID-19 Crisis and the Case for Online GeoParticipation in Spatial Planning (2022)
2. Song, X., Cao, M., Zhai, K., Gao, X., Wu, M., Yang, T.: The effects of spatial planning, well-being, and behavioural changes during and after the COVID-19 pandemic. Front. Sustain. Cities. **3**, 686706 (2021)
3. Dey, B.L., Al-Karaghouli, W., Muhammad, S.S.: Adoption, adaptation, use and impact of information systems during pandemic time and beyond: research and managerial implications. Inf. Syst. Manag. **37**, 298–302 (2020)
4. Constantinescu, T., Devisch, O., Liesbeth, H.: Participation, for whom? The potential of gamified participatory artefacts in uncovering power relations within urban renewal projects. ISPRS Int. J. Geo Inf. **9**, 319 (2020). https://doi.org/10.3390/ijgi9050319
5. Mayer, I.: The gaming of policy and the politics of gaming: a review. Simul. Gaming **40**, 825–862 (2009)
6. Ampatzidou, C., Constantinescu, T., Berger, M., Jauschneg, M., Gugerell, K., Devisch, O.: All work and no play? Facilitating serious games and gamified applications in participatory urban planning and governance. Urban Plan. **3**, 34–46 (2018)
7. Constantinescu, T., Devisch, O., Kostov, G.: City makers: Insights on the development of a serious game to support collective reflection and knowledge transfer in participatory processes. Int. J. E-Plan. Res. **6**, 32–57 (2017). https://doi.org/10.4018/IJEPR.2017100103
8. Constantinescu, T., Devisch, O., Kostov, G.: Game Mechanics as Thinking Mechanisms for Urban Development (2020). https://doi.org/10.4018/978-1-7998-4018-3.ch009
9. Sousa, M., Antunes, A.P., Pinto, N., Zagalo, N.: Serious games in spatial planning: strengths, limitations and support frameworks. Int. J. Serious Games **9**, 115–133 (2022). https://doi.org/10.17083/ijsg.v9i2.510
10. Tan, E.: The evolution of city gaming. In: Complexity, Cognition, Urban Planning and Design, pp. 271–292. Springer (2016)
11. Tan, E.: Play the City: Games Informing the Urban Development. Jap Sam Books (2017)
12. Tan, E.: Network of Games : An Ecology of Games Informing Integral and Inclusive City Developments. **7**, 264–277 (2022)
13. Poplin, A.: Digital serious game for urban planning:"B3—Design your Marketplace!" Environ. Plan. B Plan. Des. **41**, 493–511 (2014)
14. Poplin, A.: Playful public participation in urban planning: a case study for online serious games. Comput. Environ. Urban Syst. **36**, 195–206 (2012)
15. Sousa, M.: Cities: skylines: the digital and analog game design lessons for learning about collaborative urban planning. In: International Conference on Videogame Sciences and Arts, pp. 257–271 (2023)
16. Fullerton, T.: Game Design Workshop: A Playcentric Approach to Creating Innovative Games. AK Peters/CRC Press (2014). https://doi.org/10.1201/b16671
17. Ham, E.: Tabletop Game Design for Video Game Designers. CRC Press (2015)
18. Brathwaite, B., Schreiber, I.: Challenges for Game Designers. Nelson Education (2009)

19. Dodig, M.B., Groat, L.N.: The Routledge Companion to Games in Architecture and Urban Planning: Tools for Design, Teaching, and Research. Routledge (2019)
20. Kasprisin, R.: Play in Creative Problem-Solving for Planners and Architects. Routledge (2016)
21. Lerner, J.A.: Making Democracy Fun: How Game Design Can Empower Citizens and Transform Politics. MIT Press (2014)
22. Sousa, M., Antunes, A.P., Pinto, N., Zagalo, N.: Fast serious analogue games in planning: the role of non-player participants. Simul. Gaming **53**, 104687812110736 (2022). https://doi.org/10.1177/10468781211073645
23. Sousa, M.: A planning game over a map: playing cards and moving bits to collaboratively plan a city. Front. Comput. Sci. **2**, 37 (2020). https://doi.org/10.3389/fcomp.2020.00037
24. Sousa, M.: Modeling urban spaces with cubes: building analogue serious games for collaborative planning. Int. J. Film Media Arts. **8**, 8–35 (2023)
25. Ampatzidou, C., Gugerell, K.: Mapping game mechanics for learning in a serious game for the energy transition. Int. J. e-Plan. Res. **8**, 1–23 (2019). https://doi.org/10.4018/IJEPR.2019040101
26. Champlin, C.J., Flacke, J., Dewulf, G.P.M.R.: A game co-design method to elicit knowledge for the contextualization of spatial models. Environ. Plan B Urban Anal. City Sci. **49**, 1074 (2021)
27. Sousa, M.: Transforming Google Drawings into a game-based nudging tool for collaboration. In: First International Workshop on Digital Nudging and Digital Persuasion (2022)
28. Sousa, M.: Modding modern board games for e-learning: a collaborative planning exercise about deindustrialization. IEEE Int. Conf. Port. Soc. Eng. Educ. **5**, 10 (2021). https://doi.org/10.1109/CISPEE47794.2021.9507250
29. Mayer, I., et al.: The research and evaluation of serious games: toward a comprehensive methodology. Br. J. Educ. Technol. **45**, 502–527 (2014). https://doi.org/10.1111/bjet.12067
30. Engelstein, G., Shalev, I.: Building Blocks of Tabletop Game Design: An Encyclopedia of Mechanisms. CRC Press LLC, Boca Raton (2019). https://doi.org/10.1201/9780429430701
31. Engelstein, G.: Achievement Relocked: Loss Aversion and Game Design. MIT Press (2020)
32. Sousa, M.: The mechanics of drawing : helping planners use serious games for participatory planning Micael Sousa. plaNext. (2022). https://doi.org/10.24306/plnxt/80.NEXT
33. Sousa, M., Oliveira, P., Zagalo, N.: Mechanics or mechanisms : defining differences in analog games to support game design. In: IEEE Conference on Games 2021. IEEE (2021)
34. Silva, C., Bertolini, L., Pinto, N. (eds.): Designing Accessibility Instruments· Lessons on Their Usability for Integrated Land Use and Transport Planning Practices. Routledge, United Kingdom (2019)

A Systematic Review on Designing, Developing, and Implementing Effective Digital Game Interventions in Mental Disorders

Rebeca Mendes[1]([✉]) [iD], Ana Veloso[2] [iD], and Ana Torres[3] [iD]

[1] Department of Communication and Art and Department of Education and Psychology, University of Aveiro, Aveiro, Portugal
rebecamendes@ua.pt

[2] Department of Communication and Art, University of Aveiro, Aveiro, Portugal
aiv@ua.pt

[3] Department of Psychology and Education, University of Beira Interior, Covilhã, Portugal
ana.carla.torres@ubi.pt

Abstract. Digital games have been used as mental health resources to reduce and prevent mental problems and disorders. Despite the growing evidence, few studies have attempted to test the effectiveness and gather useful data for designing, developing, and implementing these resources. The present research aims to collect and examine evidence on digital game-based interventions that have been effective in mental disorders, focusing on mood, anxiety, substance use, and obsessive-compulsive disorders. The goals are to provide an overview of effective digital game-based interventions and to synthesize data for designing, developing, and implementing these resources. The research consists of a systematic review of the "Preferred Reporting Items for Systematic Reviews and Meta-Analyses". It collects evidence on effective digital game-based interventions in mental disorders, between January 2018 and December 2023, appealing to SCOPUS, Web of Science, and EBSCOhost databases. Includes three reviews and reviewers, with data analyzed on design, development, and implementation. The search yielded 3.751 sources, but only two were included as evidence. Two digital game-based interventions in mental disorders present effective results for depression and anxiety. An overview of the effective digital game-based interventions in mental disorders, and useful data for designing, developing, and implementing these resources are achieved, being supported and complemented by a discussion. This research achieved its goals and attests that digital game-based interventions can be effective in mental disorders, specifically in anxiety and depression. It highlights that more research is still lacking to support the design, development, and implementation of effective digital game-based interventions.

Keywords: digital games · mental disorders · effective interventions

A. Marto et al. (Eds.): VJ 2024, CCIS 2324, pp. 70–83, 2025.
https://doi.org/10.1007/978-3-031-81713-7_5

1 Introduction

This systematic review merges mental health with digital games. Mental health is the state of physical and psychological health and well-being, enabling the subjects to cope with adversities, enjoy life, and actively contribute to society [1]. Mental problems and disorders, by provoking a negative significant impact on the physical and psychological well-being of the population, have been a major concern of health authorities and a global burden for societies [1–3]. This is especially observed among adolescents and young adults, namely in anxiety, depression, substance use, and obsessive-compulsive disorders [4].

Digital games, especially, serious digital games, are powerful tools for conveying interventions in several educational and professional domains [5]. These games combine mechanic, dynamic, and aesthetical elements of digital game design [6–8] with the goals and knowledge from other domains, to provide a new form of learning and development [5].

The combination of the mechanics and dynamics of digital games [8] gives them an inherited potential to stimulate real-life situations that can help to recover or prevent neurocognitive problems and disorders [9]. Besides, aesthetic elements supported by game features – for instance, the world, characters, and narrative – play an important part in these real-life digital simulators, supporting the transference for the real world [5, 6, 10]. Simulations in digital games have been proven to be effective in promoting cognitive (advanced processes relying on neurological and cognitive structures and functioning) and socio-emotional skills (fundamental abilities, attributes, and characteristics for successful functioning and interaction) [11, 12]. The development of cognitive and socioemotional skills helps to prevent mental problems and disorders and reinforces healthy and adaptive physical and psychological [13].

In the past decade, digital games have been contributing to promoting mental health and preventing mental disorders, essentially through the stimulation of cognitive and neuropsychological functions. Recent reviews outline the diversity of resources that have been used and developed, directed to different mental problems and disorders, paying attention to the value of digital games as a support for mental health interventions [9, 11, 12]. However, clinical and health psychology appeals to evidence-based practice, and little information was found about the effectiveness of digital game-based interventions in mental problems and disorders and the useful data to produce these resources.

This systematic review aims to bridge the gap in evidence-based data for designing, developing, and implementing effective digital game-based interventions in mental disorders. It intends to provide an overview of the recent evidence on effective digital game-based interventions in anxiety, depression, substance use, and impulse control disorders, and to synthesize the data regarding the design, development, and implementation of these resources. The work expects to contribute to digital game design and development, evidence-based mental health interventions, and human-computer interaction.

2 Method

2.1 Protocol

The present research presents a systematic review following the "Preferred Reporting Items for Systematic Reviews and Meta-Analyses" (PRISMA 2020; Page et al., 2021). It is supported by mixed nature and methods to explore effective digital game-based interventions in mental disorders, focusing on anxiety, depression, substance use, and obsessive-compulsive categories of mental disorders, according to the fifth edition of the "Diagnostic and Statistical Manual of Mental Disorders" (DSM-V; APA, 2013). The research expects to provide an overview of effective digital game-based interventions in mental disorders and to synthesize data for designing, developing, and implementing these resources.

2.2 Eligibility Criteria

The observation of three inclusion criteria supports the sources of evidence eligible for the present systematic review: (i) the data regards a digital game; (ii) the digital game has the purpose of intervention in (at least) one mental disorder under the scope of this research; (iii) the digital game presents (at least) one significant positive effect on the mental disorder assessment measure.

Non-eligible evidence presents (at least) one of the following exclusion criteria: (a) not a game; (b) game not digital; (c) digital game not for intervention in target mental disorders; (d) digital game without a positive and significant effect in mental disorder assessment measure; (e) virtual or augmented reality digital game; (f) eye-tracking and biofeedback devices; (g) educational skill training; (h) exergames and exercise, motor, physical, kinematic training; (i) digital game for cognitive and neurocognitive training (neuropsychological rehabilitation); (j) protocol, review, non-concluded research, or non-randomly controlled trial (RCT); and (k) non-accessible or paid articles.

2.3 Information Sources

The present research gathers scientific evidence published from January 2018 to December 2023, appealing to three bibliographic databases: Web of Science (WOS), SCOPUS, and EBSCOhost (EBSCO).

2.4 Search Strategy

Appealing to Boolean operators, the search is conducted on three main domains of the systematic analysis – digital game, mental health, and empirical evidence – through a string of 23 keywords, using the following expression: "computer game" OR "video game" OR "videogame" OR "serious game" OR "digital game" OR "game" OR "game-based" AND "mood" OR "depression" OR "anxiety" OR "stress" OR "impulse control" OR "substance use" OR "substance abuse" OR "suicide" OR "self-harm" OR "mental health" OR "mental disorder" OR "psychotherapy" OR "intervention" OR "psychotherapeutic intervention" AND "experimental" OR "clinical trial" OR "controlled trial" OR "randomized controlled trial" OR "RCT".

The search is restricted to the following criteria: article type, concluded study, English language, random-controlled trial, published between January 2018 and December 2023. The present strategy aims to replicate the study conducted by Shah et al. (2018), to provide an updated and more specific systematic review of digital game-based interventions in mental disorders, focusing on evidence of effectiveness.

2.5 Selection of Sources and Data Collection

An expert in mental health (1st reviewer) performs the data collection and primary review (title and abstract) and categorizes sources into included, excluded, and dubious. During the second review, eliminate repeated evidence and screen the full content. In the third review, the text is fully screened again, and the dubious sources are discussed with a second expert in mental health (2nd reviewer). An expert in digital games (3rd reviewer) supports the selection, validates the sample, and helps through data collection. The data is categorized into three dimensions: (i) design – elements of digital game design and psychotherapeutic intervention; (ii) development – development features and game experience; and (iii) implementation – research, outcomes, and limitations.

Data items.

The items considered encompass the specifications of the dimensions, integrated as follows:

a) Design: game name, description, genre, goal, rules, levels, intervention (targeted disorder, therapeutical objective, support theory, and therapeutical techniques) [15–17], mechanics, dynamics, and aesthetics (world, narrative, and characters) [8–10, 18];

b) Development: system, engine, game mode, graphics, camera perspective, language, support features, and game experience (engagement, adaptability, and generalization) [6, 7, 12, 18, 19];

c) Implementation: population, research type, intervention format, assessment instruments, outcomes (intervention effect, and game usability), and limitations (research, intervention, and digital game) [20].

2.6 Study Risk of Bias Assessment

The present systematic review encompassed three reviews by three authors. The first author collected and reviewed all data, the third reviewed the data regarding the mental health dimension, and the second reviewed the data regarding the digital game dimension and items, which were estimated to have a low risk of bias assessment.

2.7 Effect Measures

The evidence must present (at least) one positive and significant effect between the pre-and post-intervention in the assessment instruments directly evaluating the mental disorder symptomatology.

2.8 Synthesis Methods

This systematic review presents the synthesized data on effective digital game-based interventions in mental disorders by the dimensions and respective items. It is divided by the mental disorders, and, in case of multiple evidence for the same intervention, is presented ascendingly by year.

3   Results

3.1 Study Selection

The synthesis of the evidence's selection process is presented in the following scheme (Fig. 1).

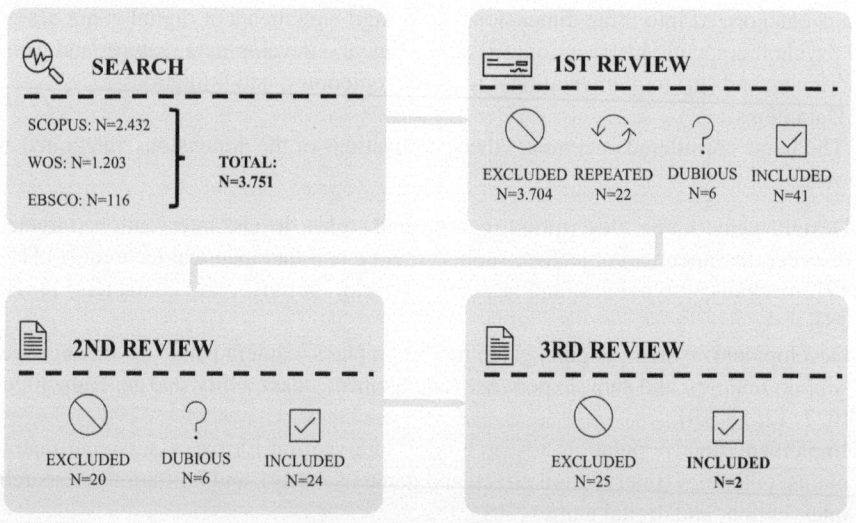

Fig. 1. Scheme of the evidence selection process.

The total number and distribution of the excluded sources of evidence by exclusion criteria are presented in the figure below (Fig. 2).

Supported by the PRISMA 2020 [14], the flow chart of the present systematic review is presented further (Fig. 3).

3.2 Study Characteristics

The sources of evidence retrieved from the present systematic review are the following:

i) Evidence 1 (E1). "CBT4Depression: A Cognitive Behaviour Therapy (CBT) Therapeutic Game to Reduce Depression Level among Adolescents." 2022. Yusof, N., Shaari, N., & Yusoff, E. International Journal of Advanced Computer Science and Applications. DOI: https://doi.org/10.14569/IJACSA.2022.0130930.

Total excluded (N=3.749)
A. Not a game (N=1.322; 35,24%)
B. Not digital (N=302; 8,05%)
C. Not for MHI (N=469;12,50%)
D. No effect (N=46; 1,23%)
E. VR/AR game (N=274; 7,30%)
F. Biomarkers (N=62; 1,65%)
G. Educational (N=372; 9,92%)
H. Exergame (N=328; 8,74%)
I. Rehabilitation (N=228; 6,08%)
J. Non RCT (N=338;9,01%)
K. Not accesible (N=10; 0,27%)

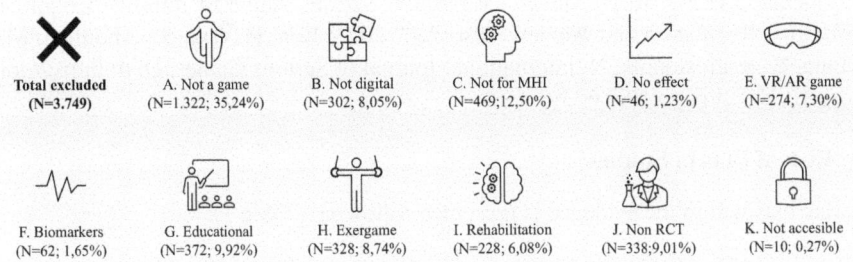

Fig. 2. Number (N) and percentage (%) of excluded sources by criteria.

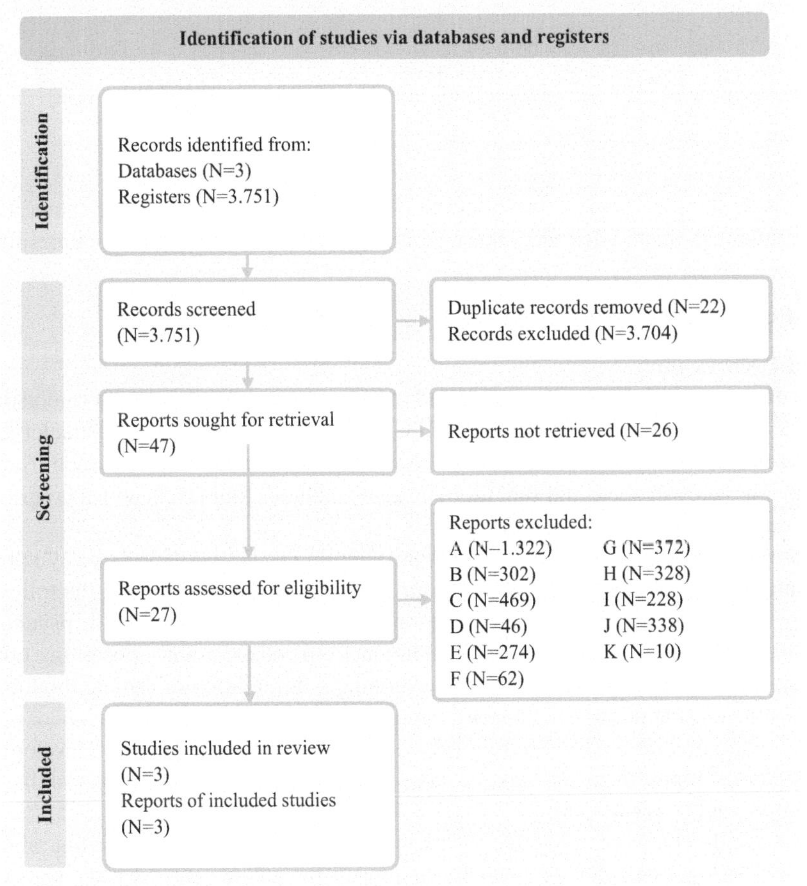

Identification of studies via databases and registers
Identification
Records identified from:
Databases (N=3)
Registers (N=3.751)
Screening
Records screened (N=3.751)
Duplicate records removed (N=22)
Records excluded (N=3.704)
Reports sought for retrieval (N=47)
Reports not retrieved (N=26)
Reports assessed for eligibility (N=27)
Reports excluded:
A (N–1.322) G (N=372)
B (N=302) H (N=328)
C (N=469) I (N=228)
D (N=46) J (N=338)
E (N=274) K (N=10)
F (N=62)
Included
Studies included in review (N=3)
Reports of included studies (N=3)

Fig. 3. Systematic review flow chart, supported by the PRISMA 2020 [14].

ii) Evidence 2 (E2). "ReWIND: A CBT-Based Serious Game to Improve Cognitive Emotion Regulation and Anxiety Disorder," 2023. Ken, H., Yan, J., Abdullah, M., Tang, Y., & Prestopnik, N. International Journal of Serious Games. DOI: https://doi.org/10.17083/ijsg.v10i3.603y.

3.3 Risk of Bias in Studies

The risk bias within the evidence is presented following (Table 1).

Table 1. Dimensions, items, and unspecified data within the evidence collected.

ID	Design															Development							Implementation											
							Intervention						Aesthetics										Experience			Intervention						Outcome		Limitations
	Name	Description	Game	Goal	Rules	Levels	Disorder	Objective	Theory	Techniques	Mechanics	Dynamics	World	Narrative	Characters	System	Engine	Mode	Graphics	Camera	Language	Support	Engagement	Adaptability	Generalization	Population	Research	Format	Instruments	Intervention	Game	Research	Intervention	Game
E1	X	X	X	X	X	X	X	X	X	X	X	X	X	X	X	X		X	X		X	X	X	X	X	X		X	X	X	X			
E2	X	X	X	X			X	X	X	X	X	X		X	X	X	X	X	X	X	X	X				X	X	X	X	X	X	X	X	X

This systematic review provides an overview of effective digital game-based interventions in mental disorders through a detailed description of the resources found.

3.4 Results of Individual Studies

"CBT4Depression"

"CBT4Depression" [21] is a digital game-based intervention in depression (mood disorders). Consists of a role-playing game (RPG) for computers. The main character needs to beat the enemy by appealing to other characters' support, to finally escape from the jungle. The rules are clear and specific. The game includes four levels with a progressive increase in difficulty.

Its therapeutic goal is to assist in identifying and reducing depressive symptoms. The intervention is supported by the cognitive-behavior theory, including the following therapeutical techniques: (i) psychoeducation – relevant content about health, psychological constructs, and psychotherapeutic techniques; (ii) observation – observing others' behaviors and reactions; (iii) Socratic questioning – stimulation of critical thinking by questioning that encourages active participation in seeking answers; (iv) behavior experiment – experimenting with different behaviors; (v) skills development – development of problem-solving, decision-making, coping, communication, and social skills; (vi) relaxation; and (vii) role-play.

Is a single-character (playable character) game control with a storyline about a teenager (protagonist) who was lost while exploring the jungle (game world). The player needed to find a small village to seek help from the villagers – non-playable characters (NPCs) – to fight the enemies. The enemies are incarnations of the protagonist's negative thoughts, appearing in various forms (e.g., friend, stepmother, or black shadow) throughout the game. The player must beat the enemies to escape from the jungle, by fighting negative thoughts and modifying them into positive ones.

The mechanics included in "CBT4Depression" are the following: exploration; survival; making predictions; taking notes; memorization; defeating the enemies; finding objects; finding information; helping; and interacting.

Regarding the dynamics, players learn techniques to strengthen skills to prevent depressive symptoms: problem-solving; decision-making; coping; communication; and social.

The game is developed in single-player mode, with 2D graphics, and from a third-person perspective. The language is Malay, the maternal language of the population.

Engagement in "CBT4Depression" is endorsed by built-in flexibility for players to explore the quests in any desired order, being also supported by realism to promote identification, facilitating the players' projection into characters and immersion in the game. Adaptability includes the design of the levels with moderate and gradually increasing difficulty, matching the players' performance to enable the learning of the skills and completion of the challenges, and providing feedback discrepancies. In matters of generalization potential, the relevant skills used by the player can be practiced outside the game context, facilitating an effective skill transfer in the player's daily life.

The digital game-based intervention was evaluated in adolescents (13 ≥ 16 years; N = 115), a non-clinical population, with pre- and post-intervention assessment embedded into the game, appealing to the Beck Depression Inventory in Malay.

The results demonstrated that the depression level of the respondents decreased from the pre- to post-intervention. It was found that depressive scores reduced significantly after using the game. "CBT4Depression" can be used as an assistive tool in mental health treatment among young patients. The therapeutic elements and game elements in this digital game-based intervention are effective in reducing depression levels among adolescents.

"ReWIND"

"ReWIND" [22] is a digital game-based intervention in anxiety (anxiety disorders). Consists of a role-playing game (RPG) for computers. The game goal is to assist quest-givers (NPCs) in resolving their problems to complete each quest in the game. The rules are not specified in the source of evidence. It includes six quests (levels), without a defined progressive order.

The therapeutic goal is to improve cognitive emotion regulation and reduce anxiety levels. The intervention is supported by the Cognitive-Behavior Theory (CBT) and the Rational-Emotive Behavior Therapy framework (REBT), focused on Ellis' ABCDE model of activating event (A), belief (B), consequence (C), disputation (D), and effect (E).

It includes the following therapeutical techniques: (i) psychoeducation; (ii) behavioral experiment; (iii) socioemotional skills development – problem-solving and decision-making; (iv) observation; (v) relaxation; (vi) role-play; (vii) cognitive restructuring – reshaping negative or faulty thoughts, evaluations, and beliefs that cause psychological disturbance into more positive and supportive ones; (viii) mindfulness practice – recognizing what is happening in the present moment and focusing our attention on the task at hand; and (ix) reinforcement – positive and negative reinforcement of thoughts and behaviors.

"ReWIND" was designed from scratch to be a digital game-based intervention in anxiety. The overarching story revolves around the main character (protagonist), who travels around the game world and encounters six quest givers, the non-playable characters (NPCs) needing help overcoming anxiety-inducing hardships. The players assist the NPCs in resolving their problems to complete each quest in the game.

The game mechanics included are the following: orientation; exploration; finding information; simulation; making choices; interaction; collecting objects; collecting information; solving mazes; and solving problems.

The dynamics within the game – summary, navigation system, quest system, decision-making dialogue, emotive bar, and incentive system – are strictly related to the theoretical model supporting the digital game-based intervention.

The digital game-based intervention was developed using the Unity 2019 LTS engine, in single-player mode, with 3D graphics in a third-person perspective, in English.

"ReWIND" focuses on enjoyable and engaging game mechanics, integrating CBT techniques into the gameplay. The 3D virtual world affords a greater level of immersion.

Enables players to think more critically. The game's structural flow and storyline attempt to simulate real-life situations and lead players to think more clearly and thoroughly whenever they encounter similar occurrences in the real context. Provides practicing ways for developing positive thought patterns towards the issues faced under negative stimuli, facilitating effective cognitive emotion regulation skills that help improve anxiety disorders in real life. Leveraged decision-making elements to guide players with anxiety disorders to change the way they perceive their surroundings, using functional and rational means.

The digital game-based intervention was evaluated in young adults ($\geq$18 years; N = 40), a non-clinical population from institutes of higher learning, including 17 males (47.5%) and 23 females (57.5%). A randomized controlled trial was performed with two groups (intervention and control), integrating pre- and post-intervention assessment. The intervention format included a total of 3 sessions, each lasting from thirty minutes to one hour, with a maximum of two game scenarios in each session (6 scenarios total). The assessment instruments included related to mental health were the following: Beck Anxiety Inventory, State-Trace Anxiety Inventory, Cognitive-Emotion Regulation Questionnaire, and Anxiety Control Questionnaire. Additionally, the digital game usability was also assessed.

"ReWIND" significantly reduced the severity level of anxiety symptoms and trait anxiety levels and increased perceived control of anxiety better than the non-game task (control group). Results indicate a greater decrease in the severity level of anxiety symptoms and trait anxiety level, and a greater increase in perceived control of anxiety compared to the non-game task.

The limitations of the research included the short duration of the intervention, the small size of the sample (non-representative), and the use of self-reported measures (bias risk). Attending to the intervention limitations, more research needs to be developed to validate "ReWIND" for a wider variety of anxiety disorders.

3.5 Results of Syntheses

The present systematic review supports the effectiveness of digital game-based interventions in the most prevalent mental disorders among young adults. Despite the reduced amount of evidence in the sample, the quality of the sources provided relevant data, covering two categories of mental disorders – mood and anxiety. The data retrieved for designing effective digital game-based interventions in anxiety and depression is synthesized following (Fig. 4).

DESIGN

- RPG genre [21,22]
- Adaptative game difficulty [21,22]
- Well-designed goals and therapeutic objectives [21]
- CBT theoretical support [21,22]
- CBT practical support [21,22]
- Original and realistic aesthetical elements [21,22]
- Relevant stories [21,22]
- Interactive dialogues [21,22]
- Non-inclusion of time dynamics [21]
- Good playability [21,22]

- Inclusion of relevant mechanics [21,22]
- Add complexity and support through dynamics [21,22]
- Simulations of real-life situations in non-harmful contexts [21,22]
- Development of cognitive, emotional, and social skills [21,22]
- Correct application of interventive structure and techniques [21,22]
- Relevant mental health content and practice [21,22]
- Self-awareness increment [1,2]
- Practice of adaptative behaviors and healthier habits [21,22]
- High realism aesthetics (situations, characters, and world) [21,22]

"CBT4Depression" (Yusof et al., 2022)[21] and "ReWIND" (Ken et al., 2023)[22]

Fig. 4. Synthesis of design elements in effective digital game-based interventions in anxiety and depression.

The development and implementation data for effective digital game-based interventions are merged into the following figure (Fig. 5).

DEVELOPMENT

- Computer game [21,22]
- Developed in powerful engines – Unity 2019 LTS [22]
- Single-player mode [21,22]
- 3D graphics for a greater sense of immersion [22]
- Third-person camera view [21,22]
- Use of the native language [21,22]
- Aesthetic elements contributing to engagement [21,22]
- System adaptability to the player performance [21,22]

IMPLEMENTATION

- Large audience [22]
- Adequate research type [21,22]
- A follow-up assessment [22]
- Intervention format capable of producing results [21,22]
- Adequate and validated assessment measures [21,22]
- Intervention duration, sample size and diversity, and the use of self-reported measures are considered potential limitations [22]

"CBT4Depression" (Yusof et al., 2022)[21] and "ReWIND" (Ken et al., 2023)[22]

Fig. 5. Synthesis of development and implementation features in effective digital game-based interventions in anxiety and depression.

4 Discussion

Despite the volume of sources retrieved (N = 3.751) and the latest body of research merging digital games and mental health [9, 11, 12], the effectiveness of digital game-based interventions in the populations' most prevalent disorders is still scarce.

This comes as surprising and curious, given that commercial off-the-shelf (COTS) digital games – not designed, developed, and implemented for mental intervention purposes – have been positively impacting mental health, helping players to manage anxiety and depression symptoms by themselves, and attesting the potential of digital games in the field [9]. Supporting previous results, this research found and excluded one piece of evidence reporting a positive significant effect in a group playing a COTS digital game when compared to other two groups, one playing the digital game-based intervention under the scope of the study and another receiving the treatment as usual [23].

This lack of scientific evidence within the research might be related to several factors; however, two evidence-exclusion criteria raised our attention: non-effective digital game-based interventions (D); and non-randomly controlled trials (J). The sources within these criteria were excluded mostly due to research design problems, conditioning the potential of finished interventions. These patterns reinforce the need for adequate scientific research, fundamental to the evidence-based practice supporting clinical psychology and an adequate intervention in mental disorders.

Previous studies have surpassed these problems through an extended length of research and less restrictive criteria: (i) Ferrari et al., 2022 appealed to any research (quantitative, qualitative, or mixed) with no effect measures required, and (ii) Shah et al., 2018 focused on empirical studies with a pre-and post-intervention statistical outcome on a psychological self-report measure (quantitative), not necessarily significant. We believe the problem must be addressed properly, so the scope of our study was decided to be maintained.

Attending to the results achieved within the study dimensions, most of the design, development, and implementation characteristics are in line with previous literature.

RPG genre facilitates the application of psychotherapeutic techniques [11, 12, 21, 22]. The rules clear and specific are fundamental for playability [5, 6, 8, 21, 24]. The game goals are different from the characterizing goals (therapeutic objectives) which, by definition, corresponds to serious digital games, despite the lack of reference to serious game theories [5]. Therapeutical objectives should be objective, feasible, integrated into the design, and directed to the mental disorder under intervention [12], aligning with the principles of digital and serious games [5, 8, 21, 22]. The support of cognitive-behavior theory [21, 22], besides being an evidence-based psychotherapy practice [25, 26], is supported by most of the research merging digital games and mental health interventions [11, 12].

Aesthetics should be designed from scratch [11, 21, 22], supported by design techniques to facilitate engagement, likeliness, and realistic representation [6, 10], enriched with relevant content within significant narratives [5, 21, 22]. Given the sensitive content embedded into these resources, the design should attend to the several forms of stigma and discrimination in digital games [6, 10, 11, 18], despite these matters not being addressed within the evidence. The game narrative is fundamental to support the intervention techniques conveyed, providing learning content, promoting interaction, and potentially changing the story course [5, 21, 22]. The mechanics included are designed to achieve the game goal and the therapeutic objective, and dynamics not only enable the mechanics and playability but can also add layers of complexity to the game system, increasing fun and engagement [7, 8, 21, 22].

One piece of excluded evidence in the present research, directed to depression symptoms and psychological distress, resulted in higher anxiety, disgust, and arousal ratings [27]. This draws our attention to the need for a quality of the design to a digital game to effectively convey a digital game-based intervention for mental disorders, and the calculation of potential side effects when considering the elements to include or exclude in the game.

Regarding the development, previous literature and present evidence suggest that the use of tridimensional graphics allows a greater sense of immersion [6, 22], and the camera perspective in the third person triggers a higher sense of identification, facilitating the players' projection [18, 19]. The system adaptability provides the adjustment of difficulty to the player's performance, ensuring the player can master the required skills and successfully progress [12, 21, 22]. Generalization achievement is fundamental in effective digital game-based interventions, as it ensures the knowledge transfer to the players' real life, enabling the effect on mental health [12, 21, 22].

Finally, focusing on implementation, effective digital game-based interventions should follow the best practices in mental health [25, 26], appeal to evidence-based practices in clinical and health psychology, and use adequate mental health assessment instruments for assessment [21, 22, 26]. No considerations were found regarding the potential negative impact of digital games – such as potential additions (internet or digital gaming) – and digital game-based interventions – such as side effects – but our research considers those relevant to address in the future.

In summary, the present systematic review attests that digital game-based interventions can be effective in mental disorders [21, 22], namely anxiety and depression categories of mental disorders (DSM-V; APA, 2013) but highlights that more quality research is still lacking, reinforcing the value of this paper. Despite not being found the volume and diversity of effective digital game-based interventions expected – lacking interventions for substance use and impulse control disorders – the two interventions for different categories analyzed – anxiety and mood disorders – provide relevant recommendations in line with previous evidence. The results show the need to explore and give more attention to the relationship between digital games and mental health interventions through human-computer interaction. Limitations of this research include the reduced sources of evidence responding to the restrictive criteria applied, and the calculated low bias eventually interfering minimally with the results achieved. Future work should focus on evidence-based interventions in anxiety, depression, substance use, and impulse control disorders, supported by CBT techniques and conveyed through digital games, with an adequate research design. These could include the design, development, and implementation of new interventions or the application of proper research design and processes with interventions already created. It is also suggested that future research should include clinical populations to increase the validity of digital game-based interventions in mental disorders.

Acknowledgments. The present paper was supported by a research grant established between the DigiMedia – Digital Media and Interaction Centre, Department of Communication and Art, University of Aveiro, Portugal, and the Foundation for Science and Technology, Portugal, with the reference UI/BD/154591/2022.

References

1. WHO: Comprehensive Mental Health Action Plan 2013–2030, p. 40. World Health Organization (2013)
2. WHO: World Mental Health Report: Transforming Mental for All (2022). https://www.who.int/teams/mental-health-and-substance-use/world-mental-health-report
3. WHO: The Mental Health Coalition: a WHO/Europe Flagship Initiative (2020)
4. WHO: 10 Facts on mental health. World Health Organization (2022). https://www.who.int/news-room/facts-in-pictures/detail/mental-health
5. Dörner, R., Göbel, S., Effelsberg, W., Wiemeyer, J.: Serious Games: Foundations, Concepts and Practice. Sringer Link (2016)
6. Adams, E.: Fundamentals of Game Design, 3rd edn. New Riders (2014)
7. Fullerton, T.: Game Design Workshop: A Playcentric Approach to Creating Innovative Games, vol. 3. Elsevier, Morgan Kaufmann (2014)
8. Hunicke, R., LeBlanc, M., Zubek, R.: MDA: a formal approach to game design and game research. In: Game Developers Conference (2004)
9. Mendes, R., Ribeiro, T., Veloso, A.I.: Mental health and digital games: a comprehensive qualitative review. In: 24th GAME-ON'2023 Conference Digital Gamified Systems. Carlow, Ireland (2023)
10. Isbister, K.: Better game characters by design : a psychological approach. Better Game Characters by Design (2006). https://doi.org/10.1201/9780367807641
11. Ferrari, M., et al.: Gaming my way to recovery: a systematic scoping review of digital game interventions for young people's mental health treatment and promotion. Front. Digit. Health **4**, 814248 (2022)
12. Shah, A., Kraemer, K.R., Rong Won, C., Black, S., Hasenbein, W.: Developing digital intervention games for mental disorders: a review. Games Health J. Res. Dev. Clin. Appl. **7**, 213 (2018)
13. OECD: Beyond Academic Learning: First Results from the Survey of Social and Emotional Skills (2021). https://doi.org/10.1787/92a11084-en
14. Page, M.J., et al.: The PRISMA 2020 statement: an updated guideline for reporting systematic reviews. BMJ **372**, 1 (2021)
15. APA: Diagnostic and Statistical Manual of Mental Disorders: DSM-5^{TM}, 5th edn. American Psychiatric Association, DSM-5 Task Force (2013)
16. Beck, J.: Cognitive Behavior Theory: Basics and Beyond (2021)
17. Ferguson, K.E., O'Donohue, W.: Behavior therapy: the second and third waves. In: International Encyclopedia of the Social & Behavioral Sciences, pp. 431–436. Elsevier (2015). https://doi.org/10.1016/B978-0-08-097086-8.21090-8
18. Ribeiro, T., Mendes, R., Veloso, A.: Playable characters in digital games: aesthetics and gender identity in digital game player's preferences. In: 25th International Conference in Human-Computer Interaction. Copenhagen (2023). https://doi.org/10.1007/978-3-031-35979-8_23
19. Denisova, A., Cairns, P.: First person vs. third person perspective in digital games: Do player preferences affect immersion? In: Conference on Human Factors in Computing Systems - Proceedings 2015-April, pp. 145–148 (2015)

20. Gray, D.E.: Doing Research in the Real World. SAGE Publications Ltd (2013)
21. Yusof, N., Shaari, N.A.M., Yusoff, E.H.: CBT4Depression: a cognitive behaviour therapy (CBT) therapeutic game to reduce depression level among adolescents. Int. J. Adv. Comput. Sci. Appl. **13**, 257 (2022)
22. Ken, H.Y., Yan, J.S., Abdullah, Bin, M.F.I.L., Tang, Y., Prestopnik, N.: ReWIND: a CBT-based serious game to improve cognitive emotion regulation and anxiety disorder. Int. J. Serious Games (2023). https://doi.org/10.17083/ijsg.v10i3.603y
23. Bergmann, M., et al.: Effects of a video game intervention on symptoms, training motivation, and visuo-spatial memory in depression. Front Psychiatry **14**, 1173652 (2023)
24. Huizinga, J.: Homo Ludens. (1938)
25. APA: APA GUIDELINES on Evidence-Based Psychological Practice in Health Care. American Psychological Association (2021). https://www.apa.org/about/policy/psychological-practice-health-care.pdf
26. APA: APA Clinical Practice Guideline Development. American Psychological Association (2020). https://www.apa.org/about/offices/directorates/guidelines/clinical-practice
27. Haberkamp, A., et al.: Testing a gamified spider app to reduce spider fear and avoidance. J. Anxiety Disord.Disord. **77**, 102331 (2021)

"Caught in a PICL": Proposal of a Game to Support Language Therapy for Pre-school Children

Carolina Simões Araújo[1,2](✉)(iD), Ana Rita Valente[3,5,6], Liliana Vale Costa[1,4], and Samuel Silva[2,3]

[1] Department of Communication and Art, University of Aveiro, Aveiro, Portugal
{carolina.araujo00,lilianavale}@ua.pt
[2] Department of Electronics, Telecommunication and Informatics,
University of Aveiro, Aveiro, Portugal
sss@ua.pt
[3] Institute of Electronics and Informatics Engineering of Aveiro (IEETA),
Aveiro, Portugal
rita.valente@ua.pt
[4] Digital Media and Interaction (DigiMedia), Aveiro, Portugal
[5] iStutter Center, Lisbon, Portugal
[6] School of Health Sciences, Polytechnic of Leiria, Leiria, Portugal

Abstract. This paper presents the concept of a digital game designed to support language therapy for pre-school children (ages 3–6) with language disorders, which affect language acquisition and lead to communication difficulties. The game is based on the Language Intervention Programme (PICL), a scientifically validated tool used by speech and language therapists (SLTs) in Portugal to address difficulties in semantics and morphosyntax. Designed and developed in close collaboration with SLTs, the project adopts an iterative design approach to ensure the game aligns with the therapeutic objectives. The game adapts the semantics activities from PICL to engage children in exercising their language skills. This approach can contribute to an increase in children's motivation during therapy sessions and provide opportunities for at-home practice, which are crucial for improvement. The paper outlines the design process, insights from SLTs, and the potential of game-based interventions to complement traditional therapy.

This ongoing work aims to study the potential of a programme-based game to support language therapy in the area of semantics, and the receptivity of this approach by therapists and patients.

Keywords: Speech and Language Therapy · Language Disorders · Pre-School Children · Digital Game Development · Serious Games

1 Introduction and Background

Communication is an active, interactive process through which humans exchange information and ideas [2,19, p. 21]. In Portugal, it is estimated that approxi-

© The Author(s), under exclusive license to Springer Nature Switzerland AG 2025
A. Marto et al. (Eds.): VJ 2024, CCIS 2324, pp. 84–92, 2025.
https://doi.org/10.1007/978-3-031-81713-7_6

mately 8–11% of children have a developmental language disorder (Côrrea, Castro and Costa, cit. in: [13]). Language disorders cause significant impairments in the acquisition, comprehension and/or production of language, therefore hindering the ability to successfully communicate. These disorders may be associated with underlying biomedical conditions (e.g., hearing loss, autism spectrum disorder, cognitive disabilities); however, when no such conditions are present, they are referred to as developmental language disorders [12].

Language disorders affect one or more areas of language, including semantics (word meaning and interpretation of word combination), syntax (sentence structure), and morphology (word structure) [3,20]. Research supports that children with language disorders face both immediate and long-term challenges, such as difficulties in school and learning, as well as an increased risk of behavioural and psychological problems during adolescence and early adulthood [4,5,11].

Speech and Language Therapy
The severity of the impacts of language disorders underscores the importance of early assessment and treatment through speech and language therapy. Research indicates that timely interventions are crucial, particularly during pre-school years (ages 3–6), when improvement is most probable [6], and before these challenges interfere with the formal education process (Wallace et al., 2015; Law et al., 2003, cit. in: Rinaldi et al. [17]).

Speech and language therapists (SLTs) are healthcare professionals who assess patients' difficulties and implement adequate strategies aimed at addressing these challenges. During sessions, SLTs also employ techniques to engage and motivate patients [13]—often drawing from the patients' interests. To achieve this, therapists create or adapt a myriad of exercises using images, books, worksheets, and both digital and non-digital games. Typically, therapy is provided once a week, with sessions lasting 30–45 minutes.

Despite the recognised importance of intervention, an international European survey of over two thousand practitioners contended that the use of external scientific evidence in intervention is uncommon [10]. Many therapists base their decision-making on personal experience rather than research-backed methods. The aforementioned study urges practitioners to afford due importance to the available scientific evidence. In Portugal, similar concerns have been raised, with researchers such as Castro et al. stressing the need for greater focus on evidence-based interventions [13].

Language Intervention Programme (PICL)
The Language Intervention Programme (Programa de Intervenção em Competências Linguísticas, PICL) is the only scientifically validated programme in Portugal for intervention in the domains of semantics and morphosyntax [14,15]. Subsequent independent case studies have been conducted, further evidencing the efficacy of PICL [7,18].

The programme is aimed at pre-school children, or children at the beginning of school age, who have language disorders. It is comprised of several image cards, scenarios, cutout figures and cubes, and a physical manual. This manual proposes activities, detailing their objective, instructions, therapeutic strategies, necessary materials, and reinforcement clues [15].

Regarding semantic development, PICL includes the following activities [14, 15]: (i) identifying and naming pictures (colours, shapes, food, animals, clothes, body parts); (ii) identifying and naming images by function; (iii) identifying and naming semantic categories; (iv) spacial notions; (v) opposites; (vi) semantic relationships; (vii) semantic judgement; and (viii) semantic absurds.

Why Game-Based Approaches Might Make Sense

Digital technology is evermore present in children's lives, and many are already familiar with the use of computers, smartphones, and tablets. This paradigm shift has influenced intervention services in Portugal, which now frequently incorporate these devices into the treatment of language disorders [1]. Nonetheless, there remains a notable gap in digital solutions tailored to support language therapy in European Portuguese, particularly in the area of semantics.

SLTs have reported that digital games offer valuable benefits in clinical practice, as they can motivate patients, gather data, and enhance interactivity during interventions (Edwards, J., Dukhovny, E., cit in: [8]). Additionally, research indicates that games can lead to improved learning outcomes, and increased retention and attention (Boyle, E. et al., cit in: [9]).

The advantages of digital tools extend beyond therapy sessions, as improvement necessitates at-home practice and repetition—a process that can become monotonous. Furthermore, pre-school children rely on their caregivers to initiate and encourage this practice, which can pose challenges. However, digital games can transform this routine into an autonomous, engaging experience through reward mechanisms and progression within the game [9].

To the extent of our knowledge, the Phonological Awareness Stimulation Programme (Programa de Estimulação da Consciência Fonológica, PECF) is the only validated speech and language therapy intervention programme in Portugal that has been adapted to a digital game [22,23]. This adaptation has demonstrated its effectiveness in promoting the development of children's phonological awareness [23]. While it serves as a valuable reference, it addresses a different area of speech and language therapy intervention.

In light of these insights, and considering that PICL is currently only available in paper format, the adaptation of part of this programme as a digital game emerges as a promising approach. Thus, this work, conducted in collaboration with speech and language therapists, aims to study the potential of a programme-based game to support language therapy in the area of semantics, and the receptivity of this approach by therapists and patients.

2   Designing a Programme-Based Game for Language Therapy

This project adopts an iterative design and development approach, actively collaborating with speech and language therapists to inform and validate each phase. Their insights have been instrumental in shaping this proposal, which is detailed in the following sections.

2.1   Understanding Patient and Clinician Needs

To gain a deeper understanding of the work conducted by speech and language therapists, the initial stages involved multiple meetings with two therapists, each with extensive experience in treating pre-school children with language disorders. Additionally, the first author attended a therapy session to observe the interactions between therapist and patient. This process allowed for a better understanding of speech and language intervention and how to best approach patients and players, as summarised in Table 1.

Table 1. Key insights regarding practices in session and approaching patients, and SLTs requirements for a system and how to approach its players.

Key insights	
Sessions and approaching patients	Sessions are not solely comprised of therapeutic exercises but also incorporate moments of play and conversation to avoid fatigue.
	Exercise objectives should be communicated clearly to the patient.
	Providing performance-based feedback is essential, but negative feedback should be avoided when responses are incorrect. Instead, patients should be encouraged to keep trying.
	Patients are motivated by competition, whether with others or concerning their own performance.
System and approaching players	Offering choices and customisation for non-essential game elements, such as characters and accessories, is beneficial.
	Current solutions are designed for use in therapy sessions and are not intended for home use. A solution for both settings should include an exercise-only mode.
	The system should be easily accessible for both children and parents.
	The system should allow therapists to select specific exercises and/or customise their content, e.g., the words and images used.
	The system should track the patient's progress, voice recordings, and usage, making this information readily available to the therapist.

These insights inform the digital game development process by highlighting the need for (i) a balance of play and therapy to maintain engagement; (ii) clear communication of exercise objectives; (iii) positive, performance-based feedback to encourage players; (iv) rewarded challenges; and (v) customisation through in-game cosmetics. Additionally, the game should support both therapy sessions and home practice, providing easy access and customisation for therapists, along with progress tracking. Given the target audience's age (3–6), the game should rely on speech sound cues and automated speech recognition rather than text-based interaction.

2.2   A First Approach

The first approach to incorporating PICL's exercises into a game was based on the children's fable The Three Little Pigs. This concept stemmed from the idea that using a familiar story could be beneficial, as players could engage with known characters rather than simply hearing about them, and eliminate the need to understand a completely new narrative universe.

Players would assume the role of the hero and be introduced to the game through a conflict between the wolf, and themselves alongside the three little pigs. Each pig would present tasks for the player to complete in order to collect the necessary materials for building their houses before nightfall, and therefore escape the wolf. For example, players would need to collect wood by chopping down a large tree; however, to use the axe, they would have to correctly answer three therapy exercises.

After assessing details with the SLT, a game design document was created, outlining the player, visual aesthetic, core mechanics, assets, and progression. A paper prototype was constructed and the game's first challenge was evaluated with a 5-year-old child undergoing language therapy intervention[1]. The insights garnered from this evaluation, along with feedback from the SLT and digital game development experts, informed the development of a digital prototype in Godot[2].

In light of a deeper analysis of the digital prototype, it became evident that the concept based on the Three Little Pigs was overly simplistic. As noted by Miller [16, p. 165], games that appear too infantile can lead children to feel frustrated and disengaged, instead of mature and represented. While this initial prototype served as a valuable proof of concept, it ultimately lacked the necessary complexity and feedback mechanisms, such as animations, sound, and vocalised instructions, to fulfil its purpose and maintain player engagement—especially for those at the upper end of the target audience. This reflection prompted a reevaluation of the game's overall design and concept.

2.3   A New Approach

To stimulate the game design process, each of PICL's semantic activities was analysed. Based on their objectives, a list of potential mini-games targeting these goals was crafted. After discussing the ideas with the SLT, the most suitable concepts were selected. Each mini-game is set in a different environment,

---

[1] Paper Prototype Testing video. Available at: https://www.youtube.com/watch?v=AL2_9002B1s.

[2] Digital Prototype Gameplay video. Available at: https://www.youtube.com/watch?v=qdTTNudJo5E.

necessitating an overarching concept to cohesively incorporate these varied scenarios.

The player assumes the role of a king or queen ruling a 2D kingdom in need of their leadership. Their companion, the court jester, introduces the player to their responsibilities—the villagers need help. The game features an interactive hub (Fig. 1a) where players can choose which villager to assist, or retreat to their castle (Fig. 1b), where they can customise their character with accessories purchased using coins they have earned.

This proof of concept focuses on naming and identifying shapes, foods, animals, and colours through small challenges that reward players with stickers and coins. For example, when the player interacts with a spaceship through the hub, they are prompted to embark on a space quest—a challenge focused on identifying and naming shapes, consisting of three parts. First, the player is introduced to shapes (square, triangle, circle, and rectangle) and their key features. To engage players in this activity, they must trace star constellations that form each shape to continue their journey (Fig. 1c). After landing on a new planet, players practice naming shapes by saying aloud the shapes of passing asteroids to destroy them (an automatic speech recognition system is used to detect their answers). Finally, to practice identifying shapes, they must click on passing galactic debris that matches an announced shape.

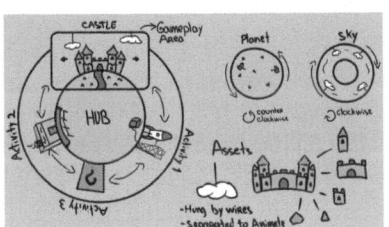

(a) The game's hub and environment.

(b) The player's castle in the hub.

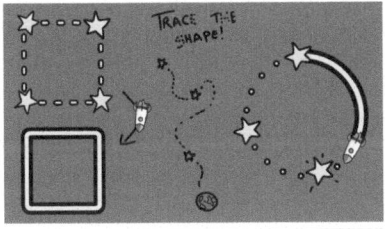

(c) The tracing shapes game.

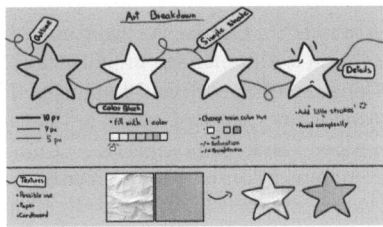

(d) Asset creation breakdown.

Fig. 1. Examples of concept art and art style breakdown.

The art direction of the game is being developed by an environment game artist with the target audience in mind, following visual design recommendations for this age group [21]. Considering young children's limited ability to process complex visuals, assets should have thick outlines and flat colour shading to ensure they stand out clearly from the backgrounds. However, hero assets can incorporate simple shading, gradients, details, and animations to highlight them (Fig. 1d). Beyond that, all elements should appear large and easy to interpret.

3 Conclusions and Future Direction

The process of game design and development is rarely straightforward, and this project exemplifies that reality. The collaboration with speech and language therapists has proven invaluable, as their expertise is essential for understanding how to best adapt therapeutic strategies for a game. Furthermore, given the inherent challenges of working with pre-school children with communication impairments, therapists act as crucial mediators, providing insights into the needs and preferences of the target audience.

As the project transitions to the development phase, it aims to create a meaningful tool to support language therapy. This phase will allow for formative evaluations during the development process to inform ongoing design improvements. A subsequent formal evaluation of the final product will be conducted to assess both therapist and patient receptivity to this tool. Ultimately, the findings from these evaluations may contribute to the academic discourse surrounding the design and development of therapeutic digital games, potentially informing best practices in the field.

Acknowledgments. We would like to acknowledge the Research Center (DigiMedia), funded by the Foundation for Science and Technology (FCT) in the context of the project [UIDB/05460/2020] - Project PLAYMUTATION, and the Institute of Electronics and Telematics Engineering of Aveiro, funded by the FCT in the context of the project [UIDB/00127/2020].

References

1. Alves, D.C., Castro, A., Coutinho, C.: COST IS1406 practitioner survey on intervention with language impairment results of Portugal. In: 10th European CPLOL Congress of Speech and Language Therapy (2018)
2. American Speech-Language-Hearing Association: Definition of Communication and Appropriate Targets. https://www.asha.org/njc/definition-of-communication-and-appropriate-targets/
3. American Speech-Language-Hearing Association: Definitions of communication disorders and variations [Relevant Paper] (1993). https://doi.org/10.1044/policy.RP1993-00208. https://www.asha.org/policy/rp1993-00208/
4. Beitchman, J., Brownlie, E.: Language Development and its Impact on Children's Psychosocial and Emotional Development (2012)
5. Bishop, D., Leonard, L.: Speech and Language Impairments in Children: Causes. Characteristics, Intervention and Outcome (2000)
6. Canadian Agency for Drugs and Technologies in Health: Screening Tools Compared to Parental Concern for Identifying Speech and Language Delays in Preschool Children: A Review of the Diagnostic Accuracy (2013)

7. Cardoso, A.P.T.: Impacto de um programa de intervenção em competências linguísticas numa criança com perturbação específica de linguagem (PEL) em idade
   escolar. Master's thesis (2018)
8. Du, Y., Salen Tekinbas, K.: Bridging the gap in mobile interaction design for children with disabilities: perspectives from a pediatric speech language pathologist.
   Int. J. Child-Comput. Interact. **23-24** (2020). https://doi.org/10.1016/j.ijcci.2019.
   100152
9. Duval, J., et al.: Designing Towards Maximum Motivation and Engagement in an
   Interactive Speech Therapy Game, p. 594 (2017). https://doi.org/10.1145/3078072.
   3084329
10. Forsythe, R., Murphy, C.A., Tulip, J., Law, J.: Why clinicians choose their lan-
    guage intervention approach: an international perspective on intervention for chil-
    dren with developmental language disorder. Folia Phoniatr. Logop. **73**(6), 537–551
    (2022). https://doi.org/10.1159/000513242
11. Fricke, S., Bowyer-Crane, C., Haley, A.J., Hulme, C., Snowling, M.J.: Efficacy of
    language intervention in the early years. J. Child Psychol. Psychiatry **54**(3), 280–
    290 (2013). https://doi.org/10.1111/jcpp.12010. https://onlinelibrary.wiley.com/
    doi/pdf/10.1111/jcpp.12010
12. Law, J., Garrett, Z., Nye, C.: The efficacy of treatment for children with devel-
    opmental speech and language delay/disorder: a meta-analysis. Journal of speech,
    language, and hearing research: JSLHR **47**(4), 924–943 (2004). https://doi.org/10.
    1044/1092-4388(2004/069)
13. Law, J., Mckean, C., Murphy, C.A., Thordardottir, E., Castro, A., Alves, D.C.,
    Departamento de Linguagem na Criança da Sociedade Portuguesa de Terapia da
    Fala: History of language impairment/developmental language disorder in Portu-
    gal. In: Managing Children with Developmental Language Disorder Theory and
    Practice Across Europe and Beyond, pp. 374–386. Routledge (2019). https://doi.
    org/10.4324/9780429455308
14. Lousada, M., Ramalho, A., Marques, C., Machado, B.: Programa de Intervenção
    em Competências Linguísticas (2017)
15. Lousada, M., Ramalho, M., Marques, C.: Effectiveness of the language intervention
    programme for preschool children. Folia Phoniatr. Logop. **68**(2), 80–85 (2016).
    https://doi.org/10.1159/000448684
16. Miller, C.H.: Digital Storytelling: A Creator's Guide to Interactive Entertainment.
    Taylor & Francis, USA (2004)
17. Rinaldi, S., et al.: Efficacy of the treatment of developmental language disor-
    der: a systematic review. Brain Sci. **11**(3), 36 (2021). https://doi.org/10.3390/
    brainsci11030407, number: 3 Publisher: Multidisciplinary Digital Publishing Insti-
    tute
18. Santos, M.R.P.d.: Aplicação de um programa de intervenção em competências
    linguísticas a uma criança com perturbação de linguagem secundária. Master's
    thesis (2017)
19. Sim-Sim, I.: Desenvolvimento da linguagem. No. 158 in Manuais, Universidade
    Aberta, Lisboa (1998)
20. Sociedade Portuguesa de Terapia da Fala: Dicionário Terminológico de Terapia da
    Fala (2020)
21. Soni, N., Aloba, A., Morga, K., Wisniewski, P., Anthony, L.: A Framework of
    Touchscreen Interaction Design Recommendations for Children (TIDRC): Charac-
    terizing the Gap between Research Evidence and Design Practice (2019). https://
    doi.org/10.1145/3311927.3323149

22. Sá, M., et al.: Development of the phonological awareness stimulation programme (PECF) - digital version. In: 2019 14th Iberian Conference on Information Systems and Technologies (CISTI), pp. 1–6 (2019). https://doi.org/10.23919/CISTI.2019. 8760963. https://ieeexplore.ieee.org/document/8760963. ISSN 2166-0727

23. Sá, M.C.V.B.d.: Análise da eficácia do programa de estimulação da consciência fonológica: versão digital. Master's thesis (2019). https://ria.ua.pt/handle/10773/ 27052

Edutainment

"Hello, Virus. Is this Cell You're Looking For?" A Virtual Reality Game for Perspective-Taking in Young Adult's Virology Education

Bernardo Marques[1], Liliana Vale Costa[2(✉)], Nuno Veloso[2], Daniel Francisco[2], Paulo Dias[1], and Beatriz Sousa Santos[1]

[1] IEETA, DETI, LASI, University of Aveiro, Aveiro, Portugal
[2] DigiMedia, DeCA, University of Aveiro, Aveiro, Portugal
lilianavale@ua.pt

Abstract. Viruses operate on a microscopic scale, and their behavior involves intricate interactions at a molecular level. The complex and dynamic nature of viral transmission pathways, host interactions, and mutation processes adds a layer of abstraction that requires students to engage in imaginative thinking. Thus, traditional approaches relying on static visuals or theoretical explanations fall short of providing a comprehensive understanding. This work proposes Virulent Odyssey, a Virtual Reality (VR) game to assist the education process for virus epidemiology and viral infections. Equipped with a VR Head-Mounted Display (HMD), students can use the controllers to navigate through a bloodstream, encountering various challenges. The game unfolds as a series of levels, using audio cues as a way-finding strategy to provide information on the virus' life cycle. The game was created following a Human-Centered Design (HCD) methodology with domain experts. In addition, an initial assessment was conducted with 22 individuals comprising distinct backgrounds to obtain first impressions regarding game narrative, interaction, and usability. Results suggest Virulent Odyssey has the potential to capture users' attention while helping to acquire new concepts.

Keywords: Education Process · Virus Epidemiology · Viral Infections · Virology · Human-Centered Design · Virtual Reality · Game Design

1 Introduction

One of the major concerns in education is finding strategies on how to establish the application of theoretical concepts in real-world scenarios while fostering critical thinking and research skills. For example, within microbiology, i.e., the study of microorganisms often imperceptible to the naked eye but wielding significant impact on our environment (e.g., bacteria, fungi, and viruses), traditional methods fall short in conveying the complete depth of phenomena. A particular example is virus epidemiology and viral infections. Understanding the dynamics of these microscopic entities holds paramount importance in the current global landscape. As viruses pose significant threats to public health and are key players in the emergence of infectious diseases, grasping the intricacies of their behavior is crucial for effective disease management and prevention [3,22]

However, traditional methods employed in teaching virus epidemiology, like static images, videos, and detailed explanations are insufficient in conveying the complexities of viral interactions within a host organism [19]. This limitation stems from the elusive nature of viruses, operating on a scale not visible to the naked eye. The intricate dynamics of viral replication, mutation processes, and interactions with host defenses require a level of spatial understanding that traditional methods struggle to provide. The limitations of static visuals become apparent when attempting to elucidate the real-time behaviors and nuanced interactions that occur at the microscopic scale, potentially hindering students' motivation, engagement, and overall learning capacity during classes [14].

To address this concern, information and communication technologies may have a key role in bridging the gap in traditional teaching to bring more inter-action and hands-on approaches [4,10,11]. Enhancing the learning experience in this field is not only essential for students' academic success but is also crucial in fostering a new generation of professionals equipped to address the ongoing and future challenges posed by viral infections.

In recent years, Virtual Reality (VR) has been gaining attention as a trans-formative technology in education, providing an immersive 3D digital environment [9,12,13,15]. VR may offer an interactive space for students and teachers to navigate, interact, and manipulate virtual objects. Its popularity stems from immersive features, technological advances, and cost reductions. The shift from a flat text-based interface to a rich, immersive 3D environment has shown promising benefits in various educational areas, spanning geography, natural sciences, mathematics, robotics, and beyond [5,6], including increased spatial visualization, information retention, innovative thinking, and problem-solving skills [21,23].

Another trending topic is the use of game-based approaches for educative purposes, which have emerged as a dynamic solution to enhance the learning process. Game-play approaches may captivate students' attention and immerse them in a compelling learning environment, which is known to foster increased interest and motivation among students. This departure from conventional classroom techniques aligns with the evolving needs of a technologically driven generation, opening new avenues of information acquisition [1,2].

One of the primary advantages of game-based learning is its ability to stimulate critical thinking and problem-solving skills. Educational games often present challenges and scenarios that require active decision-making, strategic thinking, and quick problem-solving, providing a hands-on experience that goes beyond passive learning. This dynamic engagement not only reinforces the comprehension of academic content but also nurtures skills that are invaluable in real-world applications. Moreover, the immediate feedback mechanisms incorporated into these games empower students to learn from their mistakes in a risk-free environment, promoting a cycle of continuous improvement [7,8].

Despite all the relevant research that has been disseminated in recent years, combining all the previous topics has been the target of a very reduced amount of work, although several research opportunities still exist. For instance, existing studies have explored the integration of VR within the higher education curricula [17], practical skills training with laboratory simulated exercises [20], deliver some cellular simulations [16], virtual labs [18], among others. Nevertheless, combining game elements within this VR reality and balancing fiction and reality has not been quite covered, allowing see the phenomena from different perspectives. Changes in perspective are likely to be essential to problem-solving, especially when dealing with uncertain scenarios and learning about the interactions of different microorganisms. Recognizing these challenges, there is an evident demand for innovative educational approaches that transcend conventional methods, providing a more immersive and effective means of comprehending the multifaceted complexities within the microbial world [5,24].

This work proposes Virulent Odyssey, a VR game designed to enhance the educational experience in virus epidemiology and viral infections addressed to young adult students. The game narrative, divided into various levels, delves into the concept of immersing the student in a microscopic world, assuming the role of a virus, in a first-person perspective. Equipped with a VR Head-Mounted Display (HMD), Oculus Quest 2, students must use the controllers to navigate through a bloodstream, while avoiding other cells until a designated location is reached and the infection process can begin. In this game, the player takes the perspective of a virus, and game audio is used as a way-finding strategy throughout the plot and provides information on the virus life cycle. One of the major challenges that is posited in these games is to balance reality-fiction in the process of learning and making it informative and interactive. For that, two domain experts in the area of microbiology were consulted to discuss the implications and behaviors of assuming the role of a virus. While passing each level, a survey on acquired knowledge about the virus life cycle is being considered to provide insights for continuous game refinement by the research team and domain experts. This dynamic feedback loop aims to optimize the learning experience, rendering the process of grasping virology more engagingly and effectively. Developed through a Human-Centered Design (HCD) methodology, Virulent Odyssey incorporated the perspectives of domain experts in virology during its design phase. An initial assessment involving 22 participants from diverse backgrounds was conducted to gather first impressions on aspects like game narrative, interaction, and usability.

The remnant of this paper is structured as follows. First, we describe the methodology used to understand the challenges of domain experts and target users. Next, we present a proposal for assisting with the education process of virus epidemiology and viral infections, focusing on the use of a VR game. Then, we report first impressions and propose next steps. Last, conclusions are drawn.

2   Methodology

In Fig. 1, the detailed depiction of the HCD methodology illustrates a collaborative integration of domain experts hailing from diverse fields, including Biology,
Virology, Education, Human-Computer Interaction (HCI), VR, Game Design
and Development, Computer Science, and others. The involvement of a total of
26 individuals was pivotal in comprehending the multifaceted challenges faced
by students, with a specific focus on virus epidemiologic-themed digital games
and the attitudes and behaviors of young adult learners concerning viral infections. This collective effort forms the foundation of a multidisciplinary research
project[1].

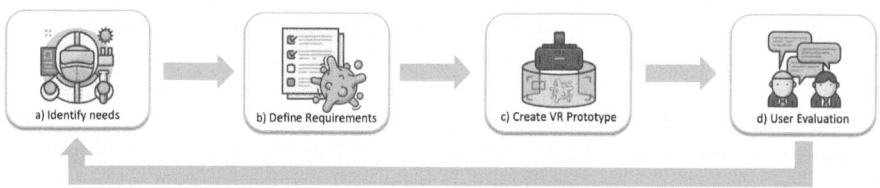

Fig. 1. Illustration of the HCD methodology employed to integrate domain experts
into the design of a VR game for helping students learning virology concepts. Assets
from iconfinder.com.

Within this comprehensive approach, the initial phase comprised a hybrid
exploration of the specific needs of the target users. This involved an extensive
series of focus group sessions and meetings, fostering an in-depth understanding
of the educational requirements and preferences crucial for effective learning.
Subsequently, the methodology progressed to establish meticulous requirements,
outlining key features and functionalities. The third step of the HCD methodology saw the actual creation of a prototype, leveraging VR technology to mate-
rialize the envisioned vision. This prototype served as a tangible manifestation
of the conceptualized design, allowing for a practical assessment of its viabil-
ity and effectiveness. Finally, the methodology culminates in a critical phase of
user evaluation, where the VR-based prototype underwent diverse testing scenarios and comprehensive feedback collection. This iterative process, guided by
the active involvement of domain experts, ensures a continuous refinement of
the serious game, aligning it precisely with the specific needs and expectations
of the intended users, ultimately striving for optimal educational impact.

Within the context of this work, we delineate a significant portion to the
proposed HCD methodology. A substantial investment of effort has been directed
towards the requirement elicitation and game development. Pertaining to the
final topic, user evaluation, it remains an ongoing work as we intend to delve
into additional user studies encompassing diverse audiences. Notwithstanding,

---

[1] https://playmutation.web.ua.pt.

we present preliminary results obtained from an initial assessment involving 22 participants. This represents the first step along the evaluation highway, serving as a prerequisite to garner initial impressions before embarking on more formal and exhaustive user studies in subsequent phases.

3   Framework Virology Learning Using a VR Game

This section presents the conceptual framework aimed at transforming the educational process of virus epidemiology and viral infection (Fig. 2). At its core lies a VR game designed to immerse students in the captivating world of virus dynamics, offering an engaging learning experience. The framework incorporates a visualization manager responsible for rendering all user interface elements, including intricate 3D models, animations, images, and text, ensuring a visually stimulating environment. Simultaneously, the interaction manager serves as the backbone of students' engagement within the virtual realm, offering intuitive control and facilitating meaningful interactions with the educational content. It orchestrates various aspects of students' navigation, manipulation, and engagement throughout their immersive educational journey.

Central to the framework is a data processing module handling the game's narrative delivery and enforcing level-specific constraints to optimize the student experience. This module ensures that students receive the game's storyline in a coherent manner while encountering appropriate challenges at each level. Additionally, a robust data storage component safeguards all pertinent game-related information, enabling students to seamlessly resume their progress and revisit levels to improve their scores and learning experience. This fosters a competitive spirit among students, motivating them to strive for excellence and outperform their peers, thus enhancing engagement and learning outcomes.

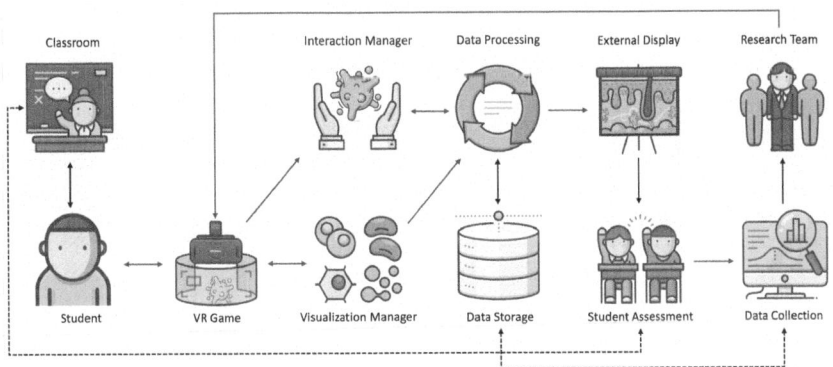

Fig. 2. Framework Overview delineating the primary modules of the envisaged solution and their interconnected dynamics. Assets from iconfinder.com

Moreover, the VR game supports external display sharing, enabling the entire class to witness and learn from the immersive experience of a single student. This innovative approach facilitates collaborative learning, where students can collectively participate in periodic assessments, providing valuable feedback on various aspects such as game narrative, usability, and interaction dynamics. Leveraging this feedback, the research team, including domain experts in virology, may iteratively refine the game design, ensuring its relevance, engagement, and alignment with educational objectives. This continuous improvement cycle ensures that the VR game evolves dynamically to meet the evolving needs and preferences of students.

3.1   Virulent Odyssey: A VR Serious Game for Virology Education

Virulent Odyssey envisions a captivating journey within the microscopic world. Placing students in the role of a virus, the game offers a mesmerizing first-person perspective, immersing them in the vibrant and dynamic environment of the bloodstream. By assuming the persona of a virus, students have an unique opportunity to delve deep into the complexities of viral interactions and transmission pathways, experiencing firsthand the challenges and obstacles encountered at the molecular level. As they navigate through the game (Fig. 3), students are tasked with maneuvering through the intricate network of blood vessels, encountering an array of challenges along the way. From evading vigilant immune cells to navigating through turbulent currents, students are presented with a series of obstacles that test their agility, strategic thinking, and problem-solving skills. Each encounter allows students to gain insights into the intricate dynamics of virus-host interactions and the complex mechanisms underlying viral infections.

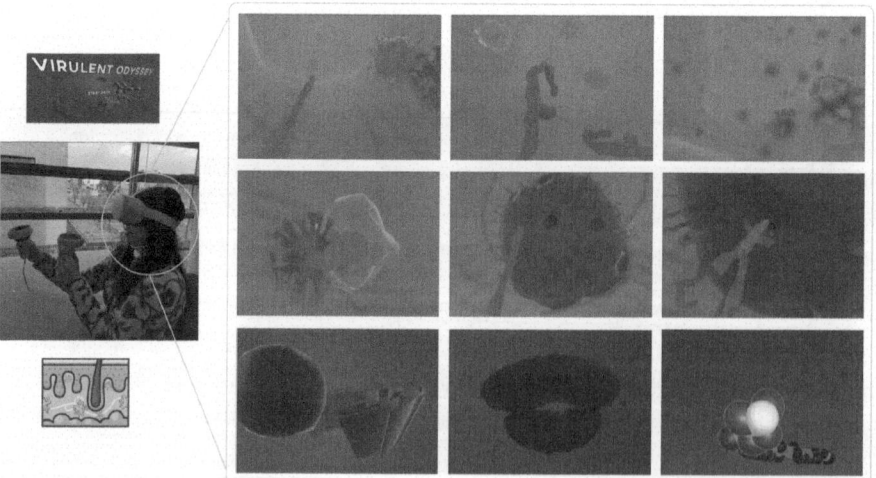

Fig. 3. Visual depiction of Virulent Odyssey game, showcasing a user equipped with a HMD navigating a microscopic world as a virus in a captivating first-person perspective.

Beyond, each level contributes to a larger storyline, gradually unveiling the mysteries of virus epidemiology and viral infections. As students progress through the game, they are not merely navigating through virtual environments but actively engaging with the underlying concepts and principles of virology. Through a combination of immersive gameplay, interactive challenges, and narrative-driven storytelling, Virulent Odyssey seeks to ignite students' curiosity, foster a deeper understanding of virus epidemiology, and inspire a passion for scientific inquiry within the captivating realm of VR.

Figure 4 illustrates how the game progression occurs. Transitioning beyond the introductory home screen, students embark on a journey through the tutorial stage, crafted to acquaint them with the diverse interactive features of the game (described below). This initial phase serves as a stepping stone, empowering students with the requisite skills and knowledge to navigate the virtual environment effectively. Through hands-on exploration and guided instruction, students are equipped to engage meaningfully with the subsequent levels and challenges presented within the game.

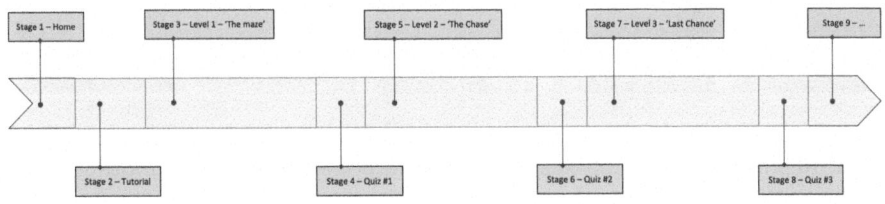

Fig. 4. Visualization of the progression within the Virulent Odyssey game.

As students progress, they unlock a series of captivating levels. Each level introduces new concepts and learning objectives, fostering a comprehensive understanding of virus-related topics while propelling students forward in their educational journey. Integral to the educational experience is the incorporation of quizzes at the conclusion of each level, serving as a checkpoint for students to assess their comprehension and retention of the virus life cycle. A diverse range of question formats are used (e.g., open-answer, true of false, as well as multiple choice) to cater to varied learning preferences and cognitive styles. By presenting a mix of question types, the quizzes aim to challenge students to apply their knowledge in different contexts, fostering critical thinking and analytical skills while reinforcing key learning objectives. These not only reinforce learning outcomes but also provide valuable feedback to both students and educators. As students answer the quizzes, their responses are logged for later analysis, providing valuable data to both students and educators alike. This data-driven approach enables educators to identify areas of strength and weakness among students, allowing for targeted interventions and personalized support where needed. Additionally, students benefit from immediate feedback on their quiz performance, facilitating self-assessment and enabling them to track their progress over time.

Figure 5 illustrates an array of the 3D models used for the game experience of Virulent Odyssey. Each model featured underwent a crafting process, that contribute to the immersive gameplay experience. The necessity for such creation stemmed from the nuanced and intricate nature of the subject matter, which demanded a thorough exploration and representation of every element within the game. The design and development team embarked on a search for visual references, sourcing inspiration from various scientific resources, expert consultations to ensure accuracy and fidelity to real-world phenomena.

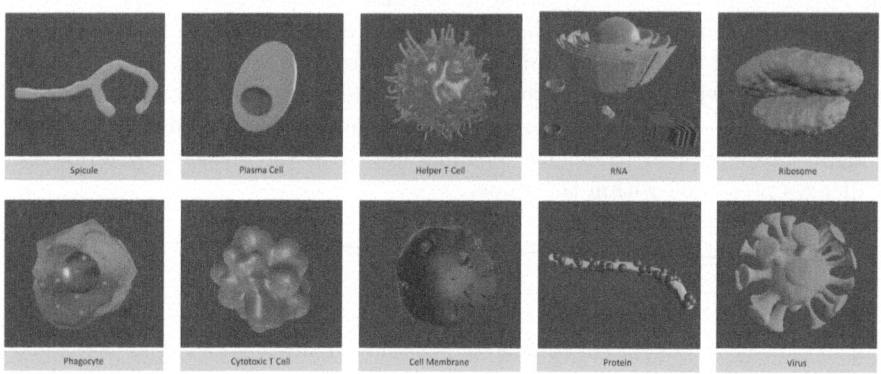

Fig. 5. Representation of a diverse 3D models incorporated into the various levels of the Virulent Odyssey game.

Once conceptual representations were gathered, the team transitioned into transforming these ideas into tangible 3D models. Each model underwent refinement to capture the essence of its real-world counterpart, from the intricate molecular structures to the dynamic interactions within the bloodstream. This attention to detail underscores the commitment to delivering a visually captivating and scientifically accurate gameplay experience. To provide concrete examples, consider the *'Spicule'* depicted in Fig. 5, which serves as a representation of the student's hands when they assume the role of the virus. Controlled via the VR headset controllers, these virtual appendages enable students to navigate through the digital space, manipulate objects, and interact with their surroundings as they progress through the game. This tactile engagement adds a layer of immersion to the learning experience, allowing students to embody the virus and actively participate in the educational journey.

Another example lies in one of the initial lessons of the game: the significance of the *'Plasma Cell'* and the *'Helper T Cell'*. Through an immersive audio narration, students are introduced to these crucial components of the immune system. For instance, as students encounter a Plasma Cell, the narration guides them, explaining its origin and function: *"In front of you, is a Plasma Cell. This organism results from the transformation of a white blood cell with the assistance of a Helper T Cell. Although appearing peaceful, if it detects an attack, it is capable of sending large quantities of antibodies, which can directly neutralize opponents or alert other protective organisms to come to aid! Therefore, do not approach too close"*. This auditory guidance not only enhances engagement but also provides valuable context, helping students understand the roles of various cellular entities within the immune response. At the outset of each level, these cues underscore the objectives, setting the stage for the learning experience. Additionally, when students encounter new elements within the environment, audio narrations provide context and explanation, laying the groundwork for future challenges and narrative developments.

A defining aspect of Virulent Odyssey is its integration of educational content seamlessly into the gameplay. For instance, students come face-to-face with *'Phagocytes'*, represented in vivid 3D detail, while receiving virtual narrations that elucidate their role in the immune response: *"over there... do you see it? Yes, a Phagocyte. This cells engage in the direct ingestion of intruders after being alerted by other organisms. Ohh no! It is coming for you! Quick, invade, invade..."*. These encounters serve as educational moments, allowing students to deepen their understanding of virology concepts in a hands-on manner.

Another interesting example is the final level, tasking students with penetrating the *'Cell Membrane'* to initiate the infection process from within. As they navigate this critical stage, students must intercept proteins on predefined paths, gaining insights into the intricate mechanisms of viral replication and cellular invasion. As before, an audio narrative is used, offering further information to assist with the learning process: *"Great, you arrived at the membrane of the cell to be infected! It is time to initiate the replication process. Start by attaching with its spike to penetrate the cell wall.... now intersect a protein to form a new virus. There is no mistake, proteins are small moving elements that look like threads, covered with tiny spheres."*. These and other lessons exist, illustrating how this dynamic representation underscores the engaging and educational nature of the Virulent Odyssey experience, offering a novel approach to learning about virology in a virtual environment.

Throughout their journey, students receive real-time feedback to enrich their learning experience. Visual cues, auditory signals, and haptic feedback mechanisms provide immediate responses to student actions, reinforcing successful interactions, highlighting key information, and alerting students to potential challenges. This feedback loop fosters an immersive and effective learning environment, ensuring that students remain engaged and motivated as they navigate the complexities of virology within the virtual realm.

Ultimately, Virulent Odyssey represents a transformative approach to education, leveraging the power of gaming technology for learning virology in an interactive and educational manner. The creation of the 3D models, essential for bringing the virtual environment to life, was executed using Blender. This powerful software allowed for the detailed crafting of each element, ensuring that the visual representations remained faithful to their real-life scientific counterparts. Through attention to detail, the 3D models accurately depict the various structures and organisms encountered within the game, enhancing the overall immersive experience for students. In parallel, the development of the VR environment itself was orchestrated using the Unity 3D game engine, a versatile platform renowned for its capabilities in creating interactive and visually stunning experiences. Leveraging C# scripts, the development team implemented various functionalities and interactions within the virtual environment, allowing students to navigate, interact, and engage with the content seamlessly. Furthermore, the Meta Quest SDK was instrumental in exporting the VR environment to the intended headset, ensuring compatibility and optimal performance for students using the Meta Quest platform.

The integration of audio cues played a pivotal role in enhancing the immersive experience and guiding students through the gameplay. Utilizing the ElevenLabs generative voice AI tool, text-to-speech technology was employed to provide students with dynamic narrations of their goals, lessons, and hints throughout the game. This innovative approach not only adds another layer of engagement but also facilitates comprehension and retention of key concepts, as students receive auditory reinforcement of the game's objectives and educational content.

4   First Impressions and Next Steps

In the initial assessment phase of the Virulent Odyssey game, the primary objective was to thoroughly analyze the reception of its innovative VR approach, placing a specific emphasis on key aspects such as the narrative, interaction dynamics, and overall usability. During the immersive journey through the game narrative, participants were accompanied by a researcher who played a dual role—offering assistance whenever needed and meticulously documenting noteworthy events that occurred during gameplay. This dual engagement aimed not only to observe participants' reactions but also to capture real-time insights into their interactions, challenges faced, and moments of engagement.

The experimental setup was an indoor environment, i.e., designated event room within our research institute, chosen for its controlled conditions and suitability for VR testing scenarios. Participants, each equipped with an Oculus Quest 2 headset, delved into the immersive experience of Virulent Odyssey. The choice of the headset, a widely acclaimed VR device, aimed to provide participants with a seamless and high-quality experience, ensuring that the technology did not hinder their engagement with the game. This meticulous setup allowed for a holistic evaluation, offering a glimpse into how participants interacted with the VR environment, their responses to the narrative elements, and their overall experience with the game. Following the gameplay session, a structured interview provided a valuable avenue for participants to express their thoughts, preferences, and any challenges encountered. Together, these elements contributed to a comprehensive understanding of the initial user experience and paved the way for iterative improvements in the game's design and functionality.

Overall, a total of 22 participants (4 females) were recruited for this assessment (Fig. 6). The age range spanned from 22 to 65 years, encompassing individuals with diverse professions such as faculty members, researchers, and Master and PhD students from various fields, including game design and development, informatics engineering, computational engineering, biomedical engineering, as well as robotics and intelligent systems. Importantly, all participants possessed prior knowledge and experience with VR technologies, ensuring a level of familiarity with the immersive medium. Every participant in the assessment also demonstrated familiarity with games across various areas of application. In contrast, none of the individuals had prior familiarity with virology concepts.

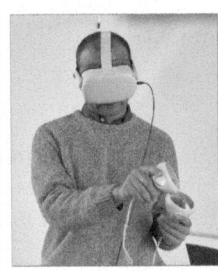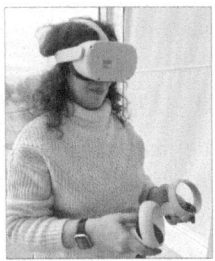

Fig. 6. Depiction of diverse individuals deeply engaged in the immersive experience of the Virulent Odyssey game.

Preliminary findings indicate that Virulent Odyssey shows promise in capturing users' attention to virology concepts, facilitating the acquisition of new knowledge. Participants emphasized that the use of VR provides an immersive and interactive environment, enhancing users' comprehension of abstract concepts and offering a deeper perception of 3D content. For example, one participant emphasized that: *"The pandemic helped me become aware that there is much about virology I do not know. Yet, this game helped me better understand the role of a virus, and how things happen. Almost like being in the front line of a guided trip"*. In fact, participants noted that such depth might not be achievable through traditional methods, particularly static image visualization. The spatial context within VR allowed for intuitive exploration and interaction, potentially fostering a more profound understanding of complex concepts compared to static representations, as highlighted by a participant: *"I pretty much loved being able to be the 'bad guy', because this helped me find out more about a topic I only learned in high-school, now with a distinct perspective"*.

Despite the potential layer of complexity from a shift to a 3D perspective, it is crucial to underscore that incorporating VR offers a more tangible learning experience, enabling a captivating experience into the intricacies of virus epidemiology. Furthermore, it was highlighted that the 3D dynamic elements of the virtual environment may likely play a role in boosting participants' motivation and engagement during the learning process. Plus, the ability to navigate, manipulate objects, and dynamically interact with content contributed to a more engaging and captivating learning experience. Some participants even commented that: *"Can this be applied to other areas of application?"* and *"This was a great experience. Do you have other levels besides virology I can play with?".*

Also relevant, participants noted a potential limitation in the VR environment, stating that: *"While I was immersed, although I was focused on the game narrative, I also notice the absence of awareness from the world around me"* and *"This would be great as a team experience. Either with other students or with a teacher. Are you planning on having a multi-player option?"* This seems to hinder their ability to establish empathy with other individuals, suggesting that exploring a collaborative setting might be beneficial in overcoming this constraint while improving users' social interaction.

4.1   Further Work

The insights garnered serve to elevate the current educational VR-based game but also establish a foundation for the evolution of this research line. Next, we explore some of the next steps we intend to delve into.

- **User Study with Target Users** - Conduct a comprehensive user study specifically targeting children to gather valuable insights into their experience with Virulent Odyssey. This would involve assessing their engagement, comprehension, and overall learning outcomes within the context of virus epidemiology and infections;
- **Expand Content and Levels** - Expanding the game's content to encompass a broader range of virology topics and related scientific concepts. Also, introduce new levels with distinct levels of complexity, each delving into different aspects of virus epidemiology, and even exploring interactions with the immune system in more detail;
- **Gamified Rewards** - Incorporate gamified elements within the game, where students earn rewards or progress based on their performance in quizzes, challenges, or problem-solving scenarios. This could add a competitive and motivating aspect to the learning experience;
- **Enhanced Audio Support** - Expand the audio support features within the game, including enhanced narration, explanations, and background music. This augmentation aims to create a more immersive and engaging auditory experience for the players;
- **Multiplayer Collaboration** - Explore multi-user settings, enabling collaboration among students and teachers within the virtual environment. This could foster teamwork, communication, and shared problem-solving, enhancing both the educational and social aspects of the learning process;
- **Multi-Role Feature** - Enable students to take on various roles by introducing a multi-role option at the beginning of the game. This will allow students to explore different perspectives within the virology context, fostering a more holistic understanding of the subject matter;
- **Incorporation of NPCs and Varied Gameplay Experiences** - Introduce Non-Playable Characters (NPCs), offering diverse gameplay experiences and histories. This will contribute to a more dynamic virtual environment, providing students with a multifaceted learning experience;

- **Teacher-Generated Quizzes** - Allow teachers to create quizzes dynamically on a separate device while enabling students to answer them within the game environment. This interactive assessment method will provide teachers with customization features;
- **Collaborative Learning** - Explore real-time collaborative features, allowing students to engage in joint problem-solving activities or group assignments (following the previous topic) within the virtual environment. This could enhance teamwork and communication skills;
- **Long-term Learning Assessment** - Implement long-term assessments to evaluate the game's impact on students' retention of virology concepts over an extended period. This could involve follow-up quizzes, or surveys conducted weeks or months after the initial gameplay to gauge the effectiveness of the educational experience.

5   Concluding Remarks

Nowadays, conventional teaching methods prove insufficient in delivering a comprehensive understanding of intricate biological processes, demanding students to engage in imaginative thinking. These limitations prompt a crucial opportunity for innovation. This work introduces Virulent Odyssey, a VR game designed to improve the educational process for virus epidemiology and infections, specially tailored for younger students. Students assume the perspective of a virus in a first-person view, navigating through a bloodstream while equipped with a VR headset and controllers. The game comprises levels with unique objectives, contributing to an overarching narrative. Beyond captivating visuals, audio cues guide students through the plot and articulate learning goals. Each level concludes with a quiz, reinforcing acquired knowledge, while data collected during gameplay facilitates continuous refinement of the game design.

Virulent Odyssey was created following an HCD methodology, in which domain experts in virology actively participated in the game's design. Furthermore, initial assessments involving 22 participants from diverse backgrounds indicate that the proposed game holds promise in capturing users' attention and aiding in concept acquisition. The incorporation of VR provides an immersive learning experience, enabling students to comprehend the complexities of virus epidemiology more engagingly.

The obtained insights serve not only to enhance the current educational VR game but also to lay the foundation for the progression of this research line. Potential avenues encompass the expansion of content and levels within the game could delve into more intricate aspects of virology, offering a more comprehensive exploration of related scientific concepts. The prospect of integrating multiplayer functionality is another promising direction, fostering collaborative learning experiences and social interaction within the virtual realm. Last, exploring the long-term effects of the educational process facilitated by the game could involve sustained assessments over extended periods, providing valuable insights into the durability and effectiveness of the acquired knowledge. To do this, it would be necessary to collaborate with educational institutions, allowing to intro-duction of Virulent Odyssey as supplementary educational material. This could involve conducting case studies within classrooms to assess the game's efficacy in real-world educational settings.

Acknowledgments. To everyone involved in brainstorming and discussion groups - especially Teresa and Margarida from INIAV, thanks for your time and expertise. This research was conducted in the scope of project PLAYMUTATION Virus Epidemiologic-themed Digital Games and Youngsters' Attitudes to Viral Infections from the Digital Media and Interaction Research Center (DigiMedia), funded by Foundation for Science and Technology (FCT) in the scope of project [UIDB/05460/2020], It was also supported by the Institute of Electronics and Telematics Engineering of Aveiro, funded through FCT in the context of the project [UIDB/00127/2020].

References

1. Beylefeld, A.A., Struwig, M.C.: A gaming approach to learning medical microbi-ology: students' experiences of flow. Med. Teach. **29**(9–10), 933–940 (2007)
2. Bowling, K.G., Klisch, Y., Wang, S., Beier, M.: Examining an online microbiology game as an effective tool for teaching the scientific process. J. Microbiol. Biol. Educ. **14**(1), 58–65 (2013)
3. Damasceno, E.F., Fernandes, L., Da Silva, A.P., Días, J.B.: A systematic map-ping study on low-cost immersive virtual reality for microbiology. IEEE Revista Iberoamericana de Tecnologias del Aprendizaje (2023)
4. Gaiz, A., Mosawy, S.: Educational technologies for online and blended learning in medical science. Acta Sci. Med. Sci. **2**(4), 50–54 (2018)
5. Hamilton, D., McKechnie, J., Edgerton, E., Wilson, C.: Immersive virtual reality as a pedagogical tool in education: a systematic literature review of quantitative learning outcomes and experimental design. J. Comput. Educ. **8**(1), 1–32 (2021)
6. Hayes, A., Daughrity, L.A., Meng, N.: Approaches to integrate virtual reality into k-16 lesson plans: an introduction for teachers. TechTrends **65**, 394–401 (2021)
7. Lameras, P., Arnab, S., De Freitas, S., Petridis, P., Dunwell, I.: Science teachers' experiences of inquiry-based learning through a serious game: a phenomenographic perspective. Smart Learn. Environ. **8**(1), 1–25 (2021)
8. Leitão, R., Maguire, M., Turner, S., Guimarães, L.: A systematic evaluation of game elements effects on students' motivation. Educ. Inf. Technol. 1–23 (2022)
9. Marks, B., Thomas, J.: Adoption of virtual reality technology in higher education: an evaluation of five teaching semesters in a purpose-designed laboratory. Educ. Inf. Technol. **27**(1), 1287–1305 (2022)
10. Marques, B., et al.: Supporting multi-user co-located training for industrial proce-dures through immersive virtual reality (VR) and a large-scale display. In: IEEE International Symposium on Mixed and Augmented Reality Adjunct (ISMAR-Adjunct), pp. 749–750 (2022)
11. Marques, B., et al.: Enhancing the educational process of geosciences through 3D reconstruction, virtual reality and a large display. In: International Conference on Mobile and Ubiquitous Multimedia, MUM, pp. 474–476 (2023)
12. Marques, B., Santos, B.S., Dias, P.: Ten years of immersive education: overview of a virtual and augmented reality course at postgraduate level. Comput. Graph. 104088 (2024)

13. Marques, B., et al.: An overview of teaching a virtual and augmented reality course at postgraduate level for ten years. In: Eurographics 2024 - Education Papers, p. 1 (2024)
14. Hernández-de Menéndez, M., Vallejo Guevara, A., Morales-Menendez, R.: Virtual reality laboratories: a review of experiences. Int. J. Interact. Des. Manuf. (IJIDeM) **13**, 947–966 (2019)
15. Peixoto, B., Pinto, R., Melo, M., Cabral, L., Bessa, M.: Immersive virtual reality for foreign language education: a prisma systematic review. IEEE Access **9**, 48952–48962 (2021)
16. Reen, F.J., et al.: Developing student codesigned immersive virtual reality simulations for teaching of challenging concepts in molecular and cellular biology. FEMS Microbiol. Lett. **369**(1) (2022)
17. Reen, F.J., et al.: The use of virtual reality in the teaching of challenging concepts in virology, cell culture and molecular biology. Front. Virtual Reality **2**, 670909 (2021)
18. Reyna, J.C., Barriga, C.D.G., Sáenz, G.A.S.: Virtual journey and virtual bio-labs: two novel strategies implemented in the pandemic to boost dynamic learning in the microbiology laboratory. In: Future of Educational Innovation-Workshop Series Data in Action, pp. 1–6 (2023)
19. de Souza, G.A.P., Queiroz, V.F., Lima, M.T., de Sousa Reis, E.V., Coelho, L.F.L., Abrahão, J.S.: Virus goes viral: an educational kit for virology classes. Virol. J. **17**(1), 1–8 (2020)
20. Tsai, H.P., Lin, C.W., Lin, Y.J., Yeh, C.S., Shan, Y.S.: Novel software for high-level virological testing: self-designed immersive virtual reality training approach. J. Med. Internet Res. **25** (2023)
21. Villena-Taranilla, R., Tirado-Olivares, S., Cozar-Gutierrez, R., González-Calero, J.A.: Effects of virtual reality on learning outcomes in k-6 education: a meta-analysis. Educ. Res. Rev. **35** (2022)
22. Xue, F., Guo, R., Yao, S., Wang, L., Ma, K.L.: From artifacts to outcomes: comparison of HMD VR, desktop, and slides lectures for food microbiology laboratory instruction. In: Proceedings of the CHI Conference on Human Factors in Computing Systems, pp. 1–17 (2023)
23. Zhang, X., Chen, Y., Hu, L., Wang, Y.: The metaverse in education: definition, framework, features, potential applications, challenges, and future research topics. Front. Psychol. **13** (2022)
24. Zhao, J., LaFemina, P., Carr, J., Sajjadi, P., Wallgrün, J.O., Klippel, A.: Learning in the field: comparison of desktop, immersive virtual reality, and actual field trips for place-based stem education. In: IEEE Conference on Virtual Reality and 3D User Interfaces (VR), pp. 893–902 (2020)

Level up! How Gamed-Based Activities Transform Learning and Alleviate Stress in Institutionalized Elderly

Elena Lacomba-Arnau[1] (iD), Anaísa Ribeiro[2] (iD), Raquel Sabino[3] (iD),
Rafael Pinheiro[3,4] (iD), Susana Lopes[4] (iD), Marisa Gaspar[3,4],
Carmen Navarro-Mateos[5](✉) (iD), Micael Sousa[6] (iD), and Marlene Rosa[3,4] (iD)

[1] Department de Psicologia, Sociologia i Treball Social, Universitat de Lleida, Lleida, Spain
[2] Faculty of Medicine and Biomedical Sciences, Algarve University, Faro, Portugal
[3] School of Health Sciences (ESSLei), Leiria, Portugal
[4] ciTechCare - Center for Innovative Care and Health Technology, Leiria, Portugal
[5] Department of Physical Education and Sports, University of Granada, Granada, Spain
carmenavarro@correo.ugr.es
[6] School of Architecture, Planning and Environmental Policy, University College Dublin, Dublin, Ireland

Abstract. Mental health issues are a critical concern for the elderly, as the inability to manage stress during stimulation activities can significantly impair their ability to accept and effectively learn new tasks, thereby affecting their performance in daily life activities. Serious games are increasingly recognized as valuable in the context of rehabilitation; however, there is a paucity of studies examining how elderly individuals manage stress and learn in regular practice using such games. In this study, 10 institutionalized elderly participants underwent 6 game-based stimulation sessions playing the serious games *Ta!Ti!* and *Mexerico*. Learning variables, including time and error rates, were assessed at baseline (T0), mid-point (T1), and the final session (T2), along with stress management indicators, specifically cortisol levels, at T0 and T2. The findings revealed that learning profiles improved throughout the program, with more pronounced gains observed initially. Additionally, stress levels decreased following each game-based session. The study identified significant relationships between stress management and learning profiles, suggesting that game-based activities can effectively enhance both learning outcomes and stress reduction in the elderly.

Keywords: Elderly · Game-Based Intervention · Serious Games · Learning

1 Introduction

Mental health in the elderly is an increasingly important topic due to its impact on the quality of life and functionality of this population. Worldwide, mental health issues among the elderly, such as anxiety, stress, and cognitive decline, affect millions of people. These disorders not only constitute a significant burden for the individuals affected but

A. Marto et al. (Eds.): VJ 2024, CCIS 2324, pp. 110–125, 2025.
https://doi.org/10.1007/978-3-031-81713-7_8

also for global and European health systems [1]. The prevalence of these disorders, including depression, anxiety, stress, and other cognitive and emotional problems, is high and expected to increase in the coming decades [1]. Although common, many of these issues are not adequately diagnosed or treated, exacerbating the problem and increasing the risk of functional decline and disability in this population [2]. Senescence, characteristic of this population, is associated with declines in cognitive functions and emotional changes that impact daily life [3, 4]. Adapting to the physical and psychosocial changes associated with aging plays a crucial role in the mental health of the elderly [2].

Mental health problems among the elderly, such as anxiety, depression, and cognitive decline, are prevalent and significantly impact their quality of life and independence [5]. These issues often lead to social withdrawal, agitation, and difficulties in performing daily tasks. Despite the high prevalence of these conditions, there is a notable lack of tailored interventions addressing the cognitive and behavioral needs of the elderly. Innovative methods, such as game-based activities, have shown promise in improving cognitive function and reducing stress, yet their integration into regular care practices remains limited. This gap highlights the need for more comprehensive and adaptive mental health strategies to better support the elderly population.

Effective stress management is crucial in learning contexts, particularly for the elderly. Games can be a very interesting motivational tools and engagement systems [6]. There is a growing need to integrate game-based activities into regular practices to enhance stress management and cognitive health in the elderly [10]. These activities provide mental stimulation, promote engagement, and create a sense of accomplishment, which can help mitigate stress levels. However, despite their potential, the incorporation of these activities into routine care and learning environments for the elderly remains limited.

The introduction of challenging activities, such as serious games, can have a significant impact on various skills in the elderly. Engaging in these activities improves cognitive functions, including memory, attention, and executive functioning, and enhances motor skills, social interaction, and emotional well-being [7, 10]. For instance, serious games that require problem-solving can stimulate cognitive processes, leading to better mental agility and delayed cognitive decline [7]. Physical games can improve coordination and physical fitness, crucial for maintaining independence [10]. Additionally, game-based activities promote social interaction, reducing loneliness and improving mood [11]. Emotional well-being is also enhanced through the sense of achievement and enjoyment derived from these activities, providing a holistic approach to improving the quality of life for the elderly [9]. Encouraging participation through tailored and accessible game-based interventions can lead to significant improvements in cognitive, physical, and emotional health, ultimately supporting healthier aging [8].

Acceptance of challenging activities, such as serious games, is closely related to stress levels in the elderly. While engaging in these activities can potentially reduce stress and improve cognitive and emotional well-being, there is limited research exploring how learning behavior during these activities relates to stress levels. Most existing studies rely on self-perceived instruments to measure stress, such as questionnaires and surveys. However, these methods may not fully capture the physiological aspects of stress. There is a notable gap in the literature regarding the use of objective indicators, such as biomarkers, to assess the relationship between learning behavior and stress levels. Biomarkers like cortisol levels could provide more accurate and reliable measures of stress, offering deeper insights into how game-based activities affect the elderly [7–9]. Further research using these physiological indicators is essential to understand the true impact of these activities on stress and learning in this population.

This study aimed to characterize learning profiles and stress levels during game-based activities in the institutionalized elderly and explore their relationships. By engaging participants in serious games, we sought to evaluate how these activities influence cognitive functions and stress responses. We propose two board games using modern board game mechanics [12], but we present them to the users as a progression of challenges. We intend to profit from the users' familiarity with analogue games and propose simplified game systems for easier replication and dissemination.

Using a combination of performance metrics and biomarkers, such as salivary cortisol, we intended to understand the physiological and behavioral effects of game-based interventions. The goal was to determine whether improvements in learning and cognitive skills were associated with reductions in stress levels, thereby providing insights into the potential benefits of integrating game-based activities into regular care practices for the elderly. This research contributes to filling the gap in existing literature by using objective measures to assess the impact of challenging activities on the mental health and cognitive well-being of institutionalized older adults.

2 Description of the Games

For this intervention, two analog board games were used: *Ta!Ti!* and *Mexerico*. Both were specifically created to enhance motor and cognitive skills in frail elderly. Additionally, they feature different levels of play to adapt to the characteristics and needs of the players.

2.1 Ta!Ti!

Ta!Ti! is a board game (Fig. 1) that promotes different cognitive and motor skills, such as: sustained attention, selective attention, coordination, dual tasking, executive functions and communication skills. The game set is composed of 1–4 boards (all different, with a random image distribution), 1–2 paper cups, 1 bell, 20 single-sided, 15 double-sided cards and 5 different position and sound cards. Among the included mechanics [12], using Browse Board Game Mechanics (from Board Game Geek [13]) as reference, are: pattern recognition, memory, dice rolling and speed matching. With the mechanics of pattern recognition and memory, the aim is to work on working memory, a cognitive function associated with the temporary storage of information and the formulation of a response. Including reaction time also allows for an emphasis on the inhibitory control of participants, as it requires managing reactions at both attentional and behavioral levels. There are two game modes: unilateral and bilateral. Various factors have also been added to increase the game's difficulty, allowing it to be tailored to the players' characteristics and needs. The first level is the unilateral mode without dice, where players only follow the instructions on the challenge card (Level 1). The second level introduces the dice, which, depending on the position and sound of the paper cup, requires the player to verbalize (Level 2). The third level is the simple bilateral mode, where only the information on the challenge card is considered (Level 3). Finally, the last level combines the bilateral mode with the card indicating both the position and sound of the paper cups (Level 4).

Fig. 1. Elements of *Ta!Ti!* (boards with different images, paper cups, bell and single-sided challenge card)

2.2 Mexerico

Mexerico is a board game that helps prevent and intervene in individuals with dementia or psychomotor frailty. Among the objectives of the game are: short-term and long-term memory, associative memory, sustained attention, selective attention and dual-task performance. The game set is composed by 4 wooden boards approximately A3 size, each with 6 spaces for placing discs, 24 discs with real-life images (to enhance the significance of the game) and 24 challenge cards (Fig. 2). Among the included gameplay mechanics [12] is memory, pattern recognition and storytelling. These mechanics were intended to impact the working memory of the participants, as it is essential at this age to work on cognitive processes such as language comprehension and reasoning, which are linked to the storage and processing of information.

In the first level, the discs are placed on the board, and a challenge card is drawn showing one of the pictures. The elderly player needs to think about the figure, touch it, and flip it over to work on fine motor skills. For the second level, two approaches can be used: the monitor selects a card, and the player has to remember where it was placed on the board; the other option is to present a memory challenge, asking the player to recall which object appeared, for example, in the second position during the first level. The third level consists of connecting different objects based on a premise. An example would be the player selecting the objects from the discs needed to prepare a meal. In addition to memory skills, this level also works on communication skills, as the players must explain why they select certain images over others.

Fig. 2. Elements of *Mexerico* (board and discs with real-life images)

3 Materials and Methods

3.1 Sample

This study involved ten participants, all recruited from two nursing assistance services. Each participant provided prior informed consent. Inclusion criteria required that participants did not have a history of moderate to severe dementia, psychiatric or neurological disorders, major medical conditions, or traumatic brain injuries that led to loss of consciousness. However, participants diagnosed with depression were eligible and included in the study. The study was approved by the ethical committee and the management of the institution ABEP (PARECER N. ° CE/IPLEIRIA/40/2023).

3.2 Procedure

The selection process began with four preliminary assessments to confirm the suitability of participants: (1) the Sociodemographic Questionnaire [14], (2) the 6-item Test for Cognitive Decline [15], (3) the Geriatric Depression Scale [16], and (4) Ryff's Psychological Well-Being Scale [17]. These assessments aimed to evaluate each participant's sociodemographic background, cognitive status, and emotional health, ensuring a suitable and homogenous sample. Following this, a test game was employed in different moments, to assess the participants' learning profiles.

The study consisted of six gaming intervention sessions, where each session involved playing a game, with two different games alternated across the sessions. Intervention sessions were performed using the analogical games *Ta!Ti!* and *Mexerico* [22], implemented alternatively per session. During *Ta!Ti!* interventions, first the unilateral version and then the bilateral version were applied [21]. The *Mexerico* was used in this study as an intervention tool, following the procedures from the first and second levels [22]. Each game-based intervention lasted approximately 20 min. These sessions were conducted in an isolated room to minimize distractions at the two nursing assistance services.

3.3 Materials

Sociodemographic and clinical data questionnaires
The sociodemographic and clinical data questionnaire [14] is a tool that aims to characterize the sample under study. It is composed of three distinct parts: (1) personal characterization; (2) professional characterization; and (3) characterization of the participant's health condition.

Six-item Cognitive Impairment Test (6CIT)
The 6CIT, adapted and validated for the Portuguese [15, 18], is a concise tool used to assess various cognitive domains, including temporal orientation, short-term memory (memory work), attention, concentration, calculation, language skills (command understanding), and executive functions. This test is particularly effective for detecting early stages of cognitive impairment [15, 18, 19]. The cognitive impairment thresholds, based on educational attainment, are defined as follows: a score ≥ 12 for those with ≤ 2 years of education; a score ≥ 10 for 3 to 6 years of schooling; and a score ≥ 4 for ≥ 7 years of education. Regardless of educational level, a score ≥ 10 also indicates cognitive impairment (exclusion criteria).

Geriatric Depression Scale (GDS)
The GDS [16, 20] is employed to screen for depression assessment of cognitive aspects of behaviors commonly altered in elderly individuals with depression. This scale is applicable to individuals with mild to moderate dementia. In this research, we utilized the Portuguese short-form version [16], which is a concise 15-question format. The scoring is interpreted as follows: 0 to 5 points suggest no depression; 6 to 10 points indicate mild depression; and 11 to 15 points suggest severe depression.

Ryff's Psychological Well-Being Scale (PWBS)

The PWBS [15, 19] assess various dimensions of psychological well-being across six core domains: (1) Self-Acceptance, (2) Autonomy, (3) Positive Relations with Others, (4) Personal Growth, (5) Mastery over Environment, and (6) Purpose in Life. In this study the short form 18-item version of scales adapted for the Portuguese population [19] is administered. This version also includes the six domains, capturing the essential content of each domain and ensuring a valid correlation between the short and original versions. Items are rated on a Likert scale from 1 (strongly disagree) to 6 (strongly agree), with some items inversely formulated and scored from 6 to 1. The total score, which ranges from 18 to 108, is calculated by summing all item scores, with higher scores indicating greater psychological well-being.

Game-based Learning Profiles
Game-based learning profiles were characterised using data from the performance evaluation in game-based intervention using the unilateral mode, level 1 of the *Ta!Ti!* [20]. Three standard plays were selected, and the execution time in seconds and the number of errors committed as a percentage (number of correct challenges/total number of challenges) were measured. The evaluation was implemented in baseline (T0), in session 3 (T1) and at the end of session 6 (T2).

Stress Levels
Cortisol samples were collected using the Salivette Collection Kit immediately before and after the gaming sessions designated for this purpose to assess the stress response biochemically. An accredited external laboratory analyzed the samples to guarantee the accuracy and reliability of the cortisol measurements [21]. Cortisol was collected at T0 (before and after the first session) and at T2 (before and after the 6th session).

3.4 Statistical Analysis

Sociodemographic and clinical characteristics (6-CIT, GDS, PWBS Ryff scale) were presented using absolute, mean, and standard deviation values. For the analysis of the PWBS Ryff scale, data were categorized into three intervals based on scores ranging from 18 to 108: Interval 1 (18–48), Interval 2 (49–78), and Interval 3 (79–108).

Since game-timed performance and error performance (in %) at T0, T1, and T2 were not normally distributed, median values and interquartile ranges (IQR) were calculated. Differences between T1 and T0, and T2 and T1, were also determined. Comparisons of these variables at different time points were made using the Wilcoxon signed-rank test, a non-parametric test for repeated measures ($p = 0.05$).

For cortisol values, mean and standard deviation were presented for the different time points (T0 before the session, T0 after the session, T2 before the session, T2 after the session). Mean differences were calculated for T0 and T2. Comparisons of cortisol levels at different time points were conducted using a paired t-test ($p = 0.05$).

To explore the relationships between stress levels (cortisol) and game-based learning profiles, a Spearman correlation test was used ($p = 0.05$). The interpretation of the Spearman correlation coefficient is as follows: 0–0.3 indicates a weak correlation, 0.4–0.6 indicates a moderate correlation, 0.7–0.9 indicates a strong correlation, and 1 indicates a perfect correlation.

4 Results

4.1 Sociodemographic and Clinical Characteristics

Ten elderly participants with no severe cognitive decline (only P17 presents a slight cognitive decline, 6-CIT > 7) living in a senior residence were involved in this study. They had a mean age of 84.2 years old (72–97; SD = 7.13). Four were female, more than 50% were widowers, and seven only had a primary education level. One participant (P20) was in the first interval of the PWBS Ryff scale, four participants (P3, P13, P17, and P18) were in the second interval, and five participants (P1, P5, P14, P15, P21) were in the third interval (Table 1).

Table 1. Sociodemographic and Clinical Characteristics of Participants

Participants	Gender	Age (years)	Marital Status	Level of education	6 CIT	GDS	Ryff PWBS
P1	M	72	Divorced	Secondary education	4	7	82
P3	M	91	Widower	Primary education	8	6	77
P5	M	89	Single	Primary education	8	1	85
P13	M	80	Widower	Primary education	0	3	74
P14	F	88	Widower	Primary education	10	0	87
P15	M	83	Widower	Primary education	2	2	103
P17	F	97	Single	Primary education	2	10	59
P18	F	79	Widower	Primary education	4	2	75
P20	F	82	Single	Secondary education	2	11	42
P21	M	81	Widower	Secondary education	0	6–0.02	84
TOTAL	6 M 4 F	84.2 ± 7.13	6 widowers 3 singles 1 divorced	7 primary education 3 secondary education	4 ± 3.35	4.80 ± 3.80	76.80 ± 16.59

Note: M, male; F, Female; 6-CIT, cognitive item test; GDS, Geriatric Depression Scale; PWBS, Psychological Well-Being Scale

4.2 Game-Based Learning Profiles

Data on game-timed performance (Fig. 3) showed a decrease over different evaluation periods: at T0, the median was 122.20 s (IQR = 93.00); at T1, the median was 32.10 s (IQR = 64.35); and at T2, the median was 25.92 s (IQR = 15.15). The reductions in game-timed performance were statistically significant between T0 and T1 (p = 0.007) and between T1 and T2 (p = 0.021). There were larger decreases in performance between T1 and T0 (−46.80; IQR = 83.10) than between T2 and T1 (−13.20; IQR = 24.15) (p = 0.028).

Data on game error performance (Fig. 3) also decreased: at T0, the median was 32.50% (IQR = 40.0); at T1, the median was 25.0% (IQR = 33.75); and at T2, the median was 0.0% (IQR = 25.0). There were no statistically significant differences between T0 and T1 (p = 0.08) or between T1 and T2 (p = 0.10). There were also no statistically significant differences between T1 and T0 (0.00; IQR = 43.75) or between T2 and T1 (−12.50; IQR = 31.25) (p = 0.81) (Table 2).

Table 2. Game-Based Learning Profiles and Performance Metrics

Participants	Game-timed performance (sec)			Mean differences		Game errors performance (%)			Mean differences	
	T0	T1	T2	T1-T0	T2-T1	T0	T1	T2	T1-T0	T2-T1
P1	123,00	32,40	20,4	−90,6	−12,0	25	25	0	0.0	−25
P3	361,80	81,00	27	−280,8	−54,0	60	60	0	0.0	−60
P5	123,00	13,20	19,8	−109,8	6,6	40	0	25	−40	25
P13	121,20	34,20	19,8	−87	−14,4	100	25	0	−75	−25
P14	123,00	84,00	34,2	−39	−49,8	80	25	25	−55	0.0
P15	34,80	23,40	23,4	−11,4	0,00	25	25	40	0.0	15
P17	138,00	90,00	74,4	−48	−15,6	40	60	25	20	−35
P18	30,60	18,00	13,2	−12,6	−4,8	0	0	0	0.0	0.0
P20	61,20	15,60	13,2	−45,6	−2,4	25	0	0	−25	0.0
P21	25,80	31,80	13,8	6	−18	25	25	0	0.0	−25
Median	122.20	32.10	25.92	−46.80	−13.20	32.50	25.00	0.00	0.0	−12.50
IQR	93.00	64.35	15.15	83.10	24.15	40.00	33.75	25.00	43.75	31.25

Note: Median values and interquartile ranges (IQR) for game-timed performance and error rates were calculated at three different time points (T0: baseline, T1: midpoint, T2: final session). Significant reductions in game-timed performance were observed between T0 and T1 (p = 0.007) and T1 and T2 (p = 0.021). No significant differences were found in game error performance between the time points

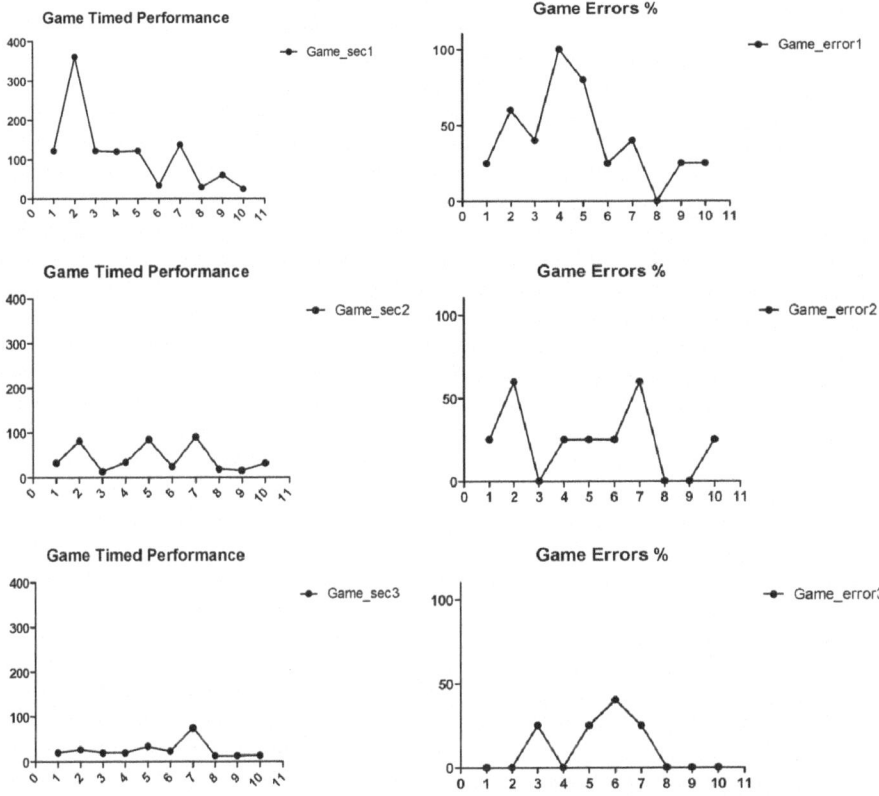

Fig. 3. Absolute values on game-timed performance and Game Errors percentage (%)

4.3 Salivary Cortisol Profiles and Relationship with Game-Based Learning Profiles

Cortisol means values significantly decreased after the game intervention (Fig. 4), both at T0 (−0.01; p [T0] = 0.004) and at T2 (−0.02; p [T2] = 0.001), with no significant differences in the magnitude of changes (p = 0.575). Cortisol values were not statically different between T0 and T2 before (p = 0.947) and after (p = 0.688) the game intervention (Table 3).

Table 3. Salivary Cortisol Profiles and Relationships with Game-Based Learning Profiles

	T0			T2		
Participants	Before	After	Diff T0	Before	After	Diff T0
P1	0,317 M	0,317M	0	0,319 M	0,301 M	−0,018
P3	0,104 A	0,141A	0,037	0,278 M	0,228 M	−0,05
P5	0,108 M	0,121M	0,013	0,165 M	0,142 M	−0,023
P13	0,214 A	0,215A	0,001	0,188 A	0,186 A	−0,002
P14	0,515 M	0,271A	−0,244	0,455 A	0,357 A	−0,098
P15	0,187 A	0,172A	−0,015	0,107 A	0,131 A	0,024
P17	0,103 A	0,134A	0,031	0,183 A	0,133 A	−0,05
P18	0,074 M	0,070A	−0,004	0,054 A	0,054 A	0
P20	0,349 M	0,299M	−0,05	0,077 A	0,062 A	−0,015
P21	0,140 M	0,226M	0,086	0,258 M	0,253 M	−0,005
Mean	0.212	0,197	−0,015	0,208	0,185	−0,024
SD	0,142	0,082	0.089	0,122	0,100	0,034

Note: M, morning reference values 0–0,736 μg/dL; A, afternoon reference values 0,069 e 0,378 μg/dL; SD, standard deviation

Values on game-timed performance at T0 were highly correlated with differences in cortisol at T2 ($S = -0.812$; $p = 0.004$). There was a borderline significant correlation between differences in cortisol at T2 and game-timed performance at T2 ($S = -0.0624$; $p = 0.054$) and between game errors difference (T2-T1) and cortisol differences at T0 ($S = -0.590$; $p = 0.072$) (Fig. 5).

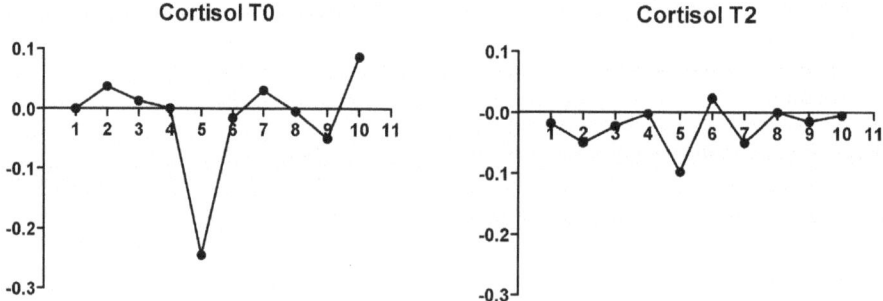

Fig. 4. Values of differences in cortisol levels before and after the first session (T0) and before and after the last session (T2)

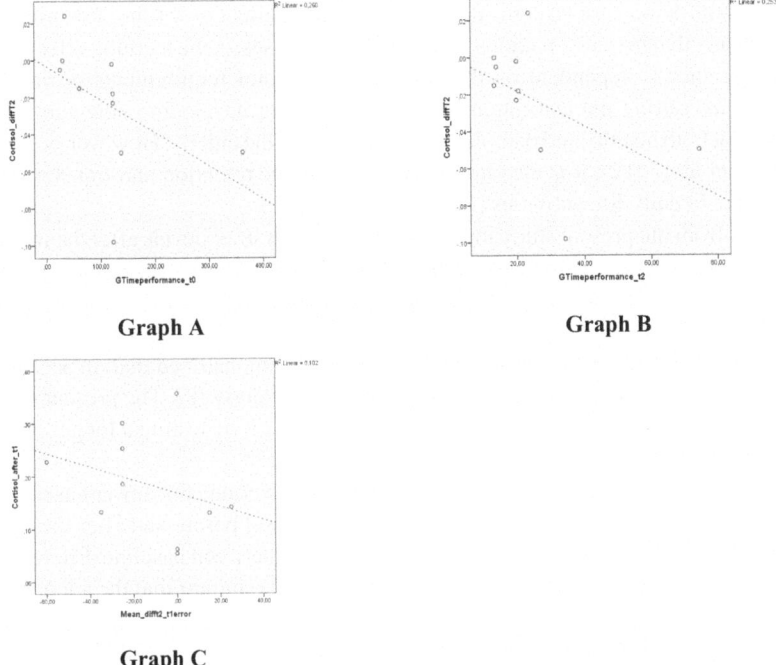

Graph A

Graph B

Graph C

Fig. 5. Scatter Plots Showing Relationships Between Cortisol Levels and Game Performance. (Graph A: Cortisol levels at T2 vs. Game-Timed Performance at T0; Graph B: Cortisol levels at T2 vs. Game-Timed Performance at T2; Graph C: Game Errors Difference (T2-T1) vs. Cortisol Differences at T0)

5 Discussion

This study finds different learning profiles during game-based activities in institutionalized elderly people according to different implementation protocol timings. It was possible to identify a significant learning process during game exposure in terms of game-timed performance but not the number of errors. The timed performance in-game improved throughout the program, with higher intensity in the first 3 sessions (T0-T1). This probably indicates that maintaining the learning effect over time in routine systems requires new and challenging adaptations for elderly people.

Reaction time during serious games has been used in other studies to distinguish elderly people with mild cognitive impairment from those with dementia. Chen et al. (2021) compared reaction times during whack-a-mole and hit-the-ball games in the elderly individuals without cognitive disability (n = 16), those with dementia (n = 11), and people with mild cognitive impairment (n = 10), finding significant differences between these groups [22]. Contrary to our results, Gee (2022) found that both game time and the frequency of errors are efficient indicators for screening impairments (especially cognition) in the elderly, using three serious games to measure vision perception coordination and psychomotor abilities [23].

Our results demonstrated some decrease in learning effect over time. According to a previous study that included patients with Parkinson's Disease, the learning effect during game experience is dependent on the cognitive and motor requirements of the tasks. Therefore, measuring the learning effect is important for developing new game-based programs for both healthy and pathological conditions in the elderly. However, according to Mendes et al. (2012), it is also important to assess the retention and transference of performance to daily life activities [24].

Results from the present study indicated a decrease in stress levels after the intervention with the game, both at the beginning and at the end of the 6-weeks of the program. In a previous study, 18 elderly participants demonstrated similar stress responses when subjected to a 3-session program based on digital games [8]. Cortisol levels did not change significantly after the game experience, which demonstrated that, in accordance with our study, game protocols are not stressful for the elderly [8]. The present study is an excellent extension of the results of Mahani et al. (2020) by testing a longer protocol (six sessions) [8].

Additionally, the results from the present study demonstrated that lower baseline performance was correlated with higher differences in cortisol before and after the session at T2. There was also a borderline significant correlation between baseline differences in cortisol and game error differences (T2-T1). These results suggest that the adaptation in cortisol levels during stimulation activities depends on the baseline performance level, or conversely, that individuals most able to adapt their baseline cortisol levels present different learning profiles. The relationships between the ability to adapt cortisol levels and learning performance have been shown in different learning domains. Specifically, Yang & Chen (2018) demonstrated that good gaming performance in elementary school children could foster the stabilization of stress levels, as measured by salivary cortisol [25]. Pireva et al. (2019) developed another study measuring affective-cognitive reactions while children play in learning contexts. Using an electroencephalography system, they compared learning using game-based strategies and pen-and-paper approaches in terms of stress, excitement, relaxation, focus and interest [26]. In their study using digital games, learners in the game-based group initially experienced less stress, indicating enhanced focus and engagement. However, the children started to lose interest and excitement during the 15 min of the protocol. In analyzing previous literature, it is important to highlight the importance of exploring game-based learning contexts and behavioral reactions in the elderly, especially due to their importance for healthy aging through rehabilitation game-based approaches.

6 Limitations

The implemented games use some of the modern game mechanics but not the way we see them on modern board games, not the games associated with the board game revolutions [27, 28]. In the future, we will go more into these designs to test other player profiles and then measure the stress levels per typified profile, which can be a very important contribution to using serious games in therapeutics and with these users.

In terms of methodological design, there are some recommendations for future related studies: (i) It will be beneficial to recruit a larger sample to explore how personal and clinical variables can influence the main variables of this study; (ii) It will be important to measure stress and learning variations over a longer period of game-based intervention, testing levels of progression in the protocol; (iii) it will be important to compare the same variables during different interventions, including game-based and other conventional therapies.

7 Conclusion

The present pilot study contributes significantly to the assumption that gamed-based strategies have a positive behavioural influence on elderly people and are potentially associated with an effective learning process. This study utilized a pre-post-test cohort methodology, which is adequate for structuring and improving the experiment. The study identified significant relationships between stress management and learning profiles, suggesting that game-based activities can effectively enhance both learning outcomes and stress reduction in the elderly.

Acknowledgments. This work has been made possible through the financial support of different key scholarships: the FI scholarship program of the Agència de Gestió d'Ajuts Universitaris i de Recerca de la Generalitat de Catalunya provided under the code "2023 FI-3 00107"; doctoral research scholarship program of the Foundation for Science and Technology of Portugal, with funding granted under the reference "2023. 02549.BDANA"; the competitive funding grants intended to finance the predoctoral hiring of research personnel in training by the agents of the Andalusian Knowledge System (code "24653").

Disclosure of Interests. There was no involvement by the company "Agilidades" in the analysis or interpretation of the data collected.

References

1. WHO, S.: Mental health of older adults. World Health Organization Internet site located at https://www.who.int/news-room/fact-sheets/detail/mental-health-of-older-adults (2017)
2. Wardhani, Y.F., et al.: The prevalence and distribution of risk factors for depression and emotional mental disorders in the elderly in Indonesia. International Journal of Social Psychiatry (2024)
3. Sequeira, T.: Depressão em idosos: O papel da família e da sociedade. In: Corotnean, T. (ed.) Estudo sobre os benefícios da estimulação cognitiva com recurso à realidade virtual em indivíduos com demência leve a moderada. Universidade Lusófona de Humanidades e Tecnologias (2010)
4. Corotnean, T.: Estudo sobre os benefícios da estimulação cognitiva com recurso à realidade virtual em indivíduos com demência leve a moderada (Master's thesis) (2019)
5. Van der Werf, M., Van Boxtel, M., Verhey, F., Jolles, J., Thewissen, V., Van Os, J.: Mild hearing impairment and psychotic experiences in a normal aging population. Schizophr. Res. **94**(1–3), 180–186 (2007)
6. Salen Tekinbas, K., Zimmerman, E.: Rules of play: Game design fundamentals (2003)
7. Anguera, J.A., Gunning, F.M., Areán, P.A.: Improving late life depression and cognitive control through the use of therapeutic video game technology: A proof-of-concept randomized trial. Depress. Anxiety **34**(6), 508–517 (2017)

8. Khalili-Mahani, N., et al.: Reflective and reflexive stress responses of older adults to three gaming experiences in relation to their cognitive abilities: mixed methods crossover study. JMIR mental health **7**(3) (2020)
9. Wiemeyer, J., Kliem, A.: Serious games in prevention and rehabilitation-a new panacea for elderly people? European Review of Aging and Physical Activity **9**, 41–50 (2012)
10. Stanmore, E.K., et al.: The effectiveness and cost-effectiveness of strength and balance Exergames to reduce falls risk for people aged 55 years and older in UK assisted living facilities: a multi-centre, cluster randomised controlled trial. BMC Med. **17**, 1–14 (2019)
11. Loos, E., Kaufman, D.: Positive impact of exergaming on older adults' mental and social well-being: in search of evidence. In: HUMAN ASPECTS OF IT FOR THE AGED POPULATION. APPLICATIONS IN HEALTH, ASSISTANCE, AND ENTERTAINMENT: 4TH INTERNATIONAL CONFERENCE, ITAP 2018. Proceedings, Part II 4, pp. 101–112. Springer International Publishing (2018)
12. Engelstein, G., Shalev, I.: Building Blocks of tabletop game design: An encyclopedia of mechanisms. CRC Press (2022)
13. Game Geek Board: Board Game Mechanics, site located at https://boardgamegeek.com/browse/boardgamemechanic
14. Dias, A., Rosa, M., Pincegher, D.: Anxiety and the Performance of Executive Functions by the Institutionalized Elderly Population - A Pilot Study Using the Matrix Instrument. In: Human Aging and Contemporaneity: Current Topics in Research, pp. 33–50. Digital Scientific Publisher (2023)
15. Apóstolo, J.L.A., Paiva, D.D.S., Silva, R.C.G.D., Santos, E.J.F.D., Schultz, T.J.: Adaptation and validation into Portuguese language of the six-item cognitive impairment test (6CIT). Aging Ment. Health **22**(9), 1190–1195 (2018)
16. Apóstolo, J.L.A., de Jesus Loureiro, L.M., dos Reis, I.A.C., da Silva, I.A.L.L., Cardoso, D.F.B., Sfetcu, R.: Contribuição para a adaptação da Geriatric Depression Scale-15 para a língua portuguesa. Revista de Enfermagem Referência **4**(3), 65–73 (2014)
17. Ryff, C.D.: Happiness is everything, or is it? Explorations on the meaning of psychological well-being. J. Person. Soc. Psychol. **57**(6) (1989)
18. Brooke, P., Bullock, R.: Validation of a 6 item cognitive impairment test with a view to primary care usage. Int. J. Geriatr. Psychiatry **14**(11), 936–940 (1999)
19. Abdel-Aziz, K., Larner, A.J.: Six-item cognitive impairment test (6CIT): pragmatic diagnostic accuracy study for dementia and MCI. Int. Psychogeriatr. **27**(6), 991–997 (2015)
20. Yesavage, J.A., Sheikh, J.I.: Geriatric depression scale (GDS) recent evidence and development of a shorter version. Clin. Gerontol. **5**(1–2), 165–173 (1986)
21. Rosa, M.N., Gordo, S., Pocinho, R., Marinho, R.: Uso de um jogo de tabuleiro na reabilitação dos membros superiores de idosos institucionalizados em Portugal: um estudo piloto quase-experimental. Revista Pesquisa Em Fisioterapia **11**(4), 657–670 (2021)
22. Wang G.Y., Simkute, D., Griskova-Bulanova, I.: Neurobiological Link between Stress and Gaming: A Scoping Review. J. Clin. Med. **12**(9) (2023)
23. Chen, Y.T., Hou, C.J., Derek, N., Huang, S.B., Huang, M.W., Wang, Y.Y.: Evaluation of the reaction time and accuracy rate in normal subjects, MCI, and dementia using serious games. Appl. Sci. **11**(2), 628 (2021)
24. Santos Mendes F.A., et al.: Motor learning, retention and transfer after virtual-reality-based training in Parkinson's disease-effect of motor and cognitive demands of games: a longitudinal, controlled clinical study. Physiotherapy **98**(3), 217–23 (2012)
25. Yang, J.C., Lin, M.Y.D., Chen, S.Y.: Effects of anxiety levels on learning performance and gaming performance in digital game-based learning. J. Comput. Assist. Learn. **34**(3), 324–334 (2018)

26. Pireva, K., Tahir, R., Imran, A.S., Chaudhary, N.: Evaluating learners' emotional states by monitoring brain waves for comparing game-based learning approach to pen-and-paper. In: 2019 IEEE FRONTIERS IN EDUCATION CONFERENCE (FIE), pp. 1–8. IEEE (2019)

27. Arnaudo, M.: The Tabletop Revolution: Gaming Reimagined in the 21st Century. McFarland (2023)

28. Sousa, M., Bernardo, E.: Back in the Game: modern board games. In: VIDEOGAME SCIENCES AND ARTS: 11TH INTERNATIONAL CONFERENCE, VJ 2019, Proceedings 11, pp. 72–85. Springer International Publishing, Aveiro, Portugal (2019)

A *Black Mirror*-Based Digital Breakout Promotes Creativity Thinking, Enjoyment and Absorption of Teachers in Training

Carmen Navarro-Mateos[✉] [iD] and Isaac J. Pérez-López[iD]

Department of Physical Education and Sports, University of Granada, Granada, Spain
carmenavarro@correo.ugr.es

Abstract. This article examines the implementation of a digital breakout inspired by the series *Black Mirror* within the Masters' in Teaching Program at the University of Granada (Spain). The primary aim was to assess its impact on students' creative thinking, enjoyment, and absorption during the course 'Learning and Teaching of Physical Education'. The methodology employed a mixed-methods approach, utilizing the GAMEX scale for quantitative analysis alongside open-ended qualitative feedback. Results indicated high mean scores in creative thinking ($\overline{X} = 4.62$), enjoyment ($\overline{X} = 4.5$), and absorption ($\overline{X} = 4.35$). Qualitative feedback highlighted how the immersive experience not only sparked creativity but also fostered problem-solving skills, linking theoretical content to practical application. While some students reported frustration, many noted the opportunity to develop emotional resilience and management skills crucial for future educators. This study contributes to the growing body of literature on game-based learning in higher education, emphasizing its role in enhancing pedagogical competencies. The results advocate for innovative teaching strategies that integrate narratives and interactive elements to foster deeper learning and engagement among pre-service teachers.

Keywords: Education · Breakout · Creative Thinking · Enjoyment · Absorption

1 Introduction

In the field of education in general, and in the university context in particular, it is essential to encourage active student participation, as it impacts the improvement of their commitment (Rodríguez-Izquierdo, 2020; Symaco and Tee, 2019). It is widely recognized that students who are actively engaged in the learning process and derive both enjoyment and practical value from their activities are more likely to achieve better learning outcomes (Kahu, 2013). The link between motivation and engagement is clear and straightforward, with motivation serving as a driver for, and often enhancing, engagement (Caruth, 2018; Lee and Reeve, 2012). In fact, there is evidence that positive motivation acts as a compensatory factor when skills and competencies are not very high, thus playing a fundamental role in the educational process (Guo et al., 2015; Majali, 2020).

© The Author(s), under exclusive license to Springer Nature Switzerland AG 2025
A. Marto et al. (Eds.): VJ 2024, CCIS 2324, pp. 126–139, 2025.
https://doi.org/10.1007/978-3-031-81713-7_9

In this context, to achieve competency development, it will be essential to focus on aspects that are significant to the current student population (Navarro-Mateos and Pérez-López 2022a, Navarro-Mateos et al., 2024). Analysing the profile of today's students, young adults are characterized by their quick decision-making ability and continuous connectivity, the inability to spend a long time doing the same thing, a preference for visual content, and a general concern about the personal image they project to society (Dauksevicuite, 2016). Considering the shifting flows and dynamics of the youth audience, such content serves as a rebranding mechanism for these entities by promoting frequent updates aligned with new trends, forms, and consumption habits of young people (Francisco-Lens and Rodríguez-Vázquez, 2020; Schanke, 2021). In such a changing and interconnected context, the effective use of Information and Communication Technologies (ICTs) is of great importance for professional development, as digital competence has become an essential requirement for personal and academic progress (Ali, 2019; Cabero-Almenara and Palacios-Rodríguez, 2020). Technology constitutes an important part of its reality and everyday life (Cilliers, 2017). On the other hand, the appeal of fiction and games, which are two of their main sources of leisure and entertainment (Del Moral-Pérez et al., 2021), can be leveraged in educational contexts.

Game-Based Learning (GBL) is one of the active methodologies that place students at the center of the teaching-learning process, emphasizing meaningfulness and engagement.This approach utilizes games to enhance cognitive processes, aligning them with curricular objectives and thereby improving student engagement and learning (Boyle et al., 2016; Tobias et al., 2014). There is evidence supporting the positive impact of GBL on cognitive development, motivation, and decision-making (Karakoç et al., 2020). According to Mora (2013), from a neuroeducational perspective, incorporating motivation and emotion into educational processes is crucial for achieving meaningful learning. GBL is a fun and interesting approach (Rotgans and Schmidt, 2011), where making a mistake is understood as something natural and part of the process (Kapur and Bielaczyc, 2012). Within GBL, escape rooms and breakouts are included (Fotaris and Mastoras, 2019). In these activities, students must achieve a specific goal within a limited time, such as escaping from a room (escape room) or opening a box (breakout). These activities provide intense, tension-filled experiences, as participants must solve various challenges (such as puzzles or enigmas) before time runs out, exploring possibilities autonomously. Such proposals may include a narrative that helps participants understand their mission and maintain interest in the game (Connelly et al., 2018). Similarly to gamification proposals, when based on a cinematic reference, it is important to recreate the same sensations and emotions experienced by the characters (Pérez-López, 2018). This contributes to increasing the credibility and immersion of the students in the experience (Navarro-Mateos et al., 2024; Pérez-López and Navarro-Mateos, 2022). As an educator, the challenge is to weave all the tasks into a narrative. This way, it can help students develop intrinsic motivation and go beyond merely focusing on extrinsic components (Yllana-Prieto et al., 2023). Moreover, such approaches can combine content work with the enhancement of skills related to soft skills, such as problem-solving, teamwork, emotional management, and critical thinking (Negre and Carrión, 2020).

The aim of this work is to describe the design and implementation of a digital breakout themed around the series *Black Mirror* in the Master's in Teaching Program, while also assessing students' perceptions and its impact on highly relevant aspects such as creative thinking, enjoyment, and absorption.

2 Context

The digital breakout was conducted in the course 'Learning and Teaching of Physical Education' during the 2023/2024 academic year, which is part of the specific module of the Master's degree in Teaching in Secondary School (Specialization in Physical Education) at the University of Granada (Spain). The purpose of this course was to develop the fundamental competencies of a Physical Education teacher, such as problem-solving ability, autonomy in learning, teamwork skills, and flexibility and adaptability to new situations.

To create a work context in which the previously mentioned competencies could be developed, it was decided to base the course on the television series *Black Mirror* (available on 'Netflix'). This allows for reflection on the impact that technology can have on our lives and the moral dilemmas behind certain advances in the digital age. This study was founded on the significance that this series holds for young individuals, as it was one of the most frequently mentioned series in a prior survey conducted to understand the participants' interests. For the course, the content was linked to different episodes of *Black Mirror*, with each session starting with a clip from an episode and a group discussion. The challenge for the instructors was to connect educational aspects with the plots and sensations of the series. Additionally, in order for students to feel like the protagonists of one of the episodes, the course recreated the application featured in the episode *Nosedive* (Season 3, Episode 1). In this episode, everyone has an app on their mobile devices through which they can rate their social interactions on a scale of 1 to 5 stars. This rating affects their daily life completely (from being able to rent a certain type of car to needing a minimum rating to book a flight). For the course 'Learning and Teaching Physical Education', each student had an app on their mobile phone (Fig. 1),

Fig. 1. Mobile application created for the subject inspired by the chapter *Nosedive*

where they could rate their classmates based on the quality of their contributions and interventions inside and outside the classroom. This created a context where developing conflict resolution and emotional management skills were very important, fundamental aspects for future educators. Through fiction, a context was thus created to address real problems with transfer to their future profession.

The breakout described below was the challenge presented to students on the first day of the course. Although there was no explicit time pressure (no countdown), it was important to complete it as quickly as possible because the time taken to complete the different stages would determine their starting score in the app. In fact, the final stage of the breakout "unlocked" access to the app.

3   Breakout Development

The digital breakout was created using 'Genially' platform, due to its interactivity and slide-locking features. It consisted of a total of 7 screens (Fig. 2), with the first one referring to the series as a whole and the remaining 6 corresponding to specific episodes from the series (one from each season). To enhance the sense of progress, a key aspect for affecting intrinsic motivation (Orbegoso, 2016), students were given the option to move from one screen to another. To access the application and complete the breakout, all screens needed to be resolved, though the order of resolution was not relevant, which

Fig. 2. Screens that comprised the digital breakout

supported the students' decision-making ability. At the bottom of each poster, there was an icon of a television remote that allowed them to navigate through the different screens. This allowed students to switch to a different screen if they were struggling with one and decide the order of the challenges.

In Table 1 below, a summary is provided of the chapters that inspired the screens, the challenges to be addressed, and their educational connection.

As can be seen in screens like the one from the *Playtest* episode, content related to the subject was included through a digital game (Fig. 3). In this case, students had to

Table 1. Connections between the chapters and the contents

Screen	Resolution	Educational Connection
Promotional Poster	Place a mirror in the middle of the mathematical operation and write it the way it appears in the reflection ($3 + 8 = 11$)	Reflection on the need to see beyond and not focus on the obvious
The Entire History of You (s.01, ep.03): An ocular device allows rewinding to any moment in the past	Locate a QR code in a store related to eyes in the city of Granada (a map of the city center appeared on the screen)	Rewind to the past of Physical Education to better understand the present image and the involved agents
The Waldo Moment (s.02, ep. 03): A cartoon begins to gain social and political impact due to its controversial interviews	Group the letters by color to find an account on the social network 'X' where the resolution code for the screen appears	Development of communication skills as a fundamental competence for future teachers
Playtest (s.03, ep.02): A company creates a chip that allows people who use it to confront their fears in a video game	Solve a digital puzzle (Fig. 3) before time runs out	Playful options in education: game-based learning, escape room/breakout, PBL and gamification
Arkangel (s.04, ep.02): A device allows parents to pixelate everything they don't want their children to see	Find out the name of the series whose poster appears pixelated	Guidelines for teaching intervention: how to address conflict resolution in the classroom
Smithereens (s.05, ep.02): A driver for a transportation company has a car accident due to using an addictive app	Locate the unique glass house featured at the end of the chapter (it is situated in a town in Granada)	Analysis of the impact of social media and its possibilities in the educational field
Joan is Awful (s.06, ep.01): Due to the information available online, a platform creates series about the lives of real people	Match the different facial fragments with the corresponding students	Possibilities of artificial intelligence for narrative creation

solve a digital puzzle that formed an image with the 'magic formula' of the Physical Education pyramid. This image included, through something as meaningful as emojis, the key aspects to consider for providing quality teaching.

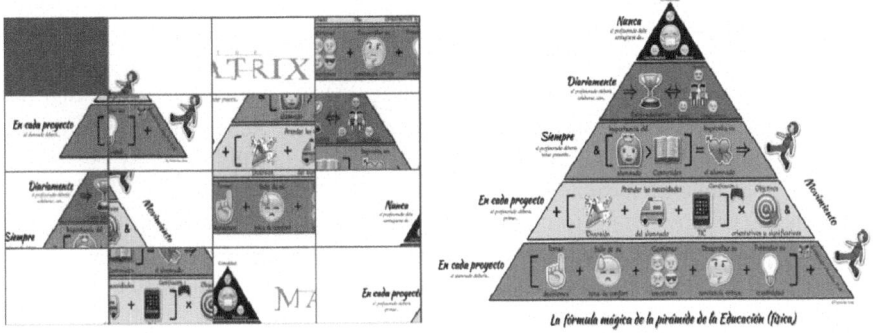

Fig. 3. Digital puzzle and resulting image

4 Methodology

To be able to know the impact of this project, the methodology used was mixed due to the impact that this approach has in the field of social sciences (Timans et al., 2019). This approach combines quantitative and qualitative tools to more comprehensively understand the research problem and obtain a more detailed picture of the studied reality (Johnson et al., 2007; Sánchez-Gómez et al., 2018). For the quantitative part, the GAMEX (Gameful Experience in Gamification) scale was used. The scale was developed and validated in English (Eppmann et al., 2018), and its reliability as a tool for collecting information in gamified experiences has been demonstrated. Particularly, the reliability of the questionnaire in the present sample showed a Cronbach's Alpha > 0.9 for each dimension. For this study, the validated Spanish version was used (Parra-González and Segura-Robles 2019). The response options were Likert-type (1 = "strongly disagree" and 5 = "strongly agree"). The scale comprised five main dimensions around the experience of participants in gamified environments:

1. Enjoyment: composed of six items and assessed the degree of the student's enjoyment with the experience. E-1: 'Playing the game was fun', E-2: 'I liked playing the game', E-3: 'I enjoyed playing the game very much', E-4: 'My game experience was pleasurable', E-5: 'I think playing the game is very entertaining', E-6: 'I would play this game for its own sake, not only when being asked to'.
2. Absorption: composed of six items and assessed the degree of absorption in the experience and evasion of the environment. A-1: 'Playing the game made me forget where I am', A-2: 'I forgot about my immediate surroundings while I played the game', A-3: 'After playing the game, I felt like coming back to the "real world" after a journey', A-4: 'Playing the game "got me away from it all"', A-5: 'While playing the game I was completely oblivious to everything around me', A-6: 'While playing the game I lost track of time'.

3. Creative thinking: composed of seven items, this dimension analysed the degree of creativity that the student perceives develops during the experience, and the confidence that the student has in himself during it. C-1: 'Playing the game sparked my imagination', C-2: 'While playing the game I felt creative', C-3: 'While playing the game I felt that I could explore things', C-4: 'While playing the game I felt adventurous', C-5: 'While playing the game I felt influential', C-6: 'While playing the game I felt autonomous', C-7: 'While playing the game I felt confident'.

4. Activation: consisted of three items and measured the degree of activity that the student considers has developed during the experience. AC-1: 'While playing the game I felt jittery', AC-2: 'While playing the game I felt frenzied', AC-3: 'While playing the game I felt excited'.

5. Absence of negative effect: formed by three items and analysed if students felt negative emotions while playing, such as frustration. AB-1: 'While playing the game I felt upset', AB-2: 'While playing the game I felt hostile', AB-3: 'While playing the game I felt frustrated'.

For the qualitative part, to better understand students' perceptions, feelings, and learnings, students were provided with a 'Google Drive' link to an open-ended question. They were invited to share their emotions, learnings, and evaluations of the breakout, all anonymously and without a limit on responses.

4.1 Sample Size and Characteristics

The breakout was implemented with group 3 of the course 'Learning and Teaching of Physical Education'. The students were allotted a two-week period to complete the breakout. The sample consisted of 26 students (24.83 ± 2.37 years), of whom 14 were men and 12 women. Following the results of an initial survey, only 46% were certain they wanted to pursue a teaching career. The rest had other professional goals, with teaching as a backup plan in case they did not achieve their main objective.

4.2 Data Analysis

For the quantitative data analysis, the statistical software 'SPSS version 25' was utilized. For the qualitative analysis, 'NVivo 11 Plus' provided essential support in the mechanical tasks of storing, organizing, and retrieving all the information collected regarding this experience, as well as for the data analysis itself.

5 Results

Next, the results of the descriptive analysis of the sample are presented (mean -$\overline{X}$- and standard deviation -σ-), addressing the dimensions included in the GAMEX: Enjoyment ($\overline{X}$: 4,5; σ: 0,46), Absorption ($\overline{X}$: : 4,35; 0,53), Creative thinking ($\overline{X}$: 4,62; σ: 0,42), Activation ($\overline{X}$: 4,1; σ: 0,61), Absence of negative effect ($\overline{X}$: 3,76; σ: 0,52).

Regarding the open question in 'Google Drive', a total of 78 responses were collected within the two-week period allowed for the breakout resolution. Starting from the initial word frequency, key categories were identified based on the participants' discourse. Following this, a detailed analysis of the most prominent concepts was conducted, establishing different categories and subcategories (Fig. 4). The most notable results will be highlighted for the items that make up each of the dimensions, accompanied by the evaluations that the students shared anonymously and voluntarily throughout the breakout.

First, it is important to highlight the dimension of 'Creative thinking', particularly the items with the highest averages: 'Playing the game sparked my imagination' ($\overline{X} = 4.9$), 'While playing the game I felt creative' ($\overline{X} = 4.9$), and 'While playing the game I felt autonomous' ($\overline{X} = 4.7$). The freedom to explore the possibilities of each screen and to try different ways to solve the puzzles allowed students to enhance their creative thinking. This idea is reflected in the following testimony: 'My brain wouldn't stop thinking of crazy ideas to solve the first screen. I never would have thought that I could come up with those solutions on my own' (N-18). Additionally, they had unlimited opportunities to test the digital password for each screen. It was impossible for them to arrive at the answer by chance, which, as one student put it, '…meant there were no limits; I could build a theory to solve the screen based on whatever my imagination came up with. It didn't matter if it wasn't the key; the process was worth it' (N-38). Such an approach creates a context in which each student can autonomously reach their own conclusions, allowing them to connect with their individual learning rhythms.

Códigos		
🔍 Buscar Proyecto		⌄
⊙ Nombre	Archivos	▲ Referencias
⊟ ○ 01_ Enjoyment	1	23
○ Motivation	1	12
○ Emotion	1	9
○ Aesthetics	1	5
⊟ ○ 02_ Creativity	1	11
○ Innovation	1	7
○ Problem_solving	1	6
⊟ ○ 03_ Immersion	1	21
○ Concentration	1	9
○ Flow_state	1	8
○ Narrative	1	6
⊟ ○ 04_ Learning	1	13
○ Digital_resources	1	7
○ Physical_Education	1	4
○ Practical_approach	1	3
⊟ ○ 05_Emotional_management	1	15
○ Time_pressure	1	9
○ Frustration	1	7

Fig. 4. Categories and subcategories extracted from students' narratives

On the other hand, analyzing the dimension of 'Absorption', the items that stand out are: 'Playing the game made me forget where I am' ($\overline{X} = 4.5$) and 'I forgot about my immediate surroundings while I played the game' ($\overline{X} = 4.5$). When an experience is immersive and the difficulty level of the challenge aligns with the participants' skill level (known as the flow state), individuals may even lose track of time while tackling the task due to high levels of focus and concentration. This state was experienced by several students: 'I was so into the breakout that I couldn't stop thinking of solutions to find the key. I looked at the clock and couldn't believe how fast the hours went by. I need more hours in the day!' (N-27) or 'It's been a while since I felt so focused on a challenge; this is testing me and keeping me completely engaged' (N-19). Additionally, the fact that it was inspired by a series that was very significant to many, along with the challenges connecting to the episodes, increased the levels of immersion: 'Black Mirror is my favorite series. This challenge has me focused, analyzing the number of connections with the show. I feel like I'm in one of the episodes!' (N-47).

In third place, it is important to analyze the dimension of 'Enjoyment', where the following items stand out: 'I liked playing the game' ($\overline{X} = 4.9$), 'I think playing the game is very entertaining' ($\overline{X} = 4.8$), and 'I would play this game for its own sake, not only when being asked to' ($\overline{X} = 4.8$). Approaches with a playful focus awaken enjoyment in participants, fostering a positive willingness to face challenges. In fact, in the narratives shared by students, it was evident that this enjoyment was closely linked to motivation: 'I'm having such a great time with this breakout. I acknowledge that it's not always easy and that I often get frustrated, but my motivation helps me overcome those moments' (N-57) or 'It had been a long time since I felt this motivated about something related to the university. When I become a teacher, I want my students to enjoy it as much as I am' (N-40). Furthermore, the fact that some tasks provided opportunities for interaction among them (like the one from the sixth season) was highly valued by the students, as illustrated in the following excerpt: 'It's really fun to observe the eyes, ears, and mouths of my classmates to solve the puzzle. This challenge allows us to interact both digitally and in person' (N-21).

Finally, regarding the two remaining dimensions, within 'Activation' the item 'While playing the game I felt frenzied' ($\overline{X} = 4.8$) stands out, and in 'Absence of negative effect', it is interesting to analyze the item 'While playing the game I felt frustrated' ($\overline{X} = 4.8$). In terms of activation, it is important to note that the breakout had many stimuli, since once they overcame the initial screen, they could access six screens with very different challenges. This made students feel '...completely overstimulated; I have to plan well to overcome the challenge' (N-4) or 'This is overwhelming; there's so much work behind this, I don't even know where to start. I love it!' (N-7). Furthermore, such proposals can generate in students the need to manage very interesting emotions from an educational perspective, such as frustration. This is why, linked to the dimension of absence of negative effect, we find comments like 'I can't take it anymore; I give up for today. This seemingly simple challenge is helping me get to know myself better. It brings out a side of me that I'm not proud of' (N-58) or 'I feel so close and yet so far. The frustration at times is enormous; I want to be one of the first to solve it' (N-12).

Additionally, after the qualitative analysis, it is evident how students mentioned the learning that such an approach was providing them. On one hand, they highlighted that, thanks to a practical approach, they experienced firsthand the educational possibilities of breakouts while also becoming more familiar with applications to create them: 'I didn't know that 'Genially' could be used for this type of activity. Plus, I'm now experiencing the rollercoaster of emotions that escape rooms and breakouts generate' (N-77). Also, during the proposal, there were students who began to see its practical application for their future teaching roles: 'As I solve screens, ideas come to me for teaching Physical Education content through a breakout. If I ever manage to solve this one, I'll start creating my own' (N-62).

6 Discussion

In recent years, there has been an increase in GBL proposals in the classroom, specifically the use of escape rooms and breakouts for educational purposes. There is evidence that demonstrates their potential in developing various soft skills and competencies with future professional transfer (Karakoç, 2020; Zhonggen, 2019). Therefore, the evidence of its implementation in the Masters' degree is less explored than in primary or secondary education (Veldkamp et al., 2020), with escape rooms being used much more frequently than breakouts, which underscores the relevance of this work.

Creative thinking is a fundamental competency for addressing the teaching challenges of the 21st century (Bellanca, 2015). There is scientific evidence linking creative thinking with critical thinking and reflective thinking, with these variables being predictors of academic achievement (Akpur, 2020). In addition, evidence suggests that creative thinking can be enhanced through enjoyment, abstraction, and the absence or effective management of negative effects (Parra-González et al., 2020). The decision to base the proposal on a series like *Black Mirror* and incorporate its aesthetic and plot connections contributed to stimulating students' creative thinking and imagination, as previously demonstrated. The results align with other interventions conducted in university contexts, where students fostered their creativity and imaginative capacity (García-Viola et al., 2019). In particular, noteworthy are the results of an escape room based on the movie *Matrix*, where the narrative respected the plots and included elements specific to the film, maintaining its aesthetic (Navarro-Mateos and Pérez-López 2022b).

Various studies have indicated that enjoyment of learning tends to decline during lower secondary school and even more so at the university level (Hagenauer, 2010; Vierhaus et al., 2016). This situation is concerning, as evidence shows that enjoyment of learning is a significant determinant of both learning and achievement (Rubach and Lazarides, 2021). In this regard, student-centred approaches, in which students are actively involved and have a voice and autonomy, positively affect the expression and development of learning enjoyment (Mötteli et al., 2023). In digital contexts, breakouts are considered tools that positively impact students' enjoyment and engagement, fostering a positive predisposition towards the task (Chandler, 2016). Furthermore, in an experience conducted with pre-service teachers, the impact of these proposals on the development of digital competencies was demonstrated, a key aspect in the current educational landscape (Llorente-Cejudo et al., 2022).

Regarding absorption, at the heart of the success of games lies an idea known as flow (Csikszentmihalyi, 1997). Achieving the flow state, or being in this zone, indicates that students are at the optimal point of motivation in the experience, fully immersed in it and away from the two most dangerous extremes: anxiety and boredom. They enter a phase of concentration in which they lose their sense of space and time. To achieve this desired state of concentration in education, Marczewski (2018) identifies several key ideas related to flow: setting clear goals and being aware of progress, having clear and immediate feedback, and balancing the perceived difficulty of the challenge with one's own skills. In this regard, it is important to highlight that the use of digital games positively impacts the flow state of students in both business and sports sciences, aiding them in the development of skills and competencies (Catalán-Gil and Martínez-Salinas, 2018; Pérez-López et al., 2023). There is evidence linking the use of GBL with the flow channel, with the level of challenge presented by the game being an important predictor.

In addition, the activation results reinforce the substantial educational opportunities it offers, consistent with research indicating strong correlations between this approach and factors related to motivation and emotional engagement (López-Belmonte et al., 2020; Sierra-Daza and Fernández-Sánchez, 2019). Lastly, it is evident that in approaches like the one developed in the present study, high levels of frustration may appear. These results align with other initiatives carried out in higher education (Navarro-Mateos and Pérez-López 2022b; Pérez-López et al., 2024). This especially relevant for students in the Master's in Teaching program (future educators), as such learning environments provide valuable opportunities to develop emotional management and resilience (Anguas-Gracia et al., 2021; Navarro-Mateos and Pérez-López 2022b).

7   Conclusion

The implementation of a digital breakout inspired by *Black Mirror* in the Master's in Teaching Program effectively engaged students and enhanced essential competencies. The high levels of creative thinking, enjoyment, and absorption observed among participants underscore the potential of GBL to create immersive educational experiences. This approach not only facilitates the acquisition of knowledge but also fosters the development of crucial soft skills, such as problem-solving and emotional management, which are vital for future educators.

Moreover, the study highlights the importance of incorporating contemporary cultural references into educational contexts, as these can significantly enhance student motivation and connection to the material. While some challenges, such as frustration, emerged, they provided valuable opportunities for students to develop resilience and adaptive skills.

Overall, this research contributes to the evolving discourse on innovative pedagogical strategies in higher education. It advocates for the broader integration of game-based learning methodologies, encouraging educators to explore creative avenues that align with the interests and needs of today's students. Future studies could further examine the long-term impacts of such interventions on teaching efficacy and student outcomes in various educational settings.

Disclosure of Interests.   The authors have no competing interests to declare that are relevant to the content of this article.

References

Akpur, U.: Critical, reflective, creative thinking and their reflections on academic achievement. Thinking Skills and Creativity **37** (2020)

Ali, W.: The efficacy of evolving technology in conceptualizing pedagogy and practice in higher education. High. Educ. Stud. **9**(2), 81–95 (2019)

Anguas-Gracia, A., et al.: An evaluation of undergraduate student nurses' gameful experience while playing an escape room game as part of a community health nursing course. Nurse Education Today **103** (2021)

Bellanca, J.A.: The focus factor: 8 essential twenty-first century thinking skills for deeper student learning. Teachers College Press (2015)

Boyle, E.A., et al.: An update to the systematic literature review of empirical evidence of the impacts and outcomes of computer games and serious games. Comput. Educ. **94**, 178–192 (2016)

Cabero-Almenara, J., Palacios-Rodríguez, A.: Marco europeo de competencia digital docente «DigCompEdu». Traducción y adaptación del cuestionario «DigCompEdu Check-In». Edmetic **9**(1), 213–234 (2020)

Caruth, G.D.: Student engagement, retention, and motivation: Assessing academic success in today's college students. Participatory Educational Research **5**(1), 17–30 (2018)

Catalán-Gil, S., Martínez-Salinas, E.: Encouraging the state of flow: the key of business simulation games. J. Manage. Bus. Educ. **1**(2), 140–159 (2018)

Cilliers, E.J.: The challenge of teaching generation Z people. PEOPLE: International Journal of Social Sciences **3**(1), 188–198 (2017)

Connelly, L., Burbach, B.E., Kennedy, C., Walters, L.: Escape room recruitment event: description and lessons learned. J. Nurs. Educ. **57**(3), 184–187 (2018)

Csikszentmihalyi, M.: Flow and education. NAMTA journal **22**(2), 2–35 (1997)

Chandler, K.: Using breakout rooms in synchronous online tutorials. Journal of Perspectives in Applied Academic Practice **4**(3), 16–23 (2016)

Dauksevicuite, I.: Unlocking the Full Potential of Digital Native Learners. McGraw Hill Education (2016)

Del Moral-Pérez, M.E., Guzmám-Duque, A.P., Bellver-Moreno, M.D.C.: Consumo y ocio de la Generación Z en la esfera digital. Prisma social **34**, 88–105 (2021)

Eppmann, R., Bekk, M., Klein, K.: Gameful experience in gamification: construction and validation of a gameful experience scale [GAMEX]. J. Interact. Mark. **43**, 98–115 (2018)

Fotaris, P., Mastoras, T.: Escape rooms for learning: A systematic review. Proceedings of the 13th European Conference on Games Based Learning, Dinamarca, pp.235–243. Universidad de Dinamarca del Sur (2019)

Francisco-Lens, N., Rodríguez-Vázquez, A.I.: La innovación de la Televisión Pública Europea en la oferta audiovisual digital: nuevas plataformas para la Generación Z. RAEIC **7**(13), 185–212 (2020)

García-Viola, A., Garrido-Molina, J.M., Márquez-Hernández, V.V., Granados-Gámez, G., Aguilera-Manrique, G., Gutiérrez-Puertas, L.: The influence of gamification on decision making in nursing students. J. Nurs. Educ. **58**(12), 718–722 (2019)

Guo, J., Parker, P.D., Marsh, H.W., Morin, A.J.S.: Achievement, motivation, and educational choices: A longitudinal study of expectancy and value using a multiplicative perspective. Dev. Psychol. **51**(8), 1163–1176 (2015)

Hagenauer, G., Hascher, T.: Learning enjoyment in early adolescence. Educ. Res. Eval. **16**(6), 495–516 (2010)

Johnson, R.B., Onwuegbuzie, A.J., Turner, L.A.: Toward a definition of mixed methods research. J. Mixed Methods Res. **1**(2), 112–133 (2007)

Kahu, E.R.: Framing student engagement in higher education. Stud. High. Educ. **38**(5), 758–773 (2013)

Kapur, M., Bielaczyc, K.: Designing for productive failure. J. Learn. Sci. **21**(1), 45–83 (2012). https://doi.org/10.1080/10508406.2011.591717

Karakoç, B., Eryılmaz, K., Turan Özpolat, E., Yıldırım, İ: The effect of game-based learning on student achievement: A meta-analysis study. Technol. Knowl. Learn. **27**(1), 207–222 (2020)

Lee, W., Reeve, J.: Teachers' estimates of their students' motivation and engagement: being in synch with students. Educ. Psychol. **32**(6), 727–747 (2012)

López-Belmonte, J., Segura-Robles, A., Fuentes-Cabrera, A., Parra-González, M.E.: Evaluating activation and absence of negative effect: Gamification and escape rooms for learning. Int. J. Environm. Res. Pub. Health **17**(7) (2020)

Llorente-Cejudo, M.D.C., Palacios Rodríguez, A.D.P., Fernández Scagliusi, V.: Learning landscapes and educational Breakout for the development of digital skills of teachers in training. Interaction Design and Architecture Journal **53**, 176–190 (2022)

Majali, S.A.: Positive Anxiety and its Role in Motivation and Achievements among University Students. Int. J. Instr. **13**(4), 975–986 (2020)

Marczewski, A.: Even ninja monkeys like to play: Unicorn edition. Gamified UK (2018)

Mora, F.N.: Solo se puede aprender aquello que se ama. Alianza Editorial, Madrid (2013)

Mötteli, C., Grob, U., Pauli, C., Reusser, K., Stebler, R.: The influence of personalized learning on the development of learning enjoyment. Int. J. Educ. Res. Open 5 (2023)

Navarro-Mateos, C., Pérez-López, I.J., Trigueros-Cervantes, C.: Análisis del rol docente en una propuesta de gamificación en el máster de profesorado. Revista de educación **405**, 275–301 (2024)

Navarro-Mateos, C., Pérez-López, I.J.: Una app móvil potencia la motivación del alumnado en una experiencia de gamificación universitaria. ALTERIDAD. Revista de Educación **17**(1), 64–74 (2022)

Navarro-Mateos, C., Pérez-López, I.J.: El escape room como estrategia didáctica en el Máster de Profesorado. Retos: nuevas tendencias en educación física, deporte y recreación **44**, 221–231 (2022b)

Negre, C., Carrión, S.: Desafío en el aula. Paidós Educación (2020)

Orbegoso, A.: La motivación intrínseca según Ryan & Deci y algunas recomendaciones para maestros. Educare, Revista Científica de Educação **2**(1), 75–93 (2016)

Parra-González, M.E., Segura-Robles, A., Morales-Cevallos, M.B., López-Meneses, E.J.: Relación de los factores asociados en el desarrollo de experiencias gamificadas. Campus virtuales: revista científica iberoamericana de tecnología educativa **9**(1), 113–123 (2020)

Parra-González, M.E., Segura-Robles, A.: Traducción y Validación de la Escala de Evaluación de Experiencias Gamificadas (GAMEX). Bordón **71**(4), 87–99 (2019)

Pérez-López, I.J., Navarro-Mateos, C., Mora-González, J.: El impacto de un doble breakout digital en un proyecto de gamificación. Retos **50**, 761–768 (2023)

Pérez-López, I. J., Navarro-Mateos, C., Mora-Gonzalez, J.: Impact of a digital serious game on emotional variables of students of the master's degree in teaching. Innovations in Education and Teaching International, 1–13 (2024)

Pérez-López, I. J., Navarro-Mateos, C.: Guía para gamificar: construye tu propia aventura. Copideporte S.L. (2022)

Pérez-López, I.J.: La docencia es un juego donde gana el que más disfruta. Habilidad motriz **50**, 2–3 (2018)

Rodríguez-Izquierdo, R.M.: Aprendizaje servicio y compromiso académico en educación superior. Revista de Psicodidáctica **25**(1), 45–51 (2020)

Rotgans, J.I., Schmidt, H.G.: Situational interest and academic achievement in the active-learning classroom. Learn. Instr. **21**(1), 58–67 (2011)

Rubach, C., Lazarides, R.: Emotionen in Schule und Unterricht: Bedingungen und Auswirkungen von Emotionen bei Lehrkräften und Lernenden. Verlag Barbara Budrich (2021)

Sánchez-Gómez, M.C., Rodrigues, A.I., Costa, A.P.: Desde los métodos cualitativos hacia los modelos mixtos: tendencia actual de investigación en ciencias sociales. Revista Ibérica de Sistemas e Tecnologias de Informacao **28**, 9–12 (2018)

Schanke, V.: 'Youthification' of drama through real-time storytelling: a production study of blank and the legacy of SKAM. Critical Studies in Television: The International Journal of Television Studies **16**(2), 145–162 (2021)

Sierra-Daza, M.C., Fernández-Sánchez, M.R.: Gamificando el aula universitaria. Análisis de una experiencia de Escape Room en educación superior. Revista de Estudios y Experiencias en Educación **18**(36), 105–115 (2019)

Symaco, L.P., Tee, M.Y.: Social responsibility and engagement in higher education: Case of the ASEAN. Int. J. Educ. Dev. **66**, 184–192 (2019)

Timans, R., Wouters, P., Heilbron, J.: Mixed methods research: what it is and what it could be. Theory and Society **48**, 193–216 (2019)

Tobias, S., Fletcher, J.D., Wind, A.P.: Game-BasedLearning. In: Spector, J. M., Merril, M.D., Elen, J., Bishop, M.J. (eds.) Handbook of Research on Educational Communications and Technology, pp. 485–503. Springer (2014)

Veldkamp, A., Van de Grint, L., Knippels, M.C.P., Van Joolingen, W.R.: Escape education: A systematic review on escape rooms in education. Educational Research Review **31** (2020)

Vierhaus, M., Lohaus, A., Wild, E.: The development of achievement emotions and coping/emotion regulation from primary to secondary school. Learn. Instr. **42**, 12–21 (2016)

Yllana-Prieto, F., Jeong, J.S., González-Gómez, D.: An online-based edu-escape room: A comparison study of a multidimensional domain of PSTs with flipped sustainability-stem contents. Sustainability **13**(3) (2023)

Zhonggen, Y.: A meta-analysis of use of serious games in education over a decade. Int. J. Comp. Games Technol. 1–8 (2019)

Leveraging Tabletop Games as Education Scenarios for Enhancing Media Literacy Skills

Ruth S. Contreras-Espinosa[1](✉) and Jose Luis Eguia-Gomez[2]

[1] University of Vic-Central University of Catalonia, 08500 Barcelona, Spain
ruth.contreras@uvic.cat
[2] Polytechnic University of Catalonia, 08028 Barcelona, Spain

Abstract. This paper presents findings from an analysis of 17 tabletop games focused on media literacy and insights from 10 interviews with educators, journalists, and crisis management professionals with expertise in game design. The study addresses key questions regarding how games can effectively foster media literacy among young people, particularly in times of crisis, and identifies established best practices. The research is part of the xxx project x. The project aims to explore how games can enhance media literacy among youth, responding to an urgent need in our information-saturated society. By examining the potential of games as educational tools, this work highlights their effectiveness in promoting media literacy and combating misinformation. The xx initiative is structured in phases, and this paper focuses on the first phase, which involves mapping tabletop games related to media literacy and conducting interviews in Spain. Through this analysis, the paper contributes to a deeper understanding of the characteristics and mechanics of media literacy games, as well as the perspectives of educators on how such games can be utilized to equip young people with critical skills necessary for navigating the complexities of media consumption in contemporary society. The findings aim to inform the development of innovative game designs that foster critical thinking and responsible media engagement among the younger generation.

Keywords: Tabletop games · Media Literacy · Disinformation · Fake News

1 Introduction

The rise of fake news in both traditional and social media has become a growing concern in recent years [1, 2]. These false narratives, presented as legitimate information, are spread through various media channels—especially online—with the intent to mislead and manipulate public opinion [3, 4]. As the amount of online content continues to expand, coupled with the increasing ease of information access via the Internet [5], the ability to discern between factual and misleading information becomes ever more crucial [6].

Fake news has evolved into a political weapon, used to sway elections, shape public opinion, and promote disinformation [7]. The repercussions for society are severe, particularly in areas such as health, politics, and security. Combating these issues necessitates a widespread effort to promote media literacy and critical thinking skills among the population [8, 9], empowering individuals to identify and challenge fake news and other forms of misinformation [10].

Traditionally, teachers have introduced students to the basics of journalism, newswriting, and the distinctions between different news genres [11]. While school curriculums touch on these concepts, they are often presented within a limited scope, and many classroom interventions stem from individual teachers' dedication [1]. However, given the already overloaded academic curriculum, simply adding more content on media literacy is insufficient. What is needed are new spaces for reflection, allowing for deeper analysis and discussion of misinformation [9]. With the growing importance of developing journalistic literacy among young people in an era of widespread misinformation [12], it is essential to rethink how media literacy is taught and integrated into education [11]. Games present a promising strategy for this, as they can seamlessly integrate educational content (e.g., fake news, digital privacy, personal media habits) into their design, supporting the development of practical skills that can be applied in real life [13, 14].

Glas et al. [15] highlight that while many games aim to teach topics like fake news or digital privacy, their effectiveness as educational tools for promoting media literacy remains uncertain. Existing research often focuses on individual games or specific aspects of media literacy, leaving a significant gap in studies exploring the practical application of these games in real educational settings [16]. Although game-based learning offers numerous potential benefits, it is important to recognize the challenges educators face when attempting to integrate games into their classrooms [17].

This paper will present findings from a comprehensive analysis of 17 tabletop games centered on media literacy. Additionally, it will draw insights from 10 interviews with teachers, educators, journalists with backgrounds in crisis management and game design. Key questions include: How can games effectively foster media literacy among young people during times of crisis? Are there established examples of best practices?

This research marks the initial phase of the YO-MEDIA project (Youngsters' Media Literacy in Times of Crisis), funded by the Calouste Gulbenkian Foundation (269094). The project aims to investigate how games can enhance media literacy among young people, addressing a critical need in today's information-saturated society. By exploring the potential of games as educational tools, this work seeks to highlight their effectiveness in promoting media literacy and combating misinformation. YO-MEDIA is structured into phases, and this paper focuses on the first phase: mapping tabletop games related to media literacy and conducting interviews in Spain.

2   Materials and Methods

We selected a sample of tabletop games through Google searches using the keywords "fake news," "media literacy," and "media." Games that did not specifically focus on media literacy, the use of fake news, or directly combat misinformation were excluded.

Priority was given to games that offered a deep, meaningful approach, suitable for educational purposes, and provided an immersive, educational experience. Ultimately, we analyzed 17 tabletop games to better understand their nature and characteristics. Of these, 16 were card-based games, with one additional game that also included a board (Table 1).

Table 1. List of games analyzed.

No.	Game	Date	Community
G1	Truth Be Told	2009	12+
G2	The Resistance	2009	13+
G3	The Big Bang Theory: Fact or Fiction Card Game (2012)	2012	12+
G4	Fact or Fiction Card Game	2013	12+
G5	Fake news	2017	12+
G6	Fake news or Not?	2017	12+
G7	Fake News Game: Kanye Edition	2018	12+
G8	The news game Fact from fiction/ political edition)	2018	-
G9	Mischief Media: The Game of Fake News	2018	10+
G10	LAMBOOZLED!	2019	11+
G11	Fake News: The Improv Game!	2019	11+
G12	¿Verdad o bulo? Coronavirus Edition	2020	14+
G13	P1: Pandemonium	2020	-
G14	Trump News Game	2021	10+
G15	Make Fake News Great Again	2021	16+
G16	Liar Liar: The Game of Truths and Lies	2022	14+
G17	Hoax Hunters Vol 1 Murder	2012	-

The interviews focus on key areas: the role of media literacy during crises as a framework (How does media literacy empower individuals to critically assess information and avoid misinformation during crises?); strategies for developing media-related skills in adolescents (What effective classroom activities can educators use to enhance adolescents' media literacy skills?), including best practices from the field; teacher training in media literacy (What professional development programs are most effective in preparing teachers to teach media literacy?); and the role of games in promoting media literacy among young people (How can interactive games be designed to effectively teach media literacy concepts to young audiences?). The Spanish team conducted 10 interviews (Table 2): 6 with teachers involved in various media literacy projects, 1 with an educator working with youth in diverse contexts, and 3 with journalists knowledgeable about games and media literacy.

Table 2. Interviewees in Spain.

No.	Role	Institution/Project
T1	Teacher/tabletop game developer	Sant Andreu de la Barca High School Quixotic Games
T2	Teacher/researcher	José Manuel Blecua High School
T3	Teacher/researcher	Camps Blancs Sant Boi de Llobregat High School
T4	Teacher/artist	Zaurín High School
T5	Teacher	Escolàpies de Sant Martí School
T6	Teacher	Maristes Sant Pere Chanel School
E1	Educator	Center for Innovation and Education in Teaching Uvic-UCC; Catalan Society of Pedagogy
J1	Journalist	El 9 Nou
J2	Journalist	Social.cat
J3	Journalist/researcher	Transmedia Literacy projects H2020; Guadalajara Press Society

Teachers were chosen for their expertise in teaching young people aged 12 to 20 and their experience in producing various audiovisual media, such as games, blogs, videos, animations, and graphics. Their teaching spans a diverse range of educational institutions within the Spanish education system, including public and private schools in cities and rural areas. Interviewee T2 also works at a government-supported resource center in Aragón that focuses on media literacy training. E1 is the director of a center for innovation and education that fosters collaboration among educators. The interviewed journalists, involved in Journalism degrees, bring both journalistic and educational perspectives. J3 participates in a national working group supported by the Spanish government to combat misinformation, J1 and J2 collaborated with students on podcasts and journalism workshops, emphasizing the critical role of digital media in promoting information literacy and the importance of reliable information.

3 Results

3.1 Tabletop Games

The analysis of these 17 tabletop games offers valuable insights into the diverse mechanics employed in media literacy games, highlighting their key features and characteristics. Several card games centered on fake news (G1, G3, G4, G6, G7), developed between 2009 and 2017, utilize a straightforward mechanic focused on distinguishing true from false statements. The primary objective of these games is to enhance awareness of misinformation and foster critical skills for discerning facts from fakes.

Post-2018, we observe a shift toward games with more complex mechanics. These newer titles not only tackle disinformation but also allow players to assume various roles, managing aspects of creating and disseminating fake news. This intricate gameplay design creates a more immersive and educational experience, enabling players to understand the complexities and societal implications of misinformation in contemporary contexts.

A common theme across all games is the emphasis on recognizing credible information sources. Thirteen games delve into the motives behind the spread of misinformation aimed at manipulation (G2, G9, G10, G12), for example Make Fake News Great Again allows players to engage in a satirical take on media manipulation, providing insights into how fake news can be intentionally crafted and spread for personal or political gain. While twelve games illustrate the processes involved in producing fake news (G9, G10, G12). Nine games facilitate an understanding of the societal and individual consequences of creating and sharing false information (G2, G9, G10, G12). Furthermore, seven games explore how specific political or commercial interests attempt to influence online behavior (G8, G10), and six games promote responsible social media usage (G9, G10, G12). Four games briefly address how algorithms shape online visibility, providing a foundational understanding (G9). Another four games offer strategies for verifying information and navigating the broader media landscape (G9, G10, G12), while three games give advice on defending against social media threats and risks (G9).

In terms of gameplay mechanics, all games present players with a series of news articles—some true and others false—requiring them to differentiate and select the authentic news. For instance, LAMBOOZLED! The Media Literacy Card Game engages players by presenting various news headlines and requiring them to determine whether they are true, false, or satirical. The game encourages discussion about why certain headlines may be misleading, prompting players to critically evaluate the sources and content of information. Similarly, Trump News Game immerses players in the chaotic world of political news, where they must discern fact from fiction in statements made by public figures, thereby highlighting the challenges of misinformation in political discourse. Nine games enable players to create and disseminate their own fake news (G2, G9, G10, G12) for example, Liar Liar: The Game of Truths and Lies encourages players to create their own deceptive headlines, fostering a deeper understanding of how easy it is to manipulate information and the potential consequences of doing so.

While three games involve researching the presented news to assess its truthfulness (G10, G13). Additionally, nine games challenge players to spread fake news via virtual social media platforms within the game (G9) Mischief Media: The Game of Fake News, and four games highlight the consequences of spreading fake news, such as potential damage to the player's reputation (G12) the game ¿Verdad o bulo? Coronavirus Edition. Eleven games incorporate questions that test players' knowledge on current topics (G5, G6, G8, G14, G15).

Among the analyzed games, five employ storytelling as a persuasive strategy, weaving narratives that emotionally engage players. Nine games integrate emotional language into their content (G2, G5, G6, G9, G10, G13, G14, G15, G16). Eleven games adopt a conductive educational approach, while six embrace a constructivist method (G9, G10) Mischief Media: The Game of Fake News and LAMBOOZLED! The Media Literacy

Card Game, focusing on active knowledge construction and interactive experiences. Only three games (G2, G9, and G10) center on fictional narratives, while the remainder addresses real-world issues.

Notably, among the fiction-based games, G10 LAMBOOZLED! The Media Literacy Card Game (Fig. 1), distinguishes itself by offering additional resources to help educators adapt the game for their curricula. This feature equips teachers with tools to effectively integrate the game into lesson plans, thereby enhancing the educational experience and aligning it with pedagogical objectives. Such support facilitates implementation in educational settings, maximizing both the utility of the game and the benefits for students.

In summary, the analysis highlights the evolving landscape of media literacy games, illustrating their diverse elements and mechanics. This foundation provides valuable insights for the future development of games aimed at improving media literacy among young audiences. By incorporating a mix of straightforward and complex mechanics, as well as emotional engagement strategies, developers can create more effective and impactful educational tools.

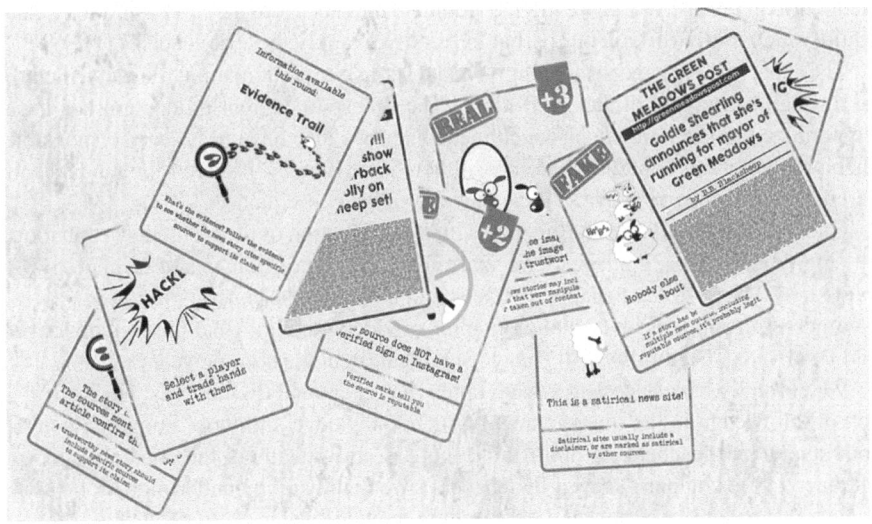

Fig. 1. G10. LAMBOOZLED! The Media literacy card game.

3.2 Interviews

According to educators in the field of journalism, students are competent consumers but lack critical skills in consumption and audiovisual production. Developing skills and competencies can help young people become critical consumers and creators of media and prepare them to navigate the complex and constantly changing media landscape of today's digital age. The crucial skills and competencies that young people need to develop to become media literate are:

Think critically about media messages and evaluate their credibility and bias: "What all teachers strive for is that they acquire critical thinking. That is how it should be" (T2).

Evaluate, locate, and use information effectively and ethically: "It is very important that students support all their contributions and positions with references they have read or in connection with other topics. Some have tried it, and some have learned it" (E1).

Use digital tools and platforms to create, communicate, and collaborate. Considering technology is evolving at a great pace, preparing students to navigate the constantly-changing media landscape of today's digital age is beneficial. The ability to gain new knowledge will be increasingly valuable: "When I started teaching in 2005... we learned to use email, later Google, and then came social media. Everything started to change with Facebook, Twitter, smartphones... And today, with artificial intelligence" (J2).

Create and share media content in various formats, including text, images, audio, video, or games: "Nowadays, one thing that is very useful (referring to teaching in schools), and that is increasingly growing, is people who explain topics in videos, YouTubers, video tutorials, etc., and kids really enjoy that" (T5).

Communicate effectively and ethically in various contexts, including online and offline: "The ethical spirit... The critical spirit of students has significantly experienced a reduction in the last years. We often encounter students, from the last promotions, that require much more work on our part at university to apply these concepts" (J1).

The capacity to engage in meaningful and responsible participation in society, including through media and digital platforms: "The access to content production has been democratized to such an extent which means that we also have to be very responsible when producing this content. Therefore, if there is no literacy, if there is no guidance, if there is no critical perspective... it can have negative consequences" (E1).

Understanding artificial intelligence with a combination of media literacy education, critical thinking skills, guidance from educators, and responsible use of technology: "I don't know what is going to happen because, for example, the expert opinions of judges, or witnesses in court, used to be photographic or videographic. All of this will no longer hold legal value. Credibility will change significantly in the coming years" (J2).

We currently have access to a wide range of content and digital tools, but there is a lack of knowledge in the management of such knowledge and tools among all actors, from students to teachers: "I don't think there is an issue about the content itself (in reference to the content received by teachers), we could significantly improve how we manage the use of these tools and the teaching methods we employ" (E1).

When it comes to promoting critical thinking, only two interviewees mentioned fact-checking tools (J1, J3). There is a significant lack of awareness outside the journalistic field about websites for news verification. These tools are commonly used in data journalism but are less known and utilized beyond that domain.

Teachers encounter foundational problems when trying to foster critical thinking. First, students often struggle to receive criticism constructively, and there is a lack of solidified logical thinking. Without establishing this foundation, it becomes challenging to cultivate a critical mindset within society to counter attempts at manipulating messages. Students are competent consumers, but they do not have command over audiovisual production, an issue often highlighted by teachers in the field of journalism.

Regarding the question "How do students consume the latest news about the war in Ukraine?", several interviewees (T1, T3, T4, T5, T6) highlighted a general ignorance among the students, which seems to have increased during the pandemic, and a generalized apathy towards acquiring general culture.

"The best way to do so (referring to how to approach students with content) would be to make them see that the issue affects them somehow, to make them see that it is useful for their daily lives because this is one of the few and one of the best approaches that can encourage them to learn and do anything" (T5).

The interest of students focuses on familiar things and for a short time span. They (T1, T3, T4, T5, T6) insist that, beyond promoting a critical spirit, it is crucial to recover a culture that promotes effort. All the teachers and educators interviewed consider that there has been progress in the knowledge of tools but, regardless of the technology students have at their disposal, they highlight a lack of maturity and low acceptance of criticism at all levels as compared to previous years. This, together with a generalized lack of effort in deepening the knowledge about the content provided, entails a big risk: in the future, those who control the message may easily manipulate them.

"Everybody has access to content and everybody can generate it. But you cannot know where that content comes from and whether it is reliable. Some people know it and they take advantage of it. Content creation is designed to capture your attention from the first line, the first image. It is supposed to catch your eye because it is attractive, but they don't look at the content in-depth" (referring to the students in general) (T5).

In light of this, some teachers have taken the initiative to create content in the format and channels students best accept, or search for content in such format.

"One thing that truly works nowadays, and that is becoming increasingly popular, is for teachers to make videos explaining topics, or becoming YouTubers, make video tutorials, etc. Kids really like this, because they are very used to watching such type of content. So it is about taking the same model, and changing the content" (T5).

Public centers provide one hour of mentoring weekly that tackles several issues from an ethical and moral perspective, and some teachers use these spaces to bring up the topic of media literacy with examples (T1, T2, T3, T4): "During these mentoring hours (one hour weekly per level) in secondary schools, we tackle a great number of transversal topics in relation to issues of harmony, sexual education, etc." (T1).

This space also serves to approach questions beyond the center's scope regarding media consumption and the critical spirit that end up being more relevant than the education provided by the center: "During the majority of cases, it is difficult to reach the kids because social issues beyond the center are typically more relevant than what we can do at school. But we do work on these issues in reality; it is a part that influences the creation of a critical consciousness, a critical capacity" (T4).

All the interviewees consider working on the ethics and morals of the use of information an essential factor. Teachers highlight that a lack of ethics and morals, together with the ease to produce content and access to more distribution channels has led to an increase in insults and abuse among students, with cases of bullying, and even identity theft crimes: "On one hand, we encounter the typical scenarios, which we experience on a daily basis, of insults and slander, threats, and such, that they direct at each other through social media. And well, we have a very recent case now of a student pretending to be another through an email account from which fake offensive emails directed to teachers in the name of the latter were sent" (T4).

Another key factor in which the interviewees agree regarding the use of information is knowing about and respecting intellectual property rights, they highlight that students have a hard time accepting authorship of what they find on the Internet: "They do not place due value on references. They should be more mindful and not grab everything and appropriate it; they tend to think that everything online belongs to them" (T3).

Nowadays, the lack of presence of games inside the classroom is not due to a negative view from the school management, the parents, or society in general. We identified two main reasons: (1) lack of knowledge, for instance, of fact-checking tools, and (2) lack of time, especially in the highest educational levels before university access: "I don't use games during this History course because, as compared to last year, we have one hour less per week. Instead of three, now we have two, and I need to provide the same content as before. I don't think games are a waste of time, because they aren't, but if I apply games in the methodology, we just won't make it. That's the main reason" (T5).

Games usage experience in the classroom is normally limited to Kahoot, despite some teachers question whether it is useful at all: "It comes down to gamification and Kahoot is a good example of that. Some platforms advertise that they gamify education, but it turns out it is just another Kahoot" (T2).

The majority of the interviewees agree that project-oriented subjects, very present in the classrooms nowadays, would enable a more natural implementation of games (T1, T2, T3, T4, T5, T6). Referring to a media literacy game: "We would include games within a project, as there is a tendency now to include project-oriented learning methodologies. For instance, in secondary school, there is a subject called Culture and Values, which is a complement to the mentoring hours" (T3).

The teachers and the educator insist on the need for the content and the consequences of the actions in the games to be familiar and transferable to their daily life: "It is necessary to use examples of concrete situations… to encourage thinking and discerning in a critical manner which options would they choose" (T5).

It is necessary to take into account that students are very used to constant feedback and to consuming information in very short periods of time, which hinders deepening the content rigorously. The interviewees (T1, T2, T6) prefer games with an ethical component, as an example of mechanics for a possible media literacy game, but they recognize the difficulty to attract an audience that typically chooses competition-dominated games: "In This War of Mine, the theme is surprising, and treats more ethical topics, such as in Papers, Please. Ethical components are included and so are the implications of the decisions taken. But, what type of games reach the youth? Massive games. Games that include war topics, such as Call of Duty. We can use something that connects with them such as, for instance, the game Among Us during the pandemic" (T1).

All the teachers highlighted to a certain extent the importance of adapting any content, including games, to the curriculum they taught, but they also accepted their responsibility of adapting the games to the curriculum in certain instances (T5, T6). The possibility of the teachers themselves adapting games to their needs is in line with the methodologies observed in subsidized schools, that are less constrained by timing.

Teachers think it is feasible to use fiction in games to motivate students, as they tend to be more attracted to fiction, which can be then used to indirectly treat topics linked to reality. On the other hand, in the journalism field fictive worlds tend to be rejected. There is a tendency to fictionalize reality to make news more attractive in podcasts: "At university, they highlight that it is crucial to distinguish between opinion and information. It is also important for the professional to be able to make this distinction" (J1).

Teachers are worried about political correctness and prefer fictive worlds that may elicit inner conflict within the individual as a catalyst for reflection, but avoid altogether possible distractions such as political division. When using reality in the story, immersive reality tools may be very motivating for students, as well as the metaverse when using fictive worlds. Finally, when promoting a possible game focused on media literacy, it is crucial to count on the recommendation of teachers working on a similar subject, rather than supported by the figure of the educator or a government agency.

4   Conclusions

The analysis of the tabletop games has provided a comprehensive understanding of the game mechanics involved and has allowed for the identification of key characteristics that stand out in these educational products. Through this evaluation, it has been observed that older games, particularly those focused on identifying fake news, share a simple mechanic aimed at raising awareness about misinformation. However, since 2008, a shift towards more complex game designs has been noted. This transformation in game design suggests a recognition of the complexities of information in today's society, providing players with more immersive and challenging educational experiences [16].

A significant finding is that most of the analyzed games emphasize the importance of recognizing credible information sources and understanding the reasons behind the spread of misinformation. This is crucial for developing critical competencies that enable players to discern facts from falsehoods. Furthermore, games that include persuasive narratives and emotional language prove effective in engaging players, highlighting the need for educational content to resonate emotionally with users.

Interviews with educators have revealed that, while students are competent consumers of media content, they lack the critical skills necessary to evaluate and create audiovisual content. The importance of fostering critical thinking in students is emphasized, helping them assess the credibility and bias of media messages. There is also a need to prepare young people for responsible participation in an ever-changing media landscape, where ethics in content production and consumption become essential. The combination of media literacy with critical thinking skills is fundamental in shaping responsible and engaged citizens in today's digital society.

However, educators face significant challenges in promoting critical thinking and media ethics. The general apathy among students towards current issues, as well as their resistance to constructive criticism, hinders the effective teaching of these competencies. Despite the democratized access to content production tools, the lack of maturity and low acceptance of criticism reflect a concerning trend that could lead to message manipulation in the future.

The underutilization of games in classrooms, despite their potential benefits, is attributed primarily to a lack of time and knowledge about fact-checking tools. It is evident that, for games to be effectively integrated into education, there needs to be greater awareness and training among educators. The implementation of project-oriented learning methodologies presents an opportunity to incorporate media literacy games more naturally into the curriculum.

Finally, adapting game content to the realities of students is crucial for educational effectiveness. Educators must work together to create learning experiences that are relevant and connect with students' daily lives. Fiction can be a powerful tool to capture students' attention, but it should be used in a way that fosters critical reflection and classroom discussion. Collaboration among educators in the design and implementation of games can maximize their potential and ultimately contribute to shaping a more aware and critical generation in the face of misinformation in the digital age.

Here are three key ideas that emerge from the connection between the analysis of the games and the interviews, which can influence the design of a new media literacy game:

The analysis of the games revealed that more recent games utilize complex mechanics that allow players to create and disseminate fake news. This feature, combined with the incorporation of persuasive narratives and emotional language, can make the new game not only interactive but also immersive. Including stories that highlight the consequences of misinformation can foster greater empathy and critical reflection in players, facilitating the understanding of complex issues.

Interviews with educators emphasized the importance of developing critical competencies in students, such as evaluating the credibility and bias of media messages. A new game could incorporate collaborative dynamics that promote teamwork, allowing players to discuss and analyze the presented information together. This would not only strengthen critical thinking skills but could also address the observed lack of maturity and resistance to criticism among students.

Finally, both the game analysis and the interviews highlighted the need for game content to be relevant and relatable to students' everyday lives. Designing a game that reflects real-world situations and uses concrete examples could increase players' emotional connection and interest. Additionally, including resources for educators to adapt the game's content to the existing curriculum would facilitate its implementation in the classroom, maximizing its educational impact.

Disclosure of Interests.   The authors have no competing interests to declare that are relevant to the content of this article.

References

1. Adams, Z., Osman, M., Bechlivanidis, C., Meder, B.: (Why) is misinformation a problem? Perspect. Psychol. Sci. **18**(6), 1436–1463 (2023)
2. Shehata, A.M.: The problem of misinformation and fake news. In: Al-Suqri, M., Alsalmi, J., Al-Shaqsi, O. (eds.) Mass Communications and the Influence of Information During Times of Crises. IGI Global, Hershey, PA, USA (2022)

3. Terian, S.: What is fake news: a new definition. Transilvania **11–12**, 112–120 (2021)
4. Jaster, R., Lanius, D.: Speaking of Fake News: Definitions and Dimensions 2021. In The Epistemology of Fake News; Oxford University Press, Oxford, UK (2021)
5. Majerczak, P., Strzelecki, A.: Trust, media credibility, social ties, and the intention to share towards information verification in an age of fake news. Behavioral Scienes **12**(2), 51 (2022)
6. Pennycook, G., Rand, D.G.: Who falls for fake news? the roles of bullshit receptivity, overclaiming, familiarity, and analytic thinking. J. Pers. **88**, 185–200 (2019)
7. Miró-Llinares, F., Aguerri, J.C.: Misinformation about fake news: a systematic critical review of empirical studies on the phenomenon and its status as a 'threat.' Eur. J. Criminol. **20**(1), 356–374 (2023)
8. Jones-Jang, S.M., Mortensen, T., Liu, J.: Does media literacy help identification of fake news? Information literacy helps, but other literacies don't. Am. Behav. Sci. **65**(2), 371–388 (2021)
9. Kellner, D., Share, J.: The Critical Media Literacy Guide: Engaging Media and Transforming Education. Brill, Leiden, The Netherlands (2019)
10. Gaozhao, D.: Flagging fake news on social media: An experimental study of media consumers' identification of fake news. Gov. Inf. Q. **38**(3), 101591 (2021)
11. Frey, N., Fisher, D.: Junior journalists: reading and writing news in the primary grades. In: Moss, B., Lapp, D., Roser, N., Fuhrken, C., Dybdahl, C., (eds.) Teaching New Literacies in Grades K-3. Resources for 21st-Century Classrooms. Guilford Press, New York, NY, USA (2010)
12. Roozenbeek, J., Van der Linden, S.: Fake news game confers psychological resistance against online misinformation. Palgrave Communications **5**, 65 (2019)
13. Basol, M., Roozenbeek, J., Van der Linden, S.: Good news about bad news: gamified inoculation boosts confidence and cognitive immunity against fake news. J. Cogn. **3**(1), 1–9 (2020)
14. Molnar, A., Kostkova, P.: On effective integration of educational content in serious games: Text vs. game mechanics. In: Proceedings of the 2013 IEEE 13th International Conference on Advanced Learning Technologies. Beijing, China (2013)
15. Glas, R., van Vught, J., Fluitsma, T., De La Hera, T., Gómez-García, S.: Literacy at play: an analysis of media literacy games used to foster media literacy competencies. Frontiers in Communication **8**, 1155840 (2023)
16. Contreras-Espinosa, R., Eguia-Gomez, J.L.: Evaluating video games as tools for education on fake news and misinformation. Computers **12**(9), 188 (2023). https://doi.org/10.3390/com puters12090188
17. Diehl, C., et al.: Exploring current challenges and opportunities in Media Literacy skills for youth: Stakeholders' Perspectives. In: Proceedings of the 18th European Conference on Game-Based Learning: ECGBL´24 (2024)

Game Design and Development

Exploring Asymmetry of Information in Cooperative Games

Daniel Reis[1]([envelope]) [ID], Pedro Pais[1] [ID], David Gonçalves[1] [ID], Kathrin Gerling[2] [ID],
and André Rodrigues[1] [ID]

[1] LASIGE, Faculdade de Ciências da Universidade de Lisboa, Lisbon, Portugal
{dsreis,pgpais,dmgoncalves,afrodrigues}@ciencias.ulisboa.pt
[2] Karlsruhe Institute of Technology, Karlsruhe, Germany
kathrin.gerling@kit.edu

Abstract. Digital gaming has the potential to foster togetherness through shared, challenging, and immersive activities. In cooperative games, asymmetry of information impacts social interactions, yet its effects on player experience are not yet fully understood. To this end, we propose a framework for studying asymmetry of information in game design, expanded from previous work on asymmetry, and apply it to the design and development of a cooperative game prototype. In a study with ten pairs of players, we examined how asymmetry of information affected the player experience, through outcomes such as connectedness, communication, fun, and challenge, through a mixed-methods approach. Findings showed that players perceived differences in the distribution of information, influencing their interactions during gameplay. Future research could focus on enhancing the framework to capture the granularity of information dynamics and investigating the asymmetry of information in different scenarios other than two-player cooperative games, offering deeper insights into gaming dynamics.

Keywords: gaming · asymmetry · cooperative · information · framework

1 Introduction

Games designed to promote social interaction have demonstrated positive effects on players, including improvements to social skills, facilitation of relationship formation and maintenance, and contributions to psychological well-being [7,8,20]. Multiplayer gameplay, regardless of the medium - whether played co-located (e.g., split-screen game modes), in the real-world (e.g., location-based games), or online - can foster distinct social experiences. Moreover, multiplayer games, even those not explicitly designed to foster social interaction, naturally involve some level of social interplay by virtue of their design - whether it is team-based competition, cooperative story-driven gameplay, or one versus one adversarial matches -, these different types elicit socialisation both in cooperative

© The Author(s), under exclusive license to Springer Nature Switzerland AG 2025
A. Marto et al. (Eds.): VJ 2024, CCIS 2324, pp. 155–170, 2025.
https://doi.org/10.1007/978-3-031-81713-7_11

and competitive settings. In the context of cooperative games, players are often encouraged to collaborate and communicate to achieve shared goals, which can have benefits [13,17] such as the development of prosocial behaviour [6,12,24], trust [9], and the reinforcement of interpersonal bonds [8,25]. Moreover, studies indicate that player interdependence, meaning the phenomenon where players are dependent on each other in some way, influences communication dynamics during gameplay, promoting social interaction and enhancing the overall social experience [8,10].

To fully harness these social benefits, it is important to understand the underlying game dynamics that potentiate them. Previous research has focused on formalising frameworks and conceptual models to capture these game dynamics: game design fundamentals [11,19,29], game design frameworks [18,28], and player motivations and typologies [3,21,31]. These studies emphasise the importance of understanding not only individual player behaviour, but also the dynamics that emerge from player interaction within structured game environments. One of these dynamics includes cooperation. Past work [26,30] shows there are many design patterns that can elicit cooperation, such as synergies, resource sharing, etc. Among these patterns is the use of asymmetry, for which Harris et al. [16] introduced a framework, identifying six types of asymmetry: ability, challenge, interface, information, investment, and goal/responsibility. This notion has since been refined further in other related work: introducing the concept of degrees of interdependence [15]; further extending the framework through analysis of related work on asymmetric gameplay in multiplayer virtual reality games, incorporating dimensions of patterns of shared control and social asymmetries [27]; and separating asymmetry into two categories: endogenous and exogenous [22,23]. Specifically, asymmetry of information inherently impacts social interactions in games. However, the impact of information asymmetry on the player experience is not yet fully understood.

In this work, we leverage the framework by Harris et al. [16], focusing specifically on the asymmetry of information, *"where one player knows something other players do not"*, since asymmetry of information naturally fosters social interaction. By doing so, we propose a framework to capture asymmetry of information and the different ways it can be implemented in digital games. Next, we developed a digital cooperative game prototype with a variety of asymmetry of information implementations. Finally, in a controlled laboratory study involving ten player pairs, we explore the player's perceptions of asymmetry of information, reflecting on their experience in terms of connectedness, communication, fun, and sense of challenge. We leveraged a mixed-methods approach aiming to answer the following research questions:

- **RQ1:** How do players perceive asymmetry of information in a cooperative game?
- **RQ2:** How do different types of asymmetry of information shape social interaction between players in a cooperative game?

The user study revealed that even if not explicitly aware, players intuitively perceived the existence of asymmetry of information and its subcategories. This asymmetry influenced cooperation and progression, with some challenges being negatively received due to their one-sided nature or the perception that both players had access to the same information. Furthermore, unexpected combinations of framework values led to either confusion or excitement. Some combinations created challenging and positively received puzzles, while others caused frustration. These findings provide insights into using asymmetry of information as a tool in game design to create diverse and engaging experiences.

2 Defining Asymmetry of Information

During the conceptualisation of asymmetry of information, we leveraged Harris et al. [16] as a starting point. The authors define asymmetry of information as the situation *"where one player knows something other players do not"* [16]. Through multiple iterations, taking into account related work [15,16,22,23,27], game mechanics in commercial games that are built on asymmetry, and our own experience with digital and analogue games, we propose a framework to capture asymmetry of information and the different ways it can be implemented in digital games. Throughout this process, the research team met regularly and iterated over the framework's definitions and categories through a shared document. This led to the identification of two categories characterising information asymmetry, *Implementation* and *Awareness*. In this sense, the notion of "information" can be split into how it is implemented and perceived in terms of what it is for, where it is used, who has access to it, and what does it lead to. In turn, all of these are also "information" since, for example, knowing who has access to information is also "information". Below, we present each category's definition, subcategories and possible values. Table 1 provides an overview of the framework.

Table 1. Framework for asymmetry of information.

ASYMMETRY OF INFORMATION							
Implementation							
Possession				**Utility**			
Single	Combined		Split	Single	Multiple		None
Awareness							
Possession				**Utility**			
Single	Combined	Split	None	Single	Combined	Split	None
Location				**Consequence**			
Single	Combined	Split	None	Single	Combined	Split	None

Implementation. Refers to who has the information and who will have to use it. This category comprises the following subcategories: *Implementation of Possession* (i.e., who has the information or who has access to it) and *Implementation of Utility* (i.e., who uses or needs the information).

Awareness. Refers to who knows about different aspects related to the information. This category comprises the following subcategories: *Awareness of Possession* (i.e., who knows who has the information), *Awareness of Utility* (i.e., who knows who needs the information), *Awareness of Location* (i.e., who knows where the information is used), and *Awareness of Consequence* (i.e., who knows the consequence of using the information).

Values. For the *Implementation* category, the values for *Implementation of Possession* can be *Single* (i.e., only one player has or has access to the information), *Combined* (i.e., multiple players have or have access to overlapping parts of the information), or *Split* (i.e., multiple players have or have access to non-overlapping parts of the information). For *Implementation of Utility*, the values can be *Single* (i.e., only one player uses or needs the information), *Multiple* (i.e., multiple players use or need the information), or *None* (i.e., no player uses or needs the information). In the *Awareness* category, the values for *Awareness of Possession*, *Awareness of Utility*, *Awareness of Location*, and *Awareness of Consequence* follow a similar pattern: *Single* (i.e., only one player is aware), *Combined* (i.e., multiple players are aware of overlapping parts), *Split* (i.e., multiple players are aware of non-overlapping parts), or *None* (i.e., no player is aware). For example, in a game scenario where players possess pieces of a treasure map, *Split* would mean each player holds a unique piece, while *Combined* would mean some or all pieces were shared by players.

3   Designing for Asymmetry of Information

In order to operationalise the framework proposed and ensure greater control over how the asymmetry of information was implemented and manipulated, we designed and developed a proof-of-concept digital two-person cooperative top-down dungeon exploration experience, *Parallel Realms: Asymmetry United*. This design selection was informed by the game genre's broad appeal for cooperative play and its scalability potential, aiming to provide a gaming experience close to a real-world scenario. In this game, players explore a dungeon composed of four floors, where each floor corresponds to a different information-based asymmetric challenge puzzle. Both players are expected to traverse and explore each of the dungeon's floors, defeating enemies, collecting loot, levelling up through experience gathered, and interacting with the environment. On each floor, players may encounter unidentified information rooms and clues for the solution to the given floor's puzzle. Moreover, the final room of each floor always contains a puzzle challenge that players must overcome by articulating the knowledge they have gathered. An example of the Weapon Challenge progression can be seen in Fig. 1. Supplementary material, including a gameplay sample, is available online[1]. A downloadable version of the prototype is also available online[2].

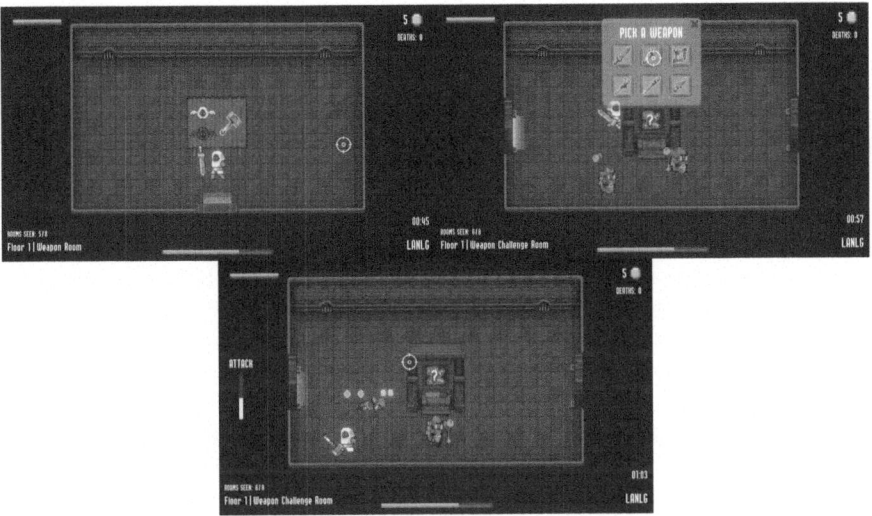

Fig. 1. Gameplay screenshots of the player completing the Weapon Challenge: top-left shows the player interpreting an informational sign, top-right depicts selecting the correct weapon, and the bottom-centre illustrates using it to defeat enemies.

We explore a subset of the possible combinations of values for each of the previously presented subcategories of the framework (Table 2), recognising the infeasibility of implementing and analysing all possible combinations of values in only one study. To ease and expedite the development process, certain subcategory values were chosen. The *Awareness of Location* subcategory was fixed on the *Combined* value since players, a priori to experiencing the game, already knew that the challenges would be found in the final room of each floor. The *Awareness of Consequence* subcategory was fixed on the *None* value, since players did not know what possible abilities they might receive by completing a challenge. If players failed to complete any of the challenges, they would be killed and forced to respawn back at the floor's initial room. This design choice discourages participants from resorting to a trial-and-error strategy in solving the challenges. Finally, the *Awareness of Possession* and *Awareness of Utility* subcategories were paired, meaning that the player who knew who had the information would also know who needed the information.

---

[1] The supplementary material can be accessed at https://osf.io/ndwv2/?view_only=f05e5f3942554c779e34d1a160fe0dcb.

[2] A downloadable version of the prototype can be accessed at https://techpeople.itch.io/parallel-realms-asymmetry-united.

Table 2. Combinations of values from the asymmetry of information framework considered.

Combination	ASYMMETRY OF INFORMATION					
	Implementation		Awareness			
	Possession	Utility	Possession	Utility	Location	Consequence
Skull Challenge	Single (Player A)	Single (Player B)	Single (Player A)	Single (Player A)	Combined	None
Chest Challenge	Single (Player A)	Single (Player B)	Single (Player B)	Single (Player B)	Combined	None
Weapon Challenge	Single (Player A)	Multiple	Single (Player A)	Single (Player A)	Combined	None
Pots Challenge	Single (Player A)	Multiple	Single (Player B)	Single (Player B)	Combined	None
Wolf Challenge	Split	Single (Player B)	Single (Player A)	Single (Player A)	Combined	None
Sequence Challenge	Split	Single (Player B)	Single (Player B)	Single (Player B)	Combined	None

As an example, in the Weapon Challenge (seen in Fig. 1) both players individually encounter the same challenge room, which contains an interactable station and two invulnerable enemies. When interacted with, the station prompts the player to choose a weapon out of six different weapons. Picking the right weapon allows the player to defeat the enemies and advance to the next floor; otherwise, the player is killed and respawns back at the floor's initial room. The correct weapon is the same for both players (*Implementation of Utility* is *Multiple*) but only one player has access to the information room revealing it (*Implementation of Possession* is *Single*). The player who encounters the information room understands they have key information needed by both players, as the sign displays both player's symbols and the answer to the puzzle (this player thus has *Awareness of Possession* and *Utility*).

Both type A and B players encounter the Weapon Challenge on the first floor. Type A players face the Skull Challenge, and type B players face the Chest Challenge on the second floor. Both types then encounter the Pots Challenge on the third floor. Type A players face the Wolf Challenge, and type B players face the Sequence Challenge on the fourth floor.

The game was developed for PC using the *Unity Game Engine* [1]. Defeating enemies awards players with experience, allowing them to level up and choose one of three random abilities. Similarly, upon completing a dungeon floor, players are presented with a choice of one of three random special abilities to further strengthen their character.

4   User Study

A controlled laboratory study was conducted with ten player pairs, who responded to questionnaires before and after playing the game. A mixed-methods approach was applied to the data analysis, aiming to answer our previously mentioned research questions.

Participants. We recruited 20 participants, aged 19 to 42 years (M = 23.9, SD = 4.79), primarily through university social media, mailing lists, and word-of-mouth. Most participants were university students in Portugal with varying levels of gaming experience, with only one (D2) reporting no prior experience with video games. Participants were encouraged to enrol in the study in pairs, thus attempting to guarantee some level of acquaintance, however, some player pairs had to be matched by the researchers. All but one pair (pair D) knew each other prior to the experiment. We will refer to participants by pair letter and number (e.g. C2).

Procedure. The study followed the ethical requirements imposed by our institution. First, participants were informed about the study procedure, and required to fill in a consent form regarding their willingness to participate and share their data, and a demographics form. The latter prompted participants on their name, e-mail, age, self-perceived digital game experience (playing frequency and duration), self-reported investment and competitive profiles, preferred game type and preferred affinity level with play partners. The study took place in a room at our university, with participant pairs seated opposite to each other, each on a laptop. Participant pairs were then given a brief explanation of the game's structure and theme. Participants were informed they could communicate with each other throughout the study and playing sessions were audio, video and screen captured, as well as game events and player actions (e.g., player death) were logged into a database. After experiencing the game, participant pairs were asked to fill out two questionnaires, the miniPXI Questionnaire [14] with a modified scale (5-point Likert scale instead of 7-point Likert scale) and a custom-made experience questionnaire with a 5-point Likert scale. The miniPXI was chosen as it is a validated scale that effectively captures player experience with only a few items. The custom questionnaire prompted participants on their perspectives regarding connection with their play partner, communication quality, fun, challenge, and satisfaction with the play partner's performance, for each of the asymmetric challenges. Moreover, the most liked and disliked asymmetric challenge was also requested. Finally, participants were asked to participate in a semi-structured group interview to understand their perspective on the cooperation throughout the game (i.e., communication needs, missing information, how the game supported and prompted their cooperation), their perspective on the various asymmetric challenges, and game elements they perceived to have influenced their communication.

Data Analysis. A descriptive and statistical analysis of the quantitative data provided by the questionnaires was conducted. The results from the demographics form were analysed in order to give an overview of the sample present in the study. We applied a Friedman (predefined significance level $\rho = 0.05$) and Wilcoxon Signed-Rank Test over the metrics of our custom questionnaire (connection, communication, fun, challenge, and satisfaction with the play partner's performance) enabling us to contextualise the data collected. The results from the miniPXI Questionnaire [14] were also analysed to get a sense of the participants' overall experience.

Regarding qualitative data, for this study, only the interviews were transcribed and subject to a mixed deductive and inductive thematic analysis [4,5]. The coding of the interviews was conducted by one of the researchers involved, and the codebook was initially populated with deductive codes informed by our research questions (e.g., asymmetry, communication) and then expanded with iterative readings of the interviews and notes taken by the researchers during the study. These codes were then discussed amongst all the researchers involved, and revised for redundancy and scope of the research. The coding of the interviews and ensuing discussions about code relationships led to the rationalisation of the overarching themes discussed in the Results section.

5   Results

Ratings resulting from the administration of the modified miniPXI Questionnaire [14] averaged M = 4.22 (SD = 0.87) for type A participants, M = 3.91 (SD = 1.05) for type B participants, and M = 4.06 (SD = 0.97) in total, which indicates an overall positive player experience that is also reflected in participant quotes: "*Genuinely, I liked it a lot. I thought it was a lot of fun.*" (B1). On average, the best-rated components were Audiovisual Appeal (M = 4.60, SD = 0.50) and Enjoyment (M = 4.50, SD = 0.51), while the worst-rated one was Mastery (M = 3.20, SD = 1.11).

There were no statistically significant differences observed in perceived fun ($\chi^2(2) = 2.348$, $\rho = 0.799$) or partner performance ($\chi^2(2) = 10.576$, $\rho = 0.060$) based on the asymmetric mechanic. However, a statistically significant difference was found in perceived communication ($\chi^2(2) = 12.204$, $\rho = 0.032$), connection ($\chi^2(2) = 17.893$, $\rho = 0.003$), and challenge ($\chi^2(2) = 17.893$, $\rho = 0.003$), depending on the asymmetric mechanic. Post hoc analysis with Wilcoxon signed-rank tests was conducted on these three measures, incorporating a Bonferroni correction with a significance level set at $\rho = 0.005$. In this case, there were no statistically significant differences in perceived communication regarding the different mechanics. However, there was a statistically significant difference in perceived connection regarding the skull and sequence mechanics ($Z = -2.871$, $\rho = 0.004$), with the sequence challenge being the highest-rated for connection (M = 4.65, SD = 0.59) and the skull challenge the lowest (M = 3.80, SD = 1.06). Conversely, a statistically significant reduction in perceived challenge was noted regarding the weapon and sequence mechanics ($Z = -3.256$, $\rho = 0.001$), with the sequence challenge rated highest for challenge (M = 4.42, SD = 0.69) and the weapon challenge lowest (M = 3.40, SD = 1.23).

Cooperative Prompting and Flow. Communication between pairs in the study was encouraged by various triggers within the game. Participants identified specific elements like signs, symbols, or interactable objects as cues prompting communication, noting that anything out of the ordinary prompted them to share information with their partner: "*It was whenever I saw a colour or a symbol, anything out of the ordinary, that I felt I had to tell them.*" (B2). For instance, one participant reported that entering specific areas, such as challenge rooms, prompted them to communicate: *"In the red rooms. [...] When you got to a room and there was nothing to kill, you had to communicate."* (G1). Participants reflected on the flow of communication, noting strategies such as delaying the transmission of information to remember clues for later discussion: *"I'd get to that room, make the connection: 'OK, I'm going to need this for later'. I'd try to remember and then when I got to the other room I'd tell them."* (D1). While this approach was effective in some cases, others found it less dynamic and organic: *"It makes it a bit... less dynamic. [...] So we keep quiet and then eventually communicate."* (B2). Frustration and stress from perceived poor performance also hindered communication flow for some participants: *"When you start performing poorly, you get frustrated and end up focusing on things other than communication."* (E2), highlighting the complex interplay between gameplay dynamics and social interaction.

Impact of Implementation and Awareness. During the interviews, some participants intuitively expressed perceptions of the varying subcategories within the proposed framework for asymmetry of information. While unable to articulate specific changes, participants noted a sense of difference in asymmetric mechanics: *"It's almost the same thing, but it's not."* (H1). This was evident due to their acknowledgment that certain challenges would influence their interdependence and communication. Some subtly identified the fixed, and therefore predictable, nature of the **Awareness of Location** sub-category, recognising that it helped guide their actions: *"I didn't necessarily need the key to know that I had to go to the chest. If there was another puzzle, then I needed an indication of which room I had to go to with that answer."* (E2), indicating that clear spatial indications can reduce confusion, especially when facing multiple potential tasks. Others perceived the **Implementation of Possession**, noting that certain puzzle elements were exclusive to one player's side: *"Unless I misunderstood, the key to solving the problem wasn't on both sides, it was only on their side."* (I1). Participants also alluded to certain values of sub-categories, such as **Multiple** in the **Implementation of Utility** or the concept of **Combined** values: *"For example, we might both have to choose the same thing in some cases."* (J2); *"Put them together, and they would represent something in the world."* (I1). Overall, managing the asymmetry of information relied heavily on participants' dependence on their play partner's perspective of the game, often requiring prompts for information exchange to overcome challenges: *"We always had to get information from our partner."* (F1). However, complexity resulting from **splitting information** or **communication requirements** occasionally led to confusion: *"The information wasn't directly on either side, so I was a bit confused as to what I was seeing related to what [D2] was seeing and what the answer was."* (D1). Nonetheless, participants reported a learning curve in the game, where, as they became more accustomed to the asymmetry of information, their performance and effectiveness improved: *"If something didn't go well, the next one I have to pay more attention to."* (H1).

Information Distribution and Enjoyment. Participants expressed a desire for increased interdependence with their play partners beyond situations where progress was blocked. One participant noted that the game's structure involved barriers where both players had to contribute to complete challenges: "*I realised that the game was more oriented towards... Barriers? [...] you both have to pass the puzzles to be able to do it.*" (I2). However, this was perceived negatively as it blocked progress, with another participant suggesting the inclusion of optional information that could aid players without being essential for progress: "*I think I could also use a bit more... Something where the other person's information helps but isn't necessary for progress.*" (I2). This desire for greater interdependence extended to exploration aspects of the game, with participants noting challenges that required **information solely possessed by one player** as hindering cooperation: "*Well, the problem is that either they get all the information and give it to me, or I give it to them.*" (B2). While some compared these challenges to similar experiences in games like *We Were Here* [2], where the exchange of information between players is necessary and beneficial: "*In one sense there was information that one person didn't have and that the other didn't have, and they had to complement the information they had.*" (I2), others found **one-way transactions of information** acceptable only if both sides did not require the same information. However, when **both sides needed information and only one possessed it**, participants found this lack of reciprocity less enjoyable: "*If it's just a transaction, that is, I'm giving them information on a one-off basis, that's fine, but if you both need information, and it's only on one side, I think...*" (I1).

Managing Individual and Shared Progress. Some participants acknowledged the need to wait for their play partner to catch up while traversing the dungeon, often finding themselves in situations where they had to wait without knowing what to do next: "*I often waited for them to finish and didn't know what to do.*" (B1). This suggests that achieving interdependence is difficult when players move at different speeds, possibly due to varying skill levels. Conversely, participants who were still engaged in dungeon activities while their partner waited reported feeling a sense of obligation to **provide the necessary information promptly**, leading to frustration when unable to do so immediately: "*The thing is, I had to kill all the beasties to be able to give you the information. It was frustrating me.*" (E1). Participants also reported doubting their previous explorations when struggling with a challenge, sometimes backtracking to check previously seen rooms: "*Because I was always like, I'm missing something, I was going back and forth, back and forth.*" (B1). For instance, on the second floor, type B players averaged 5.20 backtracks (SD = 2.90), compared to type A players who averaged 3.50 (SD = 2.17). This suggests that **splitting information**, especially when introducing high requirements of **awareness** (e.g., covert information), reduces the **discoverability factor of certain information**. On the other hand, relevant information was located at most in two rooms, sometimes leading players to lose interest in exploring the dungeon floor when they realised this and found the information: *"If I'd already found the altar room and the other room with information, I'd completely lose the motivation to explore the other rooms that I know still exist."* (F1). **Splitting and dividing information** when players **know they have all the necessary information** to overcome an obstacle could make the rest of the experience irrelevant.

Factors Influencing Difficulty in Conveying Information. The perceived impact of genre familiarity on expertise and performance was highlighted, with more experienced players being better able to pinpoint relevant cooperation prompts. Another significant contributor to the complexity was the **splitting of information**, requiring players to articulate their knowledge with their partner to progress, leading to confusion and insecurity about **conveying relevant information effectively**. This was reflected in the concern of some players about missing important details: *"I was always thinking... I was missing something and that I wasn't telling them what I was supposed to."* (B1). Participants often struggled to understand how the information they shared would impact their partner's side: *"My biggest difficulty was describing things, and I couldn't understand how it would have an impact on the other side."* (G2). For example, on the second floor, failed attempts at challenges were the most common, particularly for type B players who averaged 3.00 failures (SD = 1.49). Challenges involving **communication requirements**, such as describing information accurately or making covert associations, further added to the complexity. One participant noted: *"We should have communicated better. Some parts. There was something missing there. A better explanation of what we saw and so on. And then that makes it very difficult to move on."* (B2). Participants struggled with conveying the **importance of certain information**, particularly colours or symbols, which hindered effective communication: *"The only problem I had was the difficulty in explaining how important the colours were."* (I2).

Coherence Expectations. Participants had expectations for coherence in the game's challenge configuration, anticipating consistent mechanics and difficulty progression. However, deviations from these expectations sometimes led to confusion. For example, participants expected information to be scattered throughout the dungeon: *"When I joined the game, I expected that each room would have a little piece of information hidden in a corner."* (F2), and when these expectations were not met, confusion ensued. Additionally, participants anticipated consistency in the mechanics of the game after encountering their first asymmetric challenge: *"The one with the weapon because it was the first one, and I realised more or less what the game was about."* (C1), which occasionally caused confusion when the game deviated from these expectations. Despite this, participants appreciated the coherence in the game's difficulty progression, finding that challenges became progressively more difficult: *"Each of these challenges gets harder. So the first two weren't much of a struggle and the other two were more difficult."* (D2). For example, on the first floor, both player types had a 100% success rate in completing the floor's challenge, while on the last floor type A players had 90% and type B players, 60%. They enjoyed drawing on past experiences to overcome obstacles and found satisfaction in applying learned concepts to more complex scenarios, such as utilising knowledge gained from earlier challenges to solve later ones: "*I also liked the fact that you start with something simple and then use it later with a more complicated puzzle.*" (E2).

Player Dynamics and Communication. Participants highlighted the relevance of familiarity and trust between play partners in shaping communication dynamics. For instance, in situations where participants did not know each other beforehand, the level of connection influenced their ability to decipher each other's intentions and convey information effectively: "*We didn't already know each other, so I think it's different from being with someone you already know and who, as [D1] said, you already know, you already know how they react... It's different.*" (D2). Trust, both in oneself and in the play partner, emerged as a significant factor affecting communication: "*Then I was afraid whether the communication I was giving was the right one or not. So it affected my communication and it also affected [A1]'s.*" (A2). Some participants expressed worry about conveying accurate information, while others felt confident in their attempts to assist their partners: "*I was at least trying to help them with what they needed to do.*" (B1). However, there were instances of hesitancy, particularly after negative outcomes following communicated actions, which affected participants' willingness to trust their partners' instructions: "*When you then said yellow [...] I click on yellow, then I die. Then I'm afraid, based on that communication, to do anything else.*" (A2).

6   Discussion

In this discussion, we review the feedback from participants, delving into our findings regarding their views on the asymmetry of information in the game. We also explore how this asymmetry affects social interactions among participants, connecting our observations to the proposed framework.

Perceptions of Asymmetry of Information. Participants intuitively recognised aspects of asymmetry of information, especially those familiar with similar games (e.g., *We Were Here* [2]). Even without explicit identification, the challenges impacted their cooperation and progression. For example, the mixed reception of **one-sided information** and the need for **shared information** indicate an inherent understanding of subcategories such as *Implementation of Possession* and *Implementation of Utility*. Each challenge's unique combination of subcategory values sometimes disrupted these expectations, leading to varied responses, ranging from confusion to positive engagement. For instance, puzzles where one player unknowingly held crucial information and the other knew that, were often well-received, which is supported by the quantitative results in which, on average, the Sequence Challenge was considered the most challenging (M = 4.42, SD = 0.69) but also the highest-rated for connection (M = 4.65, SD = 0.59). These observations highlight the framework's potential as a design tool. By intentionally disrupting expectations and strategically combining subcategory values, designers can enhance curiosity, experimentation, and replayability.

Impact of Asymmetry of Information on Social Interactions. The asymmetry of information served as a catalyst for social interaction in the cooperative experience provided. The interdependence required for overcoming obstacles naturally fostered communication and cooperation. For example, participants identified game elements such as signs, symbols, or special rooms as prompts for cooperation. Moreover, strategies such as **delaying information exchange** or prioritising exploration emerged as players navigated the game, either to strategise or due to reluctance to engage with their partner when not necessary. While individual characteristics, player connection, trust, and performance influenced these social dynamics, asymmetry of information played a key role in shaping communication flow. However, different configurations of asymmetry of information elicited varied responses. For instance, challenges with **one-sided information exchange** or **minimal input from both participants** were often viewed negatively, resulting in reduced interaction. This is supported by the quantitative results, where the Skull Challenge, an example of such cases, was the lowest-rated for perceived connection on average (M = 3.80, SD = 1.06). In contrast, **covert information** (e.g., clothing items' colours) prompted more significant cooperation due to its complexity, which was for example the case for the Sequence Challenge. On the other hand, the ability to complete a challenge and the quality of interactions, influenced by complexity, affected perceived communication quality, reflected in ratings of challenge of these mechanics. Overall, despite the innate complexity of the designed mechanics, the complexity added by different types of asymmetry of information and by the need of players to learn and improve communication skills with their play partner has an impact on the number of interactions between players as well as the quality or depth of these interactions.

Limitations and Future Work. The study presents an initial framework for understanding the implementation of asymmetry of information in games. The findings and implications are based on specific values within the framework and may not apply to all game types and contexts, acknowledging the infeasibility of covering all possible design permutations and contexts. The user study was conducted with 20 participants with specific game profiles which were not controlled, and with a specific game genre and mechanics which may not generalise to other game contexts and populations. Furthermore, the order in which participants engaged with the different implementations of asymmetry of information was static, which could have influenced perceptions of difficulty and social interaction. However, the research highlights the significant impact of asymmetry of information on social interactions in a two-player cooperative game setting.

Future studies could explore additional combinations of values and their effects on players' perspectives and interactions. There is potential for further exploration of asymmetry of information in different types of gameplay and across various game genres. Understanding how this type of asymmetry is influenced by player dynamics (e.g., trust) and characteristics, and exploring their effects in family play and intergenerational gaming may offer valuable insights into gaming dynamics. The study recognises the limitations of its controlled laboratory approach and suggests incorporating more realistic scenarios in future research, such as conducting long-term observations in natural settings.

7   Conclusion

Addressing a gap in understanding nuanced variations of asymmetry of information, our study proposes a framework to analyse, ideate, design, and discuss asymmetry of information in gaming. We developed a digital cooperative game, exploring some of its dimensions, where players engage with information-based challenge puzzles. Through a user study involving ten pairs of players, we examined how various types of asymmetry of information influenced social interaction and player perspectives. Participants perceived asymmetry, influencing cooperation and progression. Asymmetry drove social interaction, shaping communication patterns and player engagement. Different configurations of asymmetry elicited distinct reactions, highlighting its impact on communication effectiveness and gameplay dynamics. Our study underscores the importance of understanding and leveraging dynamics such as these for crafting engaging and socially enriching gaming experiences. Future work could further explore how to capture the granularity of information dynamics and exploring asymmetry of information in various contexts beyond two-player cooperative games.

Acknowledgments. We thank the anonymous reviewers for their valuable feedback. This work was supported by FCT through project *Plug n' Play: Exploring Asymmetry and Modularity for Inclusive Game Design*, ref. 2022.08895.PTDC (http://doi. org/10.54499/2022.08895.PTDC), scholarships, ref. UI/BD/151178/2021 (https://doi. org/10.54499/UI/BD/151178/2021), ref. 2022.12448.BD, and ref. 2023.03373.BD, and the LASIGE Research Unit, ref. UIDB/00408/2020 (https://doi.org/10.54499/UIDB/ 00408/2020) and ref. UIDP/00408/2020 (https://doi.org/10.54499/UIDP/00408/ 2020).

Disclosure of Interests. The authors have no competing interests to declare that are relevant to the content of this article.

References

1. Unity Real-Time Development Platform | 3D, 2D, VR & AR Engine (2024). https://unity.com/. Accessed 14 Mar 2024
2. We Were Here - First Person Coop Puzzle Solving Adventure (2024). https:// totalmayhemgames.com/games/we-were-here/. Accessed 05 Mar 2024

3. Bartle, R.: Hearts, clubs, diamonds, spades: players who suit muds. J. MUD Res. **1**(1), 19 (1996)
4. Braun, V., Clarke, V.: Using thematic analysis in psychology. Qual. Res. Psychol. **3**(2), 77–101 (2006). https://doi.org/10.1191/1478088706qp063oa
5. Braun, V., Clarke, V.: Thematic analysis. In: Cooper, H., Camic, P.M., Long, D.L., Panter, A.T., Rindskopf, D., Sher, K.J. (eds.) APA Handbook of Research Methods in Psychology, vol. 2: Research designs: Quantitative, Qualitative, Neuropsychological, and Biological, Washington, pp. 57–71. American Psychological Association (2012). https://doi.org/10.1037/13620-004
6. De Schutter, B., Vanden Abeele, V.: Designing meaningful play within the psychosocial context of older adults. In: Proceedings of the 3rd International Conference on Fun and Games, Leuven, Belgium, pp. 84–93. ACM (2010). https://doi.org/10.1145/1823818.1823827
7. Depping, A.E., Johanson, C., Mandryk, R.L.: Designing for friendship: modeling properties of play, in-game social capital, and psychological well-being. In: Proceedings of the 2018 Annual Symposium on Computer-Human Interaction in Play, Melbourne, VIC, Australia, pp. 87–100. ACM (2018). https://doi.org/10.1145/3242671.3242702
8. Depping, A.E., Mandryk, R.L.: Cooperation and interdependence: how multiplayer games increase social closeness. In: Proceedings of the Annual Symposium on Computer-Human Interaction in Play, Amsterdam, The Netherlands, pp. 449–461. ACM (2017). https://doi.org/10.1145/3116595.3116639
9. Depping, A.E., Mandryk, R.L., Johanson, C., Bowey, J.T., Thomson, S.C.: Trust me: social games are better than social icebreakers at building trust. In: Proceedings of the 2016 Annual Symposium on Computer-Human Interaction in Play, Austin, Texas, USA, pp. 116–129. ACM (2016). https://doi.org/10.1145/2967934.2968097
10. Emmerich, K., Masuch, M.: The impact of game patterns on player experience and social interaction in co-located multiplayer games. In: Proceedings of the Annual Symposium on Computer-Human Interaction in Play, Amsterdam The Netherlands, pp. 411–422. ACM (2017). https://doi.org/10.1145/3116595.3116606
11. Fullerton, T.: Game Design Workshop: A Playcentric Approach to Creating Innovative Games, 4th edn. A K Peters/CRC Press, Boca Raton (2018). https://doi.org/10.1201/b22309
12. Gentile, D.A., et al.: The effects of prosocial video games on prosocial behaviors: international evidence from correlational, longitudinal, and experimental studies. Pers. Soc. Psychol. Bull. **35**(6), 752–763 (2009). https://doi.org/10.1177/0146167209333045
13. Granic, I., Lobel, A., Engels, R.C.M.E.: The benefits of playing video games. Am. Psychol. **69**(1), 66–78 (2014). https://doi.org/10.1037/a0034857
14. Haider, A., et al.: miniPXI: development and validation of an eleven-item measure of the player experience inventory. Proc. ACM Hum.-Comput. Interact. **6**(CHI PLAY), 1–26 (2022). https://doi.org/10.1145/3549507
15. Harris, J., Hancock, M.: To asymmetry and beyond!: improving social connectedness by increasing designed interdependence in cooperative play. In: Proceedings of the 2019 CHI Conference on Human Factors in Computing Systems, Glasgow, Scotland, UK, pp. 1–12. ACM (2019). https://doi.org/10.1145/3290605.3300239
16. Harris, J., Hancock, M., Scott, S.D.: Leveraging asymmetries in multiplayer games: investigating design elements of interdependent play. In: Proceedings of the 2016 Annual Symposium on Computer-Human Interaction in Play, Austin, Texas, USA, pp. 350–361. ACM (2016). https://doi.org/10.1145/2967934.2968113

17. De la Hera, T., Loos, E., Simons, M., Blom, J.: Benefits and factors influencing the design of intergenerational digital games: a systematic literature review. Societies **7**(3), 18 (2017). https://doi.org/10.3390/soc7030018
18. Hunicke, R., LeBlanc, M., Zubek, R.: MDA: a formal approach to game design and game research. In: Proceedings of the AAAI Workshop on Challenges in Game AI, vol. 4 (2004)
19. Juul, J.: Half-Real: Video Games between Real Rules and Fictional Worlds. MIT Press, Cambridge (2005)
20. Khalis, A., Ferrari, M.A., Smit, S., Ewell, P.J., Mikami, A.Y.: You teach me and I'll teach you: the role of social interactions on positivity elicited from playing Pokémon Go. Cyberpsychol. J. Psychosoc. Res. Cyberspace **16**(4) (2022). https://doi.org/10.5817/CP2022-4-9
21. Nacke, L.E., Bateman, C., Mandryk, R.L.: Brainhex: a neurobiological gamer typology survey. Entertainment Comput. **5**(1), 55–62 (2014). https://doi.org/10.1016/j.entcom.2013.06.002
22. Neto, A., Cardoso, P., Carvalhais, M.: Asymmetric gameplay: types and perspectives. In: International Conference on Design and Digital Communication, pp. 765–778. Springer (2022). https://doi.org/10.1007/978-3-031-20364-0_64
23. Neto, A., Cardoso, P., Carvalhais, M.: Endogenous asymmetry in games: expanding the typology. In: Perspectives on Design and Digital Communication IV: Research, Innovations and Best Practices, pp. 277–294. Springer (2023). https://doi.org/10.1007/978-3-031-41770-2_16
24. Nguyen, H.T.T., Tapanainen, T., Theng, Y.L., Lundberg, S., Luimula, M.: Fostering communication between the elderly and the youth with social games. In: PACIS 2015 Proceedings (2015)
25. Oswald, D.L., Clark, E.M., Kelly, C.M.: Friendship maintenance: an analysis of individual and dyad behaviors. J. Soc. Clin. Psychol. **23**(3), 413–441 (2004). https://doi.org/10.1521/jscp.23.3.413.35460
26. Pais, P., et al.: A living framework for understanding cooperative games. In: Proceedings of the CHI Conference on Human Factors in Computing Systems, pp. 1–17 (2024). https://doi.org/10.1145/3613904.3641953
27. Rogers, K., Karaosmanoglu, S., Wolf, D., Steinicke, F., Nacke, L.E.: A best-fit framework and systematic review of asymmetric gameplay in multiplayer virtual reality games. Front. Virtual Reality **2**, 694660 (2021). https://doi.org/10.3389/frvir.2021.694660
28. Schell, J.: The Art of Game Design: A Book of Lenses, 1st edn. CRC Press (2008). https://doi.org/10.1201/9780080919171
29. Tekinbas, K.S., Zimmerman, E.: Rules of Play: Game Design Fundamentals. MIT Press, Cambridge (2003)
30. Toups Dugas, P.O., Hammer, J., Hamilton, W.A., Jarrah, A., Graves, W., Garretson, O.: A framework for cooperative communication game mechanics from grounded theory. In: Proceedings of the First ACM SIGCHI Annual Symposium on Computer-Human Interaction in Play, Toronto, Ontario, Canada, pp. 257–266. ACM (2014). https://doi.org/10.1145/2658537.2658681
31. Yee, N.: Motivations for play in online games. CyberPsychology Behav. **9**(6), 772–775 (2006). https://doi.org/10.1089/cpb.2006.9.772

NFLegacy, Towards a NFT-Based Game Ecosystem

Pedro Trincheiras[1], Pedro A. Santos[1,2], and João Dias[3,4(✉)]

[1] Instituto Superior Técnico, Universidade de Lisboa, Lisbon, Portugal
[2] INESC-ID, Lisbon, Portugal
{pedro.trincheiras,pedro.santos}@tecnico.ulisboa.pt
[3] Faculdade de Ciências e Tecnologia, Universidade do Algarve, Faro, Portugal
jmdias@ualg.pt
[4] CISCA, Oeiras, Portugal

Abstract. NFT stands for Non-fungible Token, which represents unique and non-interchangeable assets, based on blockchain technology. One of the applications where NFTs are being used is in gaming, by using NFTs as in-game assets. So far, the inclusion of NFTs in games has not led to new gameplay mechanics. Most developers opt to use the blockchain just as an asset repository ignoring other blockchain capabilities. To facilitate the development of games able to explore NFT-based game mechanics, we created NFLegacy, an NFT-based Game Ecosystem. Furthermore, we propose a game mechanic, inspired by object attachment theory, where a player shares item progression across different games. Using NFLegacy, we have developed and tested a prototype with three mini-games. To assess the improvement in the player experience, we conducted tests with a selected group of users and analyzed the results.

Keywords: Blockchain · NFT · Game Design · Game Mechanics

1 Introduction

Blockchain technology is gaining popularity and recognition for its ability to securely store and verify data [11]. While it is well-known for facilitating cryptocurrency transactions, blockchain technology is rapidly expanding to a wide range of applications [6]. One of the most notable developments in this area is the emergence of NFTs, which are unique digital assets that can be bought, sold, and traded on the blockchain. These assets are unique in the sense that even though their image and attributes can be copied, the blockchain provides authenticity for the original one, just like real-world handmade collectibles where replicas can be made but there are only a certain amount of original pieces with each one being unique, and there is an authority that can verify their authenticity.

In the last couple of years, there has been a huge increase in investment in NFT projects by industry [10]. On the gaming side, studios are looking for ways to integrate NFTs into existing games or create new games that are centered

A. Marto et al. (Eds.): VJ 2024, CCIS 2324, pp. 171–185, 2025.
https://doi.org/10.1007/978-3-031-81713-7_12

around NFTs. For example, some studios are using NFTs to create "play-to-earn" games, where players can earn new NFTs by playing the game. Examples of commercial games that employ Blockchain and NFT-based elements can be found in [2,8].

1.1   Objectives

While NFTs are currently being used in many games, those games predominantly rely on mechanics already established in other types of games, often treating NFTs as static assets. There is a notable absence of games that truly embrace and harness the unique features of NFTs, such as their inherent uniqueness and the sentimental attachment we, as humans, associate with objects. There are even negative reactions from players and companies like Steam on how NFTs and Blockchain technology are currently being incorporated into games [2,8].

The primary objective of this work is to propose new gameplay mechanics made possible by inherent characteristics of NFTs, and demonstrate their feasibility and impact on user's experience.

To that end we developed a prototype of a comprehensive NFT Ecosystem: NFLegacy, including a game, a wallet and SDK necessary to support NFT integration. This approach ensures that the NFTs are seamlessly interwoven into the gaming experience, allowing for greater innovation and a deeper connection between players and their in-game assets.

We also selected one of the gameplay mechanics proposed, the NFT-based Shared Progression mechanic, and implemented it as a proof-of-concept. This mechanic connects the progression of in-game assets to the ownership of NFTs, potentially changing how players engage with and value their virtual possessions.

2   Gameplay Mechanics Enabled by NFTs

The use of NFTs has the potential to introduce different and new game mechanics. In this section we present a set of game mechanism inspired in inherent characteristics of NFTs. These mechanics were refined taking in consideration the books "Game Mechanics: Advanced Game Design" [1], "Game Usability, Advice from the Experts for Advancing the Player Experience" [4] and "Design e Desenvolvimento de Jogos" [5].

2.1   Shared Progression

One of the inherent characteristics of NFTs is the true ownership of the assets they represent. By granting players true ownership, they will potentially establish deeper connections with in-game items, thus generating stronger emotions during gameplay and interactions. This notion is underpinned by the theory of 'Object attachment as we grow older' [3], which posits that individuals develop emotional bonds with objects, and these emotions intensify as we grow older with the object. This also intensifies when the individual has pleasant memories with the object ("The increase in sentimental thoughts may be through the use of objects to recall and reminisce about pleasant memories" [3]). One example of a pleasant memory could be when a player receives an item as a reward for beating a particularly hard challenge in a game. It can then be reinforced when they use the item to help pass a hard part of that game. Furthermore, by allowing the player to visit the NFT outside of the game (even if the game ceases to exist), we are further reinforcing the attachment to the object by recalling pleasant memories with it.

Additionally, NFTs enable transferability between contexts; they are not tied to a specific game context and can be used in different contexts. Players can then explore multiple worlds while curating a collection of NFTs that narrate their unique gaming journey.

Based on these key characteristics, we propose a game design mechanic, which we name Shared Progression (depicted in Fig. 1), where an item unlocked in a game, exists outside the game, and can be used seamlessly across multiple games.

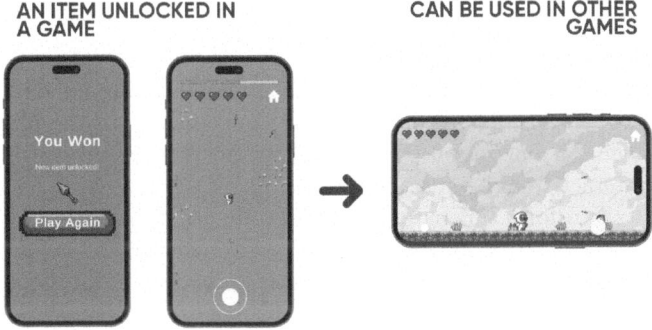

Fig. 1. Shared Progression Diagram

In an interconnected gaming ecosystem that takes advantage of this mechanic, players are no longer constrained to a single game; This aspect can further help to intensify a player's emotional connection to in-game items, given that players will more easily perceive these items as their very own possessions.

Furthermore, Shared Progression creates another opportunity for critical thinking and strategic problem-solving. When confronted with challenging scenarios, players can devise innovative solutions by leveraging their NFT inventory, or even try to advance in another game of the ecosystem to obtain new useful items.

Note that the Shared Progression mechanic is also aligned with the promise of interoperability in the Metaverse [9].

2.2   Merge NFTs

The motivation for this game mechanic is rooted in a common observation within the gaming and NFT community. Many game developers opt for static NFTs creating them with fixed images and metadata that remain unchanged throughout their lifecycle. While this ensures the uniqueness of the initial NFT mint, it often results in limited dynamism and reduced player engagement over time.

In response to this challenge, it is possible to use a game mechanic that harnesses the capabilities of smart contracts. This mechanic allows for the merging of identical NFTs, granting players the ability to create an upgraded version with mutated or enhanced attributes. The transformative nature of this mechanic not only introduces greater dynamism to the NFTs but also makes them considerably more appealing for player interaction.

With this mechanic, we aim to offer players a dynamic and evolving experience, encouraging them to actively collect and merge NFTs to enhance their in-game assets. This not only breaks away from the static nature of traditional NFTs but also provides players with a more immersive and customizable gaming experience, increasing the overall appeal and longevity of the NFTs within the gaming ecosystem.

2.3 NFTs Upgrade

NFTs are often regarded as digital assets to be held with the sole anticipation of their monetary value appreciating over time. This prevailing mindset primarily leads to NFTs being acquired for speculative purposes, with the ultimate goal of reselling them for maximum profit. This perspective limits the full potential of NFTs as interactive and engaging assets.

We seek to transform this paradigm by introducing two game mechanics that imbues NFTs with a dynamic and evolving nature, creating a new dimension of value. These mechanics center around NFTs that can evolve and mutate through in-game and out-game behaviour:

On Win Upgrade. This mechanic involves the dynamic upgrade of NFTs based on user gameplay and achievements, offering tangible benefits to players who engage with their NFTs within the game.

Upgrade per Trade. Through on-chain trading between players, players can trigger mutations that enhance the attributes or increase the in-game power of their NFTs. By encouraging users to actively trade their NFTs, we aim to disrupt the traditional notion of hoarding NFTs indefinitely, effectively bridging the gap between holding NFTs as speculative investments and actively incorporating them into the gaming experience. This not only diversifies the utility of NFTs but also fosters an interactive and dynamic experience, providing a tangible benefit to those who engage with their NFTs within the game.

3   NFLegacy

At the heart of our approach is the planning and development of new game mechanics centered around NFTs. These mechanics are designed to seamlessly integrate NFTs and blockchain capabilities into the gaming experience.

3.1 NFLegacy - a NFT-Based Game Ecosystem

One of our goals is to create an NFT-based Game Ecosystem, to facilitate the development of new games that want to explore NFT-based mechanics, and to make it easier for end-users to both play games and use out-of-game blockchain functionalities (such as accessing a wallet). To achieve this, we created a system that seamlessly connects all the components, providing users with an intuitive experience.

The proposed system consists of three key components: the wallet, the SDK (Software Development Kit), and the games. We've structured it this way to minimize code repetition and streamline the development process. Since every game requires a consistent set of tools to interact with the blockchain, these essential tools are integrated into the wallet. Additionally, we developed an SDK that simplifies the process of invoking these tools within the game. By doing so, the wallet exposes these tools to the game through Android intents.

In Fig. 2 one can see the system architecture scheme. The user interacts with both the game and the wallet app, the game has the SDK implemented to communicate with the wallet app, and the wallet app takes advantage of external services to perform blockchain-related tasks.

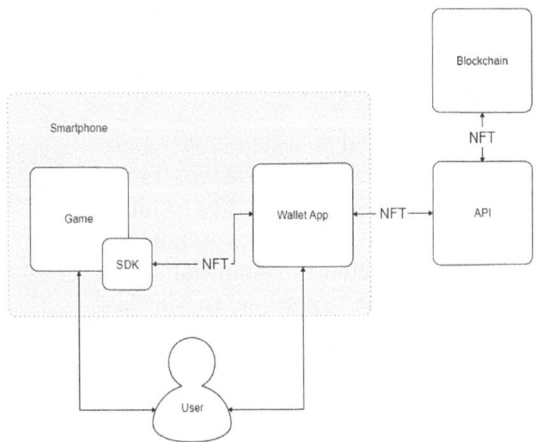

Fig. 2. NFLegacy architecture

4   The Game Trilogy, Pixy's Legacy

To showcase both the technical feasibility and the potential for NFT-based game mechanics, we implemented the Shared Progression mechanic in a set of mini-games. While the shared mechanic can be implemented within a single genre of games, to better understand its potential, we incorporated this mechanic into various game genres. Each game genre brings a unique set of challenges and opportunities that can showcase the adaptability of the mechanic and its influence on player experience and satisfaction.

We introduce thus Pixy's Legacy, a trilogy of games, each featuring its own gameplay style, to explore the feasibility of shared progression in different contexts and its effects on the player's overall experience. Although the games are different, all share the same progression method, after a certain amount of kills the player is rewarded with an NFT item, which is a weapon that can be used in all games. The only exception is the first item that is already attributed from the start. While the items are shared across the games, they are realised differently in each game, depending on gameplay and game fiction.

4.1 Topdown Pixy's Quest

We start with 'Pixy's Quest', which is a Topdown game, that draws inspiration from the well-known game called 'Survival IO'. In this game, players control a character with a joystick located at the bottom center of the screen. The character attacks are based on the cooldown of the weapon, represented by an NFT, chosen by the player at the beginning of the game and are aimed at the last direction the player was facing. The primary goal is to eliminate as many incoming enemies as possible without succumbing to defeat (Fig. 3).

4.2 Sidescroller Pixy's Adventure

For 'Pixy's Adventure' we decided on a sidescroller game experience where players control the character's movement, both left and right, and have the ability to jump. Similar to the first game, the character continuously attacks based on the NFT selected in the initial screen and aimed to the left or right according to the last player's movement. Similarly to the previous game the goal is to eliminate as many enemies as possible (Fig. 4).

4.3 Tower Defense Pixy World Defense

Finally, 'Pixy World Defense' was developed as a tower defense. Players start with three towers, and these towers utilize the weapon chosen in the in-game menu to defend against incoming meteors. The player can change the current towers at any time but can not move them. For every 15 meteors destroyed, players can install another tower. The main objective in this game is to destroy the meteors without allowing the planet to be destroyed.

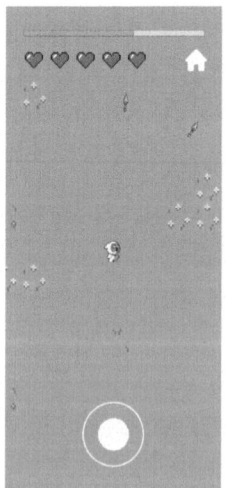

Fig. 3. Pixy's Quest

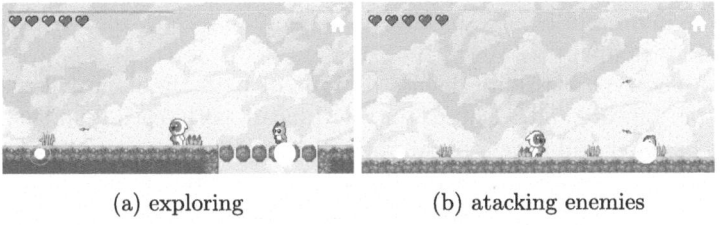

(a) exploring   (b) atacking enemies

Fig. 4. Pixy's Adventure

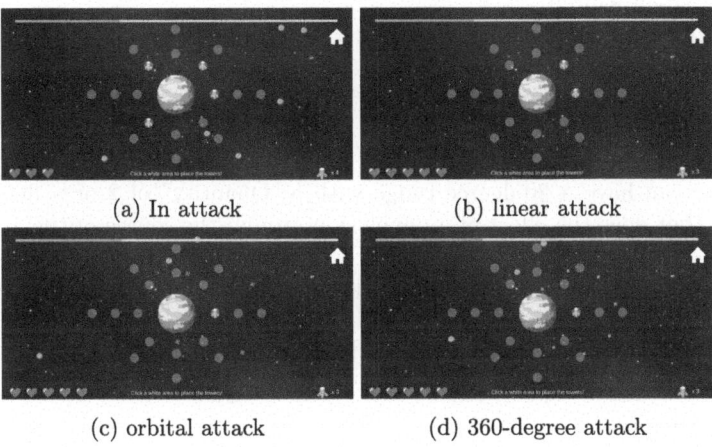

(a) In attack   (b) linear attack

(c) orbital attack   (d) 360-degree attack

Fig. 5. Pixy World Defense

4.4 NFT Attributes

Careful consideration was given to the attributes of the NFTs, ensuring they added depth and variety to the gaming experience across all three games. Each game has its unique mechanics and gameplay style and the items needed to fit all the games. For instance, in our first and second games, players throw the items, while in the third game, items are launched from a tower. Furthermore, the size and the attributes of these items can differ between games; for instance, the items in the first game are smaller than those in the second game and the velocity in the second game is just ninety percent of what it is on the first game. To accommodate these variations and ensure that the NFT items could seamlessly integrate into each game's context, we strived to abstract the items as comprehensively as possible. After several iterations, the final version of the attributes was decided to be the following:

– **Collision Type**: This attribute represents the type of collision the item will have, it could be "Single Hit" or "Pass Through". "Single Hit" means that the item will only collide with the first enemy and will be destroyed after, only being able to hit the first enemy in a line of enemies, "Pass Through" means that the items will be able to hit every enemy in its way until it exceeds its range for example.

- **Type**: The type attribute defines de type of path the item will have, this could be "Linear" and "Orbital". In the "Linear" path, each item shot by the player or tower will have a linear path starting from the source of the attack at intervals defined by the attribute "Cool Down". In the "Orbital" path the item does a circular path around the player, in this type. The number of projectiles orbiting the player is defined by the "Quantity" attribute.
- **Cool Down**: This attribute defines the interval (in seconds) between which each weapon is fired.
- **Damage**: "Damage" is the amount of health the item could take from the items when they collide.
- **Quantity**: This attribute defines the number of items the player could shoot at once.
- **Angular Range**: "Range" defines the angle that the items are shot, for example, we can have a 20-degree range with a "Quantity" of 2 or a 360-degree range that does the effect of an explosion around the source of the attack as we can see in Fig. 5(d).
- **Reach**: The "Reach" defines the maximum distance an item or projectile can travel since it is shot.
- **Velocity**: This attribute represents the velocity the item travels.

5   Evaluation

Our main goal with this study was to assess the feasibility of NFLegacy and the developed shared progression mechanic. To that end, we compared a version of the games with and without shared progression, and analysed the user experience of both versions. Additionally, we assessed the NFT Integration by examining if the inclusion of NFTs in the project introduces any difficulties or challenges for users while using the wallet and interacting with the NFT-based games.

To achieve these objectives, we adopted an evaluation approach that involved dividing our test users into two distinct groups. Each group was exposed to a different progression variant, enabling us to compare the impact of shared progression versus non-shared progression. Users in the experimental condition (With Shared Progression - WSP) experienced the shared progression mechanic with NFTs throughout their interaction with the games. Users in the control condition (No Shared Progression - NSP) followed the traditional non-shared progression mechanic, which did not involve sharing their items between games.

The evaluation process involved a sequence of actions and data collection:

1. User Interaction with Wallet: Test users commenced their engagement with the wallet by performing a predetermined set of actions. These interactions include all the implemented features in the wallet.
2. Gameplay: After interacting with the wallet, users engage in gameplay within the games. They played games that featured either the WSP or NSP progression variants.

3. Questionnaire: Following their interactions with the wallet and games, users were presented with a questionnaire. This questionnaire was designed to gather valuable feedback and insights from users regarding their experiences and perceptions.

5.1   Protocol

In our study, a protocol was put in place to evaluate the impact of the shared progression mechanic within the user experience and their interaction with the wallet. The protocol consisted of a series of carefully planned steps and considerations to comprehensively assess user interactions and perceptions.

Steps of the User Tests:

- **User Segmentation:** Our initial step involved the recruitment and categorization of participants into the two distinct conditions with fifteen (15) participants each.

  However, it's important to note that our aim was to diversify our user pool as much as possible, ensuring a broad spectrum of experiences, knowledge levels, and backgrounds. Our user pool spanned a wide range of profiles, encompassing individuals with varying degrees of expertise in blockchain and NFTs.

  In addition to this diversity in blockchain and NFT expertise, our participants covered a wide age range, with participants ranging from sixteen (16) to fifty-five (55) years old. Furthermore, our participants represented diverse educational backgrounds, from varying levels of formal education to differing degrees of gaming knowledge and interests.

  While our study aimed to diversify our user pool, we acknowledge a limitation in our participant selection process. Due to time constraints, we opted to involve individuals who were known to us. This decision may have introduced a level of familiarity and potential bias into the study. While every effort was made to maintain objectivity and impartiality throughout the evaluation process, we recognize that this factor could influence the results to some extent.

- **User Onboarding:** To ensure that participants were well-acquainted with the games, the wallet, and its functionalities, users embarked on their journey with a carefully crafted onboarding process. This introductory phase allowed them to become familiar with the wallet's features and capabilities, ensuring a smooth transition into the evaluation process.

  During the onboarding process, users were guided through various tasks designed to acquaint them with the wallet's functionality. This included actions such as viewing their NFT collection, which provided a clear understanding of the NFTs they owned. Additionally, participants were introduced to essential wallet components. This quick overview of wallet functionality served as a foundation for their subsequent interactions with the platform.

  Moreover, in the WSP version, a significant emphasis was placed on conveying the concept of shared items between games. Participants were explicitly informed about the interconnected nature of the games, emphasizing that the same item rewarded in a game was seamlessly accessible within the other games.

- **Gameplay:** A distinctive feature of our study was the provision of unlimited game time while keeping track of it. This approach encouraged participants to engage in gameplay within the NFT-based games for an extended period. The intention was to expose as much as possible the user to the version they got, enabling them to explore all the games and to jump between games without cutting their experience short but also without forcing them to play for more time than they wanted. Importantly, the players of one version of the game were unaware of the existence of the other version, ensuring that their experiences and perceptions were based solely on the specific game version they were engaging with. This "blinded" approach mitigated potential biases stemming from users' preconceived perceptions about what they are not experiencing.

- **Questionnaire:** After interacting with the wallet and games, users were presented with a structured questionnaire. This questionnaire played a pivotal role in gathering detailed feedback and insights about their experiences, preferences, challenges, and overall satisfaction. It included a series of questions and prompts to capture a comprehensive overview of user perceptions. We combined two components to create our questionnaire: the questions from the GUESS (Game User Experience Satisfaction Scale) questionnaire and our custom questions. The GUESS questionnaire is a validated tool used in game research to assess and measure user satisfaction and overall gaming experience. It includes standardized items written as affirmative statements, designed to capture various aspects of the gaming experience, such as gameplay, graphics, and overall satisfaction [7].

Participants responded to all items in the questionnaire using a Likert scale of 1 (completely disagree) to 7 (completely agree).

Additionally to the GUESS questionnaire, we also wanted to look at the usefulness and enjoyment of the specific Shared Progression mechanic. For this specific goal, we created a set of custom questionnaire items:

- I felt being stuck at a certain point in the games.
- I enjoyed unlocking an item.

For participants in condition WSP, we also asked the following questions:

- I used an item unlocked in a game in another game.
- This mechanic help me progress in the other games.
- I enjoyed using an item from a game in another game.

For participants in condition NSP, since they didn't use the Shared Progression mechanic, instead we created two different items related to the potential usefulness and enjoyment of this feature.

- It would be helpful to use an item from a game in another game.
- I Would like the items to be shared between games.

In essence, our protocol, was designed to accommodate a diverse user pool, guide them through a comprehensive onboarding process, facilitate extended gameplay experiences, and gather structured feedback through version-specific questionnaires. This methodical approach allowed us to gain a well-rounded understanding of the impact of shared progression and NFT integration on user experiences and challenges, effectively addressing our study objectives.

5.2 Results Analysis

Data was introduced into IBM SPSS and analyzed. We started by analyzing the GUESS section of the questionnaire, a component focused on game experience and satisfaction. We started by applying Cronbach's Alpha, a widely recognized measure of reliability in survey analysis. The analysis resulted in a score of 0.943 on a scale ranging from 0 to 1, signifying strong internal consistency. Consequently, we decided to amalgamate these variables into a single composite variable, User Satisfaction, representing the average of all the constituent variables.

Our analysis led us to examine the distribution of our data. To do this, we conducted the Shapiro-Wilk normality test, and the results were clear: our data did not conform to a normal distribution. This prompted us to turn to non-parametric tests.

Given the non-normal distribution of our data, we employed the Mann-Whitney U test to test for statistically significant differences in User Satisfaction across conditions. Unfortunately, the result obtained $p = .567$, indicates that there are no statistically significant differences in User Satisfaction between the two conditions. The boxplot of Fig. 6 further illustrates the lack of difference between the two conditions (although there is a wider deviation in the NSP condition). On the positive side, both versions got very high results in terms of User Satisfaction (which could be one possible explanation for no differences being found).

A similar process was applied to the other individual variables. All data was considered to be not parametric, and thus the Mann-Whitney U test was used to dependent variable results across conditions. In general, most of the dependent variables showed no statistically significant differences. However, there are three particular variables that showed close to statistically significant differences and even a statistically significant difference. The first of these we would like to report is the variable that measures whether respondents felt stuck at a certain point in the game. According to the boxplot of Fig. 6(b) there is a slight difference, as a median number of participants who played the version with No Shared Progression reported a higher level of feeling stuck. Although there are some differences, according to the Mann-Witney U test they are not strong enough to be considered statistically significant ($p = 0.148$) indicating that such differences were not substantiated.

In the second one, the dependent variable that measures if participants felt the need for a Wallet, there is also a perceivable difference between the two (see Fig. 6(c)), as most of the participants in the Shared Progression condition gave it the maximum value possible. Despite this perceivable difference, Mann-Witney U test yielded a p value of $p = .089$. Although it is not a statistically significant

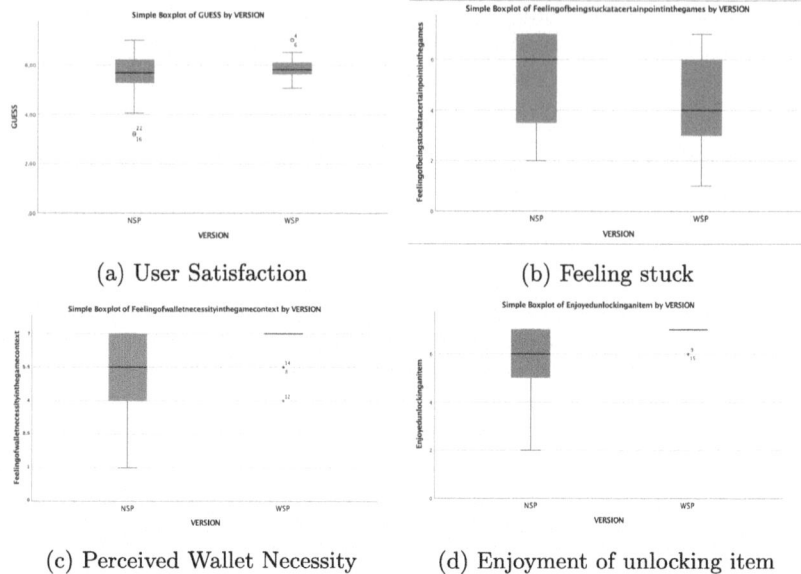

(a) User Satisfaction   (b) Feeling stuck

(c) Perceived Wallet Necessity   (d) Enjoyment of unlocking item

Fig. 6. Evaluation results

difference, it is close to being statistically significant, implying that there could be a difference here.

The third variable, measuring the "feeling of unlocking an item", presents a statistically significant difference ($p = .037$), meaning that participants in the Shared Progression felt more enjoyment when unlocking items than participants in the No Shared Progression condition. Figure 6(d) further illustrates this difference, with a median value of 6 for the NSP condition against a median value of 7 for the WSP condition (all of the participants except one rated it with the maximum value possible). This is our most important result, and it emphasizes the importance of this particular aspect in the context of the game experience and satisfaction.

Having closely observed the players during our in-person tests, we noted a behavior that repeated itself throughout the tests. It was evident that all players typically encountered early losses in the first game and switched to one of our other two games. Those using the WSP version after unlocking better items circled back to the first game for easier play. In contrast, players with the NSP version rarely returned to the first game after an initial defeat, opting to explore different gaming options instead.

5.3   Discussion

Our analysis of the questionnaire data has provided valuable insights into the user experiences and satisfaction levels across two distinct versions: the WSP and NSP. These findings shed light on the nuances of how the inclusion of shared items can influence user perception. As mentioned earlier we achieved this by evenly distributing users with diverse backgrounds between the two versions and having each user experience only one of these versions.

One notable discovery was that users of the WSP version expressed higher levels of satisfaction when it came to unlocking items compared to their NSP counterparts. A plausible explanation for this difference could be the utility of the items within each version. In the WSP version, items could be used across all three games, thus enhancing their perceived value and desirability. Conversely, in the NSP version, items remained restricted to the game where they were unlocked, mimicking the conventional model of individual game items. Users in the NSP version might have felt that their efforts in unlocking items were less rewarding compared with WSP users since the utility of these items was confined to a single gaming experience as is normal in the games people normally play.

This result underscores the importance of item integration and cross-game utility in enhancing user satisfaction. It suggests that game developers should consider the extended use of in-game items to elevate user engagement and enjoyment, a finding that holds significant implications for game design.

The analysis also revealed interesting trends in the NSP version where users reported a slightly increased sensation of feeling "stuck" in the games and awareness of the wallet necessity. While these differences were not statistically significant, it was noteworthy due to their proximity to statistical relevance. The notion of feeling "stuck" in the game can be associated with the unavailability of shared items. In the NSP version, where items remained locked within the boundaries of the specific game, users might have encountered situations where they lacked access to items that could potentially assist in progressing or overcoming obstacles. This limitation in item sharing might have contributed to the marginal difference in perceived feelings of being "stuck" between the two versions. Also, the fact that the NSP version feels more like a common game could be the reason for the users to feel less of the necessity of having a wallet in the system.

This finding suggests that the ability to share items across games can alleviate frustrations related to perceived challenges and obstacles, ultimately enhancing the overall user experience. It highlights the significance of well-thought-out game mechanics, which consider the players' ability to adapt and overcome challenges by sharing resources.

One intriguing aspect of our results was the consistency in the users' overall game experiences, which did not vary significantly between the two versions. While it is possible that user biases may have influenced their responses, given their knowledge of the researchers, it is essential to acknowledge this as a potential source of bias in the data. Users may have been inclined to provide more favorable feedback due to their familiarity with the research team. However, this consistency in-game experience may also indicate that other elements not considered in this study, such as visual appeal, for example, remained relatively stable between the versions.

In conclusion, our results highlight the influence of shared items on user satisfaction, the perceived feeling of being "stuck" in the games, and the changes in the feeling of wallet necessity. The findings emphasize the potential benefits of integrating shared items that can be utilized across multiple games, providing users with a more rewarding and engaging experience. While the bias of user familiarity cannot be disregarded, this study forms the basis for future research into the intricate relationship between shared items, user satisfaction, and the overall gaming experience. Game developers should consider these insights when designing and implementing in-game item systems to enhance user enjoyment and engagement.

6   Conclusions and Future Work

In this paper, we presented NFLegacy, an NFT-based game ecosystem to help integrate games with NFT and blockchain capabilities. We further proposed a game mechanic, Shared Progression, that helps players establish a stronger connection with game objects by allowing them to be used across different games. A set of three games were designed as a case study for the application of the Shared progression mechanic. Finally, we conducted a user study, and the preliminary results show an increased enjoyment of unlocking new items when using the Shared Progression mechanic, highlighting the potential of our approach. More interestingly, the different games can belong or not to the same company, if they agree to use the same NFTs. That way, even if a game company leaves the market, the NFTs earned by the players could still be used in all games belonging to the same ecosystem.

By introducing these incentives for using NFTs in gameplay, we aim to bridge the existing gap between holding NFTs solely for speculative purposes and actively integrating them into the gaming experience. We believe that this approach not only transforms NFTs from passive assets into interactive gaming elements but also fosters a shared experience among players, promoting more profound connections and engagement within the NFT community. As the continuation of this work, we plan to continue experimenting and evaluating additional NFT-based game mechanics.

Acknowledgements. This work was supported by national funds through Fundação para a Ciência e a Tecnologia (FCT) with reference UIDB/50021/2020.

References

1. Adams, E., Dormans, J.: Game mechanics: advanced game design. New Riders (2012)
2. Bouzid, A., Narciso, P., Wood, S.: The Gaming Industry and NFTs, pp. 85–102. Apress, Berkeley (2023). https://doi.org/10.1007/978-1-4842-9777-35
3. Dozier, M.E., Ayers, C.R.: Object attachment as we grow older. Curr. Opin. Psychol. **39**, 105–108 (2021). https://doi.org/10.1016/j.copsyc.2020.08.012

4. Mandryk, R.L., Isbister, K., Schaffer, N.: Game Usability: Advice from the Experts for Advancing the Player Experience. Elsevier/Morgan Kaufmann, Amsterdam (2008)
5. Martinho, C., Santos, P., Prada, R.: Design e desenvolvimento de jogos. FCA (2014)
6. Monrat, A.A., Schelén, O., Andersson, K.: A survey of blockchain from the perspectives of applications, challenges, and opportunities. IEEE Access **7**, 117134–117151 (2019)
7. Phan, M., Keebler, J., Chaparro, B.: The development and validation of the game user experience satisfaction scale (guess). Hum. Factors: J. Hum. Factors Ergon. Soc. **58** (2016). 10.1177/0018720816669646
8. Tavares, R., Sousa, J.P., Maganinho, B., Gomes, J.P.: Gamers' reaction to the use of NFT in AAA video games. Procedia Comput. Sci. **219**, 606–613 (2023). https://doi.org/10.1016/j.procs.2023.01.329
9. Various: Interoperability in the metaverse. Technical report, World Economic Forum (2023)
10. Vidal-Tomás, D.: The new crypto niche: NFTs, play-to-earn, and metaverse tokens. Finan. Res. Lett. 102742 (2022)
11. Zheng, X., Mukkamala, R.R., Vatrapu, R., Ordieres-Mere, J.: Blockchain-based personal health data sharing system using cloud storage. In: 2018 IEEE 20th International Conference on e-Health Networking, Applications and Services (Healthcom), pp. 1–6. IEEE (2018)

SafeARUnity: Real-Time Image Processing to Enhance Privacy Protection in LBARGs

Tiago Ribeiro📷, Anabela Marto📷, Alexandrino Gonçalves📷, Leonel Santos📷, Carlos Rabadão📷, and Rogério Luís de C. Costa(✉)📷

CIIC, ESTG, Polytechnic University of Leiria, Leiria, Portugal
{tiago.f.ribeiro,anabela.marto,alex,leonel.santos,carlos.rabadao,
rogerio.l.costa}@ipleiria.pt

Abstract. Augmented Reality applications overlay our physical world with digital components in an interactive 3D space. These applications generally capture information about the physical world around the user through cameras and sensors, which can identify user movements and interactions with objects in the real world. In recent years, Location-Based Augmented Reality Games (LBARGs) have been used in several contexts, such as entertainment, tourism, and education. However, by capturing information about the environment, AR applications can lead to failures in maintaining user and bystander privacy.

This paper addresses the identification and protection of sensitive data in LBARGs. We introduce LootAR, a location-based mobile AR game, and the SafeARUnity library, a real-time image processing middleware that acts as a layer between the AR application and the device's camera, identifying and sanitizing sensitive data prior to rendering. Implementation aspects are discussed, involving Unity Sentis, a toolkit for running machine learning models in Unity, and YOLO, a fast single-stage object detector optimized for real-time applications. We also demonstrate the integration of SafeARUnity in mobile games, using LootAR as a case study.

Keywords: Mobile Games · Augmented Reality · Privacy · Unity Sentis · YOLO

1 Introduction

In recent years, Augmented Reality (AR) systems have been adopted in several areas, such as education [1], medical training [4], and cultural heritage preservation [5,24]. These systems combine real and virtual content, enabling real-time interaction within a three-dimensional environment [3]. AR systems must scan and map the surrounding scenario and identify and track objects and actions to identify users' movements and interactions with the 3D surroundings. When scanning the environment, such applications may capture sensitive information

© The Author(s), under exclusive license to Springer Nature Switzerland AG 2025
A. Marto et al. (Eds.): VJ 2024, CCIS 2324, pp. 186–200, 2025.
https://doi.org/10.1007/978-3-031-81713-7_13

of the user or bystanders [15]. For instance, head-mounted cameras may inadvertently capture sensitive information displayed on screens, while mobile device cameras might record facial images of bystanders without their consent. Such issues are equally relevant in the context of Location-Based Augmented Reality Games (LBARGs).

In LBARGs, the available virtual content depends on the player's geographic location. The game requires information on the user's geographic location besides accessing the device's camera and other sensors, as commonly done by AR applications, thus increasing the risks of (even unintentional) privacy breaches. Indeed, several types of attacks may be issued against AR systems [34], and while some of them would lead to data integrity violations, others would lead to data breaches [27]. On the other hand, if the LBARG only tracks the necessary information to place virtual objects and doesn't identify other objects in the view, providing undetectability and unobservability, it reduces the chances that attackers or adversary may gain access to sensitive information [10].

In this work, we address the challenge of identifying and protecting sensitive data within LBARGs by implementing a specialized processing layer. This layer processes data captured by the device camera, identifies potential sensitive elements, and applies sanitization techniques to obfuscate them before rendering. This functionality is integrated within the SafeARUnity library, which uses the Unity Sentis package [35] for machine learning operations within Unity, and incorporates a YOLO ("You Only Look Once") model [28] specifically adapted for real-time instance segmentation and suited for edge computing applications.

To demonstrate and evaluate our library, we have developed a LBARG called LootAR, which serves as a practical example of SafeARUnity's implementation (Fig. 1).

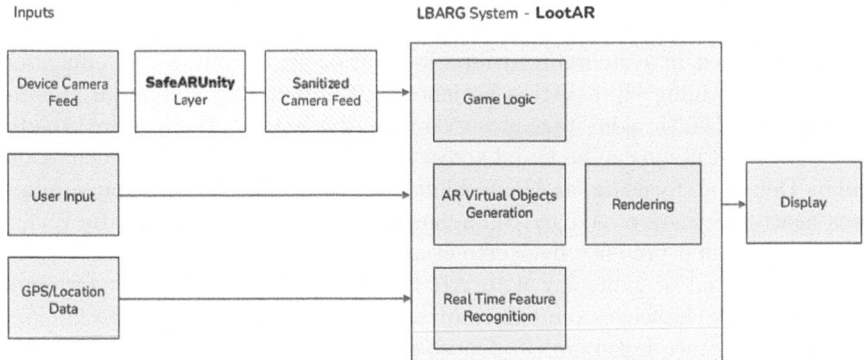

Fig. 1. SafeARUnity library integration within a LBARG. The SafeARUnity layer functions independently of the game's main logic, processing the device's camera input and applying user-defined obfuscation techniques to protect sensitive objects. The system was tested using LootAR, an LBARG developed specifically for this purpose.

The main contributions of this work include (i) the SafeARUnity library and its methods for privacy-protection (i.e., object detection, segmentation, and obfuscation), (ii) an outline of the LootAR LBARG, and (iii) a discussion on how to use the proposed privacy-protection methods in a LBARG.

The remainder of this paper is structured as follows: Sect. 2 provides background information and reviews related work. Section 3 details the SafeARUnity library and its components. Section 4 presents the LootAR and demonstrates the integration of SafeARUnity. Finally, Sect. 5 concludes the paper and outlines future research directions.

2   Background and Related Work

2.1   Location-Based AR Games

First conceived in the 1960s by Ivan Sutherland – when he created the Ultimate Display, envisioning a future computer interface that would blur the line between the digital and physical worlds – AR overlays our physical world with digital components in an interactive 3D space.

Several definitions of AR have emerged, including variations such as Mixed Reality and Extended Reality. However, this study does not seek to explore these definitions or compare their differences. Instead, it adopts Azuma's widely accepted definition, which serves as the foundation for this research. Thus, when AR technology is mentioned, it refers to any system that combines real and virtual content, enabling real-time interaction within a three-dimensional environment [3].

In recent decades AR has been advancing in both technological development and global popularity, becoming a common approach across various fields for diverse purposes, from serious contexts to purely entertainment. Numerous studies aimed at understanding the impact of AR strongly support its benefits, as demonstrated in systematic reviews focused on its advantages in education [1], medical training [4], assisting patients with autism [22], industrial maintenance [26], and cultural heritage preservation [5,24], among others. More broader research on AR usage can be found across several systematic reviews such as the one by Dey et al. focusing on AR usability studies [12], the survey that summarizes nearly 50 years of AR by Billinghurst et al. [7] or the systematic review related to AR and presence by Marto et al. [23].

The potential of gameplay with AR has been a topic of interest for some time, yet many challenges remain unsolved [6]. Most studies on this technology frequently reference Pokémon Go, focusing on various aspects such as stickiness in location-based AR games [19], player attachment [25], flow and satisfaction [19], and reasons for starting, continuing, and quitting location-based games [2], among others.

Pokémon Go has provided an interesting foundation for researchers to delve into location-based games for various purposes of study. These include cultural heritage [14], tourism [18], or education [29].

Regarding safety, the awareness of playing while moving has been a subject of study and concern [13,21,25,31]. Safety is a common theme in these studies, focusing on player safety while moving, health concerns related to extended gameplay, and issues surrounding user location tracking. Issues of privacy in AR do not seem to be a common topic of research, although some studies highlight privacy concerns as potential threats when using this technology. For instance, Chen et al. [9] demonstrated the vulnerability of AR systems by conducting experiments with password detection, showing a concerning success rate of 90% when the distance from the victim is 1.5 m.

The importance of this issue, particularly the lack of attention to privacy concerns related to image sharing through the camera when using AR features, has triggered interest in the current research and motivated the creation of a mobile AR location-based game called LootAR to address these concerns.

2.2   Object Detection and Segmentation

From a cybersecurity perspective, AR applications are particularly susceptible to threats in three main areas: the handling of input data, the management of data access and sharing, and the process of output rendering [11,30]. Indeed, the applications' abilities to capture information about the application user and bystanders lead to serious privacy risks [33], and input sanitization policies may be used as countermeasures to improve privacy protection [16].

Identifying the object to sanitize would be an initial step in video sanitization. This process may rely on, for instance, object detectors and instance segmentation. Indeed, in the last decade, machine-learning-based segmentation has quickly evolved, and the advantages of real-time instance segmentation in LBARGs include privacy-preserving gameplay, enhanced object interaction, and dynamic environment adaptation.

The field of object detection and instance segmentation has undergone a notable transformation, marked by several fundamental innovations. Initially, two-stage detectors like R-CNN and its successors, Fast R-CNN and Faster R-CNN, dominated the field [36]. These models first proposed regions of interest and then classified them, achieving considerable accuracy but at the cost of computational efficiency.

The advent of single-stage detectors, such as YOLO (You Only Look Once) and SSD (Single Shot Detector), introduced an alternative approach to object detection. These models process images in a single forward pass through the neural network, computing bounding boxes and class probabilities directly from input images. This design choice reduced computational overhead compared to two-stage detectors, making these models more suitable for real-time applications while maintaining acceptable accuracy levels [28].

The advent of single-stage detectors, such as YOLO (You Only Look Once) and SSD (Single Shot Detector), marked a paradigm shift towards real-time object detection. These models prioritized speed without significantly compromising accuracy, making them suitable for real-time applications.

Instance segmentation, which combines object detection with pixel-level segmentation, saw a breakthrough with Mask R-CNN. This model extended Faster R-CNN by adding a branch for predicting segmentation masks, effectively bridging the gap between object detection and semantic segmentation [36]. Recent advancements have further improved the speed and accuracy of instance segmentation. YOLACT (You Only Look At CoefficienTs) introduced a real-time instance segmentation approach that decomposes the task into two parallel subtasks: generating a set of prototype masks and predicting per-instance mask coefficients [8]. This innovation significantly reduced computational complexity while maintaining competitive accuracy, and gave rise to a family of YOLO models capable of instance segmentation.

The YOLO family has continued to evolve, with YOLOv8 representing one of the latest advancements [17]. It introduces several key innovations that improve its instance segmentation capabilities. It features anchor-free detection, allowing for flexible handling of object sizes without predefined anchor boxes. The decoupled head approach optimizes classification and regression tasks independently, improving overall efficiency. Its CSPDarknet53 backbone delivers superior accuracy and speed, while PANet enhances feature integration for better segmentation results.

Despite advancements, instance segmentation still faces challenges like balancing latency and accuracy, handling objects at various scales, and managing occlusions. Ongoing research aims to resolve these issues, enhancing real-time detection and segmentation for AR and other domains requiring fast, accurate visual processing [32].

2.3   Unity Sentis

Unity Sentis is a machine learning library designed for real-time applications in Unity. It enables the integration of artificial intelligence and machine learning models into the system, facilitating natural language processing, computer vision, and other functionalities. Unity Sentis can be utilized to implement algorithms that assist in obscuring user data before it is transmitted to the server. Techniques available, such as encryption or data masking, ensure that even if data is intercepted, it remains unusable. This capability is particularly crucial for data obfuscation techniques, especially given the current study's focus on privacy concerns.

Unity Sentis represents a significant advantage in terms of privacy preservation. Its support for privacy-preserving methods allows models to be trained on user data without exposing it, which is essential for maintaining confidentiality in sensitive applications.

Several factors justify the adoption of Sentis over Unity Barracuda. Unity discontinued Barracuda's development in August 2022, halting support for newer neural network models and architectures, and limiting its ability to incorporate deep learning advancements. Barracuda's general-purpose design lacks optimizations for mobile and embedded platforms, particularly in low-level hardware accelerations such as SIMD and GPU enhancements.

In contrast, Sentis introduces substantial performance and optimization improvements for Unity-supported platforms. It supports custom operators, allowing developers to implement and iterate over tailored neural network layers for real-time performance tuning. Additionally, Sentis is actively maintained and designed to accommodate emerging neural network architectures, ensuring forward compatibility. Therefore, Sentis was chosen as the preferred option for this project due to its superior capabilities and ongoing development support.

3   The SafeARUnity Library

The SafeARUnity library is a real-time image processing system designed to enhance privacy protection in AR applications. Functioning as a middleware layer between an AR application and the AR device's camera, it relies upon an object detection and instance segmentation model to identify and obfuscate objects of interest within a scene.

At its core, the library's functionality comprises several interconnected components. The process begins with real-time image acquisition from the device's camera. These captured frames undergo object detection using the YOLOv8 model, followed by instance segmentation to generate pixel-accurate masks for individual objects. Based on these masks, the system applies various obfuscation techniques such as Masking, Pixelation, or Blurring to the regions of interest. Finally, the obfuscated image is integrated back into the AR application (Fig. 2).

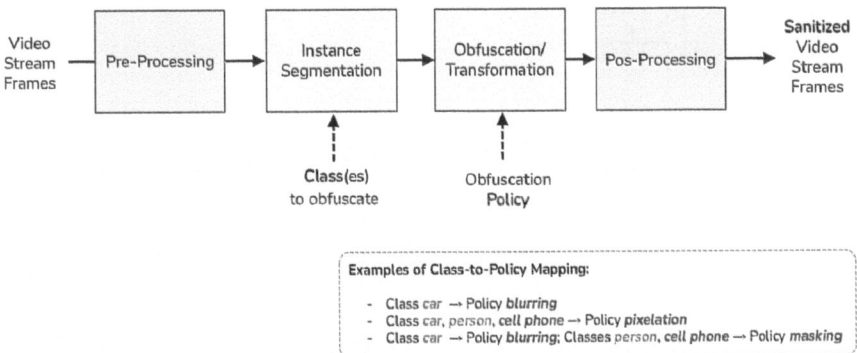

Fig. 2. High-level overview of the SafeARUnity Service Architecture. Video stream frames are processed sequentially according to user-selected object classes for obfuscation and the corresponding obfuscation policy applied to each class.

3.1   Architecture and Operation Overview

The system's architecture implements an end-to-end process for object detection, instance segmentation, and obfuscation, leveraging a YOLOv8 model as its backbone. This process is organized into four primary stages: acquisition and preprocessing, YOLOv8 processing, mask filtering and selection, and obfuscation and rendering, as schematized in Fig. 3.

In the initial stage, images are captured from the AR device's camera and converted into an RGB tensor with dimensions $3{\times}640{\times}640$, optimized for YOLOv8 input requirements. The YOLOv8 processing stage then executes the pre-trained model on this input tensor, generating bounding box coordinates, class IDs, and segmentation masks for detected instances within the frame.

To optimize computational efficiency in real-time scenarios, the third stage incorporates a similarity-based filtering mechanism. This process normalizes bounding box coordinates and dimensions, calculating Euclidean distances between corresponding boxes across consecutive frames. Prior detection results are reused if positional and dimensional changes fall below an empirically determined threshold ($\delta = 0.01$). Subsequently, relevant masks are selected, resized to match input image dimensions, and binarized for efficient application.

The final stage encompasses obfuscation and rendering. Three primary obfuscation techniques are implemented: Blurring, which applies a box blur to masked regions; Pixelation, which averages color values within specified block sizes in masked areas; and Masking, which replaces masked pixels with a predefined color. The process culminates in rendering the processed image, with detected objects effectively obfuscated.

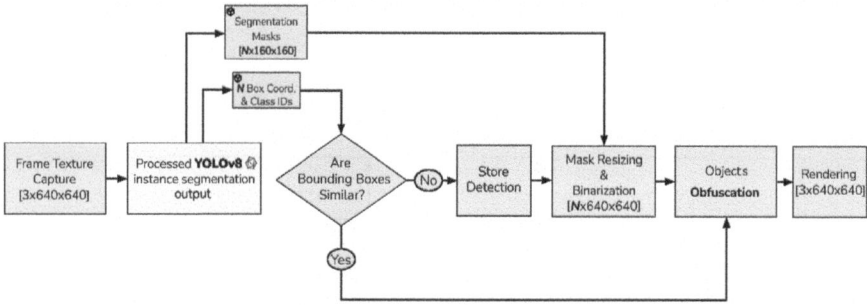

Fig. 3. YOLOv8-Based Instance Segmentation and Obfuscation Pipeline in Unity Sentis The pipeline is composed of video stream frame capture, instance segmentation via a YOLOv8 nano model, and subsequent object obfuscation. Key features include (1) optimization through inter-frame bounding box comparison, (2) mask processing involving resizing and binarization/thresholding, and (3) application of user-defined obfuscation methods (pixelation, blurring, or masking). The pipeline culminates in the rendering of obfuscated images as Unity textures.

The library's functionality can be mathematically formulated as follows: Let I represent the input image, M denote the set of masks corresponding to the objects to be obfuscated, and O represent the obfuscated image. The obfuscation process can then be expressed as:

$$O = f(I, M) \tag{1}$$

where f is the obfuscation function that applies the selected obfuscation techniques to the pixels within the regions defined by the masks.

3.2 Customization and Control

The SafeARUnity library allows user customization to meet diverse privacy protection needs. Users can specify which object classes need to be obfuscated, choosing from the 80 classes available in the COCO dataset [20]. This flexibility allows for tailored privacy solutions across various AR applications.

Furthermore, users can select from multiple obfuscation methods. Each of these methods can be associated with specific object classes, as exemplified in Table 2, providing granular control over the privacy protection strategy. The obfuscation process can be fine-tuned by adjusting parameters such as blur radius or pixelation block size, allowing users to balance visual privacy with application functionality.

It's important to note that the system's ability to detect and obfuscate specific object classes is contingent upon the capabilities of the underlying pre-train YOLOv8 nano model. While this lightweight model offers good performance on edge devices, users should be aware of its limitations in terms of detectable object categories.

3.3 Technical Implementation and Model Details

The SafeARUnity library is built upon a range of frameworks and packages, the most relevant being in Table 1. It utilizes Unity as the primary game development engine, Unity Sentis (version 1.2.0-exp.2 or greater) as the neural inference library, and ONNX as the model format for instance segmentation with YOLOv8.

Table 1. Frameworks and Packages Used in SafeARUnity

Framework/Package	Description
Unity	Game development engine
Unity Sentis	Neural inference library for Unity (v1.2.0-exp.2 or greater)
ONNX	Model format for instance segmentation with YOLOv8

The core of the system's object detection and instance segmentation capabilities lies in its implementation of the YOLOv8-seg nano model. This model, available from Ultralytics, is a lightweight version of YOLOv8 specifically developed for edge devices, making it particularly suitable for AR applications where computational resources may be limited. The nano version maintains a balance between performance and efficiency, allowing for real-time processing on a wide range of devices.

The YOLOv8-seg nano model extends the traditional YOLO architecture to incorporate mask generation capabilities, drawing inspiration from the YOLOAct architecture [8]. This extension introduces mask coefficient learning to YOLO's existing object detection pipeline. This innovation allows the model to generate masks that accurately delineate object boundaries, effectively integrating both object detection and instance segmentation into a streamlined process.

The model's workflow is structured into four primary stages: model initialization, ONNX model outputs, bounding box processing, and mask processing. In the model initialization stage, the ONNX model, pre-trained on the COCO dataset [20], is loaded via the Unity Sentis package. To ensure efficiency in real-time applications, the model's computational tasks are handled by a dedicated worker.

As shown in the diagram of Fig. 4, the ONNX model outputs stage generates raw predictions that serve as the basis for further processing. These outputs include bounding box coordinates $\mathbf{B} \in \mathbb{R}^{1 \times 4 \times 8400}$, class scores $\mathbf{C} \in \mathbb{R}^{1 \times 80 \times 8400}$, mask coefficients $\mathbf{M} \in \mathbb{R}^{1 \times 32 \times 8400}$, and prototype masks $\mathbf{P} \in \mathbb{R}^{32 \times 160 \times 160}$. The 8400 value represents the number of initial object detections, while 32 corresponds to the number of learned prototype masks.

During the bounding box processing stage, the coordinates are refined based on $\mathbf{B}$, and class scores are extracted from $\mathbf{C}$. Non-Maximum Suppression (NMS) is employed to filter out overlapping detections, preserving only the most confident predictions based on the Intersection over Union (IoU) metric.

The mask processing stage involves reshaping mask coefficients to $\mathbf{M}' \in \mathbb{R}^{8400 \times 32}$ and prototype masks to $\mathbf{P}' \in \mathbb{R}^{32 \times 25600}$. These matrices are then combined via matrix multiplication: $\mathbf{F} = \mathbf{M}'\mathbf{P}' \in \mathbb{R}^{8400 \times 25600}$. The result, $\mathbf{F}$, is reshaped to $\mathbf{F}' \in \mathbb{R}^{8400 \times 160 \times 160}$ to form mask predictions. These masks are subsequently cropped and filtered based on the surviving bounding boxes and their corresponding class IDs.

The final output of the model includes refined bounding box coordinates $\mathbf{B}_f \in \mathbb{R}^{N \times 4}$, class IDs $\mathbf{I} \in \mathbb{N}^N$, NMS indices $\mathbf{N} \in \mathbb{N}^N$, and final binary masks $\mathbf{S} \in \{0, 1\}^{N \times 160 \times 160}$ for the detected objects.

4   Privacy Protection in LootAR

In this section, we present the LootAR game and describe the required steps to integrate the SafeAR library into a LBARG.

4.1   LootAR: a Unity-Based LBARG

The LootAR prototype aims to explore the potential of AR in mobile games, providing an innovative and immersive experience for players while moving around a certain area. As a case study, the game was implemented to be used in Campus 2, School of Technology and Management of Polytechnic of Leiria, Portugal. The

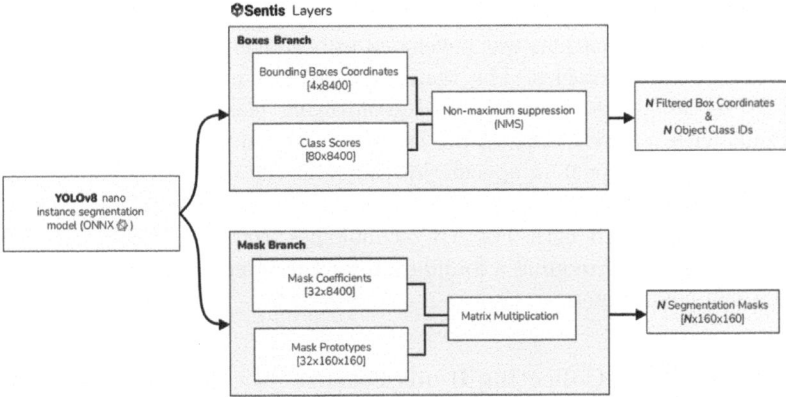

Fig. 4. The graph of the ONNX pre-trained model, loaded via the Unity Sentis package, produces tensors containing the bounding box coordinates for each detection, as well as a probability vector for each class the model was trained on. It also outputs the coefficients for the prototype masks and the prototype masks themselves. This raw output is subsequently processed using layers provided by the Sentis (Unity) package. The most notable of these is the Non-Maximum Suppression (NMS) layer, applied to the bounding boxes, while a matrix multiplication (`matmul`) operation is performed in the mask processing branch. As a result, we obtain N bounding boxes, corresponding class IDs, and N segmentation masks with a resolution of 160×160.

main objective was to develop a mobile AR location-based game where players could collect items scattered across the campus, making use of AR technology to visualize and interact with these items in the real environment. The complete source code for the LootAR game is available at https://github.com/ipleiria-ciic/loot-ar.

The game prototype is divided into three phases. The first phase is item collection, where players use their mobile devices to find and collect virtual items distributed across the campus. The location of the items is determined by GPS, and visualization and interaction are facilitated by AR technology. The user can explore the map without needing to use AR to get a general idea of where the items are located. However, when they want to collect or interact with the items, they need to activate the AR camera to visualize and pick them up. This ensures a seamless transition between navigation and the immersive AR experience for item collection.

In the second phase, players would use the collected items to craft new objects, such as weapons and armor, which would be used in battles against other players.

Finally, in the third phase, players engage in battles using the created objects, competing against each other in a virtual arena.

The current study focused on developing and testing the first phase before moving to the following phases to ensure that privacy is considered when playing LootAR.

Development of AR Using GPS. For the first phase of development, a set of experimental applications was developed to better understand the use of GPS SDKs and AR technologies. The Mapbox SDK was used for the development of GPS functionalities, which allowed geolocating players and associating virtual items with real positions on the campus. This integration was essential for enabling items to appear in specific locations on the campus. The accuracy of GPS in indoor environments and the variability in the location of items presented additional challenges. For AR technologies, the AR Foundation package for Unity was used, providing a foundation for the integration of virtual objects into the real environment.

Implementing the Collecting Items Feature. For the item collection functionality, where players can collect specific items at different locations on the campus, AR Foundation was integrated, allowing users to visualize and interact with digital items as they move around the campus. This phase involved the development of scripts and algorithms to try to correctly position the virtual items in the real environment, considering the player's location and orientation. However, significant difficulties were encountered in maintaining the correct positioning of items in AR, which compromised the user experience when interacting with them. Since the primary focus at this stage was to develop a functional prototype to test the integration with the privacy layer described later in this manuscript, the current solution was considered stable enough to move forward.

4.2   Integrating LootAR and SafeARUnity

The SafeARUnity library is built on top of the Sentis package (version 1.2.0-exp.2 or higher) from Unity and the YOLOv8 instance segmentation model in ONNX format. Implemented in C#, the system comprises several key scripts including `CameraManager.cs`, `ImageObfuscator.cs`, `ImgUtils.cs`, and `ImageWriter.cs`. These scripts work in tandem to manage the camera feed, apply obfuscation techniques, handle image processing utilities, and manage image output respectively. The complete source code for the SafeARUnity library is available at https://github.com/ipleiria-ciic/safe-ar-unity.

The integration of SafeARUnity into a LBARG involves several key steps. First, a new Unity project must be created, and the necessary SafeARUnity scripts should be imported into the project's Assets folder. Following this, the camera and user interface are configured by attaching a `RawImage` component to a camera `GameObject`, which will be responsible for displaying the video stream.

Subsequently, obfuscation is configured by adding the obfuscation script to the camera `GameObject` and importing a YOLOv8 model in ONNX format. It is crucial to ensure that the Unity Sentis package (version 1.2.0-exp.2 or higher) is installed, as it enables runtime customization through the obfuscation script.

Table 2. Example of object class to obfuscation type mapping.

Object Class ID	Obfuscation Type
0 (person)	Masking
2 (car)	Blurring
53 (pizza)	Pixelation

For obfuscation settings, a dictionary is created in the `CameraManager.cs` script that maps object class IDs to different obfuscation types (such as Masking, Blurring, and Pixelation). These settings can be tailored to specific use cases by adding or removing entries based on the object classes present in the detection model.

The corresponding C # code for the mapping in Table 2 would be:

```
obfuscationTypes = new Dictionary<int, Obfuscation.Type>
{
    { 0, Obfuscation.Type.Masking },
    { 2, Obfuscation.Type.Blurring },
    { 53, Obfuscation.Type.Pixelation }
};
```

Listing 1.1. Example obfuscation settings

Then, the system can be executed just by running the Unity project, allowing for real-time obfuscation of detected objects. The performance of the obfuscation process can be optimized by adjusting parameters and refining the model as needed. Figure 5 showcases the obfuscation results on a motorbike while using the LootAR game.

A minimal working example that demonstrates the integration of SafeARUnity into Unity and offers a starting point for implementing AR privacy protection is available at https://github.com/ipleiria-ciic/safe-ar-unity.

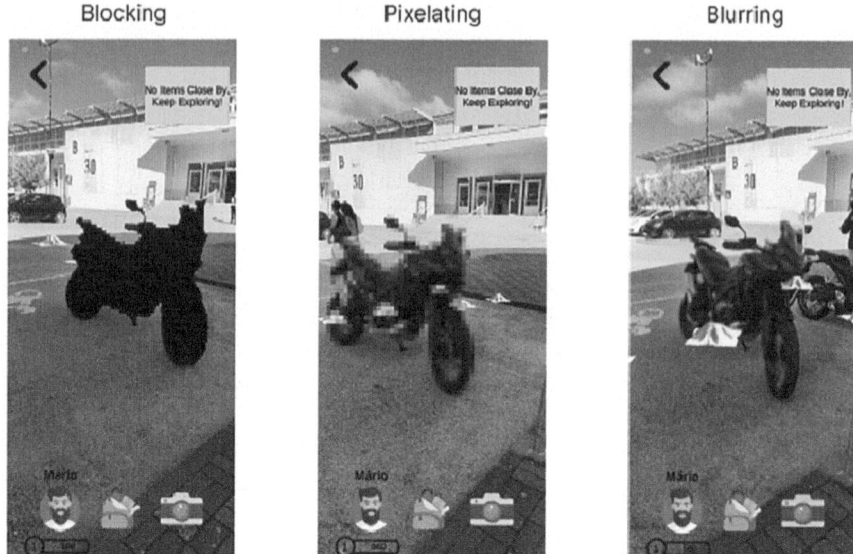

Fig. 5. Example of using different obfuscation methods (i.e., masking, pixelating, and blurring) in the LootAR game - game snapshot comprising virtual and real elements

5   Conclusions and Future Work

LBARGs have gained prominence in recent years. These games captured information from the environment to create an interaction between real and virtual. Often, sensitive data are captured as part of such process, and the processing, storage and sharing of this information can lead to privacy risks to the user and bystanders. Therefore, it is necessary to employ measures to improve privacy protection.

In this work, we presented the SafeARUnity library, which processes the data captured by the devices' cameras before it is rendered, thus providing sanitized images that may be stored or shared. The library uses Unity Sentis and relies on YOLOv8 for object detection and segmentation. We also presented LootAR, an LBARG, and described how to use SafeARUnity in an LBARG.

In future work, we intend to carry out a detailed analysis of the accuracy achieved by the segmentation model when detecting objects in usual contexts of LBARGs, a comparative analysis of the performance obtained using the library and a distributed service, and a study on obfuscation techniques more suitable for use in LBARGs.

Acknowledgements. This work is funded by national funds through FCT - Fundação para a Ciência e a Tecnologia, I.P., in the context of the projects *SafeAR - Supporting privacy and confidentiality in augmented reality contexts* - 2022.09235.PTDC and UIDB/04524/2020 and under the Scientific Employment Stimulus - Institutional Call - CEECINS/00051/2018.

References

1. Akçayır, M., Akçayır, G.: Advantages and challenges associated with augmented reality for education: a systematic review of the literature. Educ. Res. Rev. **20**, 1–11 (2017)
2. Alha, K., Koskinen, E., Paavilainen, J., Hamari, J.: Why do people play location-based augmented reality games: a study on pokémon go. Comput. Hum. Behav. **93**, 114–122 (2019)
3. Azuma, R.: A survey of augmented reality presence: teleoperators and virtual environments **6**, 355–385 (1997)
4. Barsom, E.Z., Graafland, M., Schijven, M.P.: Systematic review on the effectiveness of augmented reality applications in medical training. Surg. Endosc. **30**(10), 4174–4183 (2016)
5. Bekele, M.K., Pierdicca, R., Frontoni, E., Malinverni, E.S., Gain, J.: A survey of augmented, virtual, and mixed reality for cultural heritage. ACM J. Comput. Cultural Heritage **11**, 7 (1–36) (2018)
6. Billinghurst, M.: Grand challenges for augmented reality. Front. Virtual Reality **2**, 578080 (2021)
7. Billinghurst, M., et al.: A survey of augmented reality. Found. Trends® Hum.–Comput. Interact. **8**(2-3), 73–272 (2015)
8. Bolya, D., Zhou, C., Xiao, F., Lee, Y.J.: YOLACT++ better real-time instance segmentation. IEEE Trans. Pattern Analy. Mach. Intell. **44**(2), 1108–1121 (2022). https://doi.org/10.1109/tpami.2020.3014297
9. Chen, S., Li, Z., Dangelo, F., Gao, C., Fu, X.: A case study of security and privacy threats from augmented reality (AR). In: 2018 International Conference on Computing, Networking and Communications (ICNC), pp. 442–446. IEEE (2018)
10. De Guzman, J.A., Thilakarathna, K., Seneviratne, A.: Security and privacy approaches in mixed reality: a literature survey. ACM Comput. Surv. **52**(6) (2019). https://doi.org/10.1145/3359626
11. De Guzman, J.A., Thilakarathna, K., Seneviratne, A.: Security and privacy approaches in mixed reality: a literature survey. ACM Comput. Surv. (CSUR) **52**(6), 1–37 (2019)
12. Dey, A., Billinghurst, M., Lindeman, R.W., Swan, J.: A systematic review of 10 years of augmented reality usability studies: 2005 to 2014. Front. Robot. AI **5**, 37 (2018)
13. Guo, Y., Agrawal, S., Peeta, S., Benedyk, I.: Safety and health perceptions of location-based augmented reality gaming app and their implications. Accid. Anal. Prev. **161**, 106354 (2021)
14. Haahr, M.: Creating location-based augmented-reality games for cultural heritage. In: Serious Games: Third Joint International Conference, JCSG 2017, Valencia, Spain, 23–24 November 2017, Proceedings 3, pp. 313–318. Springer (2017)
15. Harborth, D., Pape, S.: Investigating privacy concerns related to mobile augmented reality apps - a vignette based online experiment. Comput. Hum. Behav. **122**, 106833 (2021)
16. Jana, S., Narayanan, A., Shmatikov, V.: A scanner darkly: protecting user privacy from perceptual applications. In: 2013 IEEE Symposium on Security and Privacy, pp. 349–363. IEEE (2013)
17. Jocher, G., Chaurasia, A., Qiu, J.: Ultralytics YOLOv8 (2023). https://github.com/ultralytics/ultralytics

18. Lacka, E.: Assessing the impact of full-fledged location-based augmented reality games on tourism destination visits. Curr. Issue Tour. **23**(3), 345–357 (2020)
19. Lee, C.H., Chiang, H.S., Hsiao, K.L.: What drives stickiness in location-based AR games? An examination of flow and satisfaction. Telematics Inform. **35**(7), 1958–1970 (2018)
20. Lin, T.Y., et al.: Microsoft COCO: common objects in context (2015)
21. Makhmutov, M., Asapov, T., Brown, J.A.: Safety risks in location-based augmented reality games. In: Entertainment Computing–ICEC 2021: 20th IFIP TC 14 International Conference, ICEC 2021, Coimbra, Portugal, 2–5 November 2021, Proceedings 20, pp. 457–464. Springer (2021)
22. Marto, A., Almeida, H.A., Gonçalves, A.: Using augmented reality in patients with autism: a systematic review. In: VipIMAGE 2019: Proceedings of the VII ECCOMAS Thematic Conference on Computational Vision and Medical Image Processing, 16–18 October 2019, Porto, Portugal, pp. 454–463. Springer (2019)
23. Marto, A., Gonçalves, A.: Augmented reality games and presence: a systematic review. J. Imaging **8**(4), 91 (2022)
24. Marto, A., Melo, M., Goncalves, A., Bessa, M.: Multisensory augmented reality in cultural heritage: impact of different stimuli on presence, enjoyment, knowledge and value of the experience. IEEE Access **8**, 193744–193756 (2020)
25. Oleksy, T., Wnuk, A.: Catch them all and increase your place attachment! the role of location-based augmented reality games in changing people-place relations. Comput. Hum. Behav. **76**, 3–8 (2017)
26. Palmarini, R., Erkoyuncu, J.A., Roy, R., Torabmostaedi, H.: A systematic review of augmented reality applications in maintenance. Robot. Comput.-Integr. Manuf. **49**, 215–228 (2018)
27. Qamar, S., Anwar, Z., Afzal, M.: A systematic threat analysis and defense strategies for the metaverse and extended reality systems. Comput. Secur. **128**, 103127 (2023)
28. Redmon, J., Divvala, S., Girshick, R., Farhadi, A.: You only look once: unified, real-time object detection. In: Proceedings of the IEEE Conference on Computer Vision and Pattern Recognition, pp. 779–788 (2016)
29. Ribeiro, F.R., Silva, A., Silva, A.P., Metrôlho, J.: Literature review of location-based mobile games in education: challenges, impacts and opportunities. Inform. **8**(3) (2021).https://doi.org/10.3390/informatics8030043
30. Roesner, F., Kohno, T., Molnar, D.: Security and privacy for augmented reality systems. Commun. ACM **57**(4), 88–96 (2014)
31. Shang, J., Chen, S., Wu, J., Yin, S.: ARSpy: breaking location-based multi-player augmented reality application for user location tracking. IEEE Trans. Mob. Comput. **21**(2), 433–447 (2020)
32. Sharma, R., Saqib, M., Lin, C.T., Blumenstein, M.: A survey on object instance segmentation. SN Comput. Sci. **3**(6), 499 (2022)
33. Siriwardhana, Y., Porambage, P., Liyanage, M., Ylianttila, M.: A survey on mobile augmented reality with 5G mobile edge computing: architectures, applications, and technical aspects. IEEE Commun. Surv. Tutorials **23**(2), 1160–1192 (2021)
34. Syed, T.A., et al.: In-depth review of augmented reality: tracking technologies, development tools, AR displays, collaborative AR, and security concerns. Sensors **23**(1), 146 (2022)
35. Unity Technologies: Unity Sentis: Bringing AI to Unity. https://unity.com/products/sentis (2023). Accessed 01 Oct 2024
36. Wang, Y., Ahsan, U., Li, H., Hagen, M.: A comprehensive review of modern object segmentation approaches. Found. Trends® Comput. Graph. Vis. **13**(2-3), 111–283 (2022).https://doi.org/10.1561/0600000097

A Gomoku Game-Testbed for Monte-Carlo Tree Search Algorithms

Lisa Liu[1]([✉])[ID] and Kelvin Yu[2]

[1] University of California, San Diego, La Jolla, CA 92093, USA
lil043@ucsd.edu
[2] University of Oxford, Oxfordshire, England

Abstract. Gomoku is a strategy game played on the Go board. Two players, black and white, alternate placing pieces on the intersections of the grid to form five-in-a-row. It has often been used to test tree search algorithms, including Monte-Carlo Tree Search (MCTS). Due to its simplicity, it is possible to introduce variations to the rules slightly without affecting the core mechanics and strategies for the game. In this paper, we focus on the board size (traditionally 15×15) and introducing a dynamic boundary for moves (traditionally unbounded). Both of these qualities greatly change the size of each move's action space. We contribute 9 variants of Gomoku with different combinations of board sizes and dynamic boundaries, in order to provide a set of settings ranging across action space sizes. We calculated the action space sizes per moves for each board, and implemented a new variant of MCTS that uses ancestor-based alpha-beta bounds in the selection phase. We ran this against the classical MCTS in order to demonstrate the effects of the action space changes.

Keywords: Gomoku · Monte Carlo Tree Search · Action Space

1 Introduction

Gomoku, also called "Five in a Row," is a two-player strategy board game. It is zero-sum and is a game with complete information. It's traditionally played on a Go board, and its objective is straightforward; form a straight line of five pieces in a row. There are many variations on its rules, such as not allowing six pieces in a row, or an "overline," to count as a win, or implementing various swapping rules in order to mitigate the first player advantage. Due to its simplicity and adaptability, Gomoku has often been used as a subject of research in the past, dating back to the 1993, with a paper on threat space search, a type of tree search [1]. Subsequent research on tree searches were also tested on Gomoku [2], and it is still a subject of analysis over 20 years later [8].

In this paper, we will seek to design different variations of this game in order to test out the effects of different board types on tree search algorithms. In particular, we want to focus on Monte-Carlo Tree Search (MCTS), and will demonstrate the usage of our testbed game with a new MCTS variant from 2024, which is explained in the next section.

© The Author(s), under exclusive license to Springer Nature Switzerland AG 2025
A. Marto et al. (Eds.): VJ 2024, CCIS 2324, pp. 201–211, 2025.
https://doi.org/10.1007/978-3-031-81713-7_14

2   Ancestor-Based Alpha-Beta Bounds for Monte Carlo Tree Search

2.1   Alpha-Beta Pruning

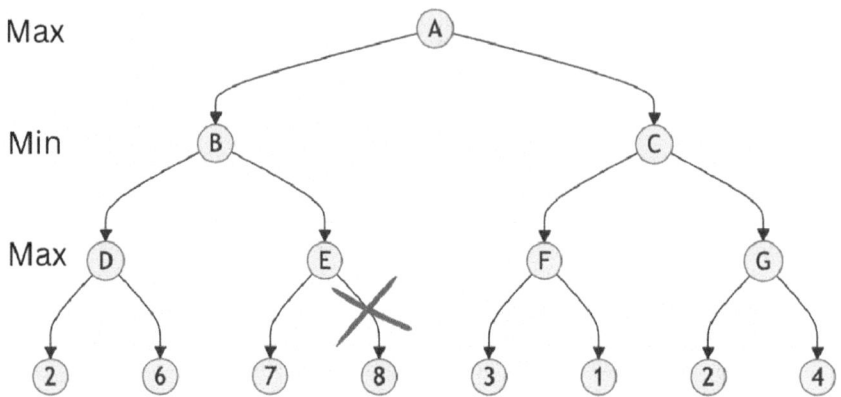

Fig. 1. Example to illustrate alpha-beta pruning on a minimax trees.

MCTS [5,9] is a tree search technique that uses random sampling and statistical evaluation to explore only the most promising areas of the search space, rather than exhaustively trying every possibility. The algorithm has four phases: 1. selection, 2. expansion, 3. simulation, and 4. back-propagation. Every node contains a current state; for Gomoku, it is the board, the placed pieces, and which player's turn it is. Its children nodes are states that are reachable in one action. The selection phase chooses the child node to explore next, either picking an untried action, or the child that returns the greatest value using the upper confidence bound (UCB) formula. In the expansion phase, all the possible actions and board states from the selected child node are added to the tree. In the simulation phase, moves are placed until the result, win or loss, is determined (a "rollout"). In back-propagation, this result is recorded up the tree back to the root node, so the root node is aware of whether this child node seems more or less promising to explore in the future. In this paper, we are focusing on the selection phase, and we use random rollouts (random moves until the end).

Alpha-beta pruning is an optimisation technique used in primarily for minimax algorithm [4]. Figure 1 shows the core concept. We start at node A and look at its first child, node B. We then look at node D. Here, the maximising player will choose the value 6 because it is the greatest, and this gets returned to B. To the minimising player here, choosing node D would result in a value of 6 so they are already guaranteed a value of 6 or less. Now we look at node E. Its first child has a value of 7, so we know the max player is guaranteed a value of 7 or more if the min player chooses node E. Because of this, the min player would never even consider node E because they can get a better value of 6 at node D. This means we did not have to look at the remaining child of node E because it would not affect the decision of the min player. In this example, only one node was pruned, but in larger trees, entire branches can be ignored which saves a lot of computational time.

When implemented, a variable "alpha" keeps track of the minimum value that the max player is guaranteed, and likewise "beta" represents the maximum value that the min player is guaranteed.

2.2   Ancestor-Based Bounds

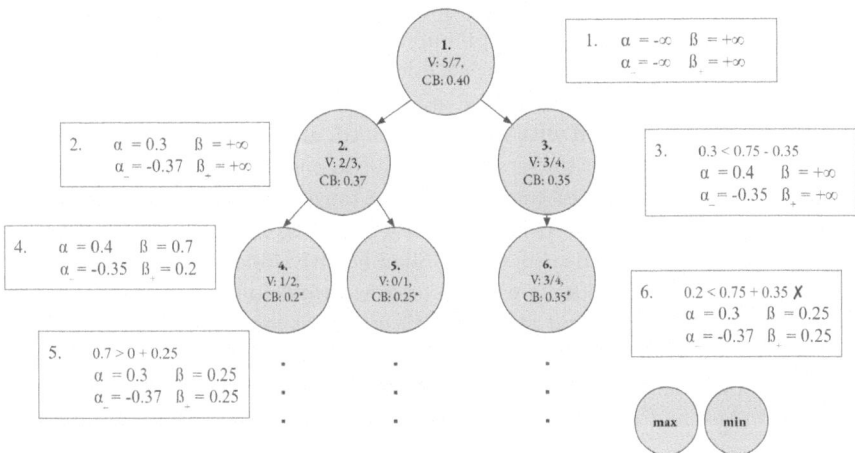

Fig. 2. Example to illustrate alpha-beta pruning on a MCTS tree. Value (V) is the number of wins/number of visits per node, while confidence bound (CB) is the c * sqrt(ln(parent visits)/node visits). We used c = 0.5. The ... represents the unexplored future moves.

In Pepels and Winands' 2024 paper [7], they describe a way to use alpha-beta pruning with MCTS. Alpha-beta pruning is used for minimax trees because minimax is a deterministic algorithm and each state will always have the same value, while MCTS relies on random rollouts so depending on the amount of rollouts so far, a state's value might change [3]. *Alpha bounds* refer to the lower bound of the values that the maximising player is guaranteed, regardless of what the minimising player does. *Beta bounds* are the opposite, being the upper bound of the value that the min player is guaranteed. Because of MCTS's random rollouts, the true alpha or beta bound might not be found. However, the ancestor node in the tree, given enough rollouts, may hope to keep a running global alpha and beta bound that can still be used to eliminate choices, even if it is not the true alpha and beta value.

Figure 2 shows the effect of alpha-beta pruning on MCTS. Each node has a value

$$V = \frac{\text{number of wins}}{\text{number of visits}}$$

and a confidence bound

$$CB = C \times \sqrt{\frac{\ln\left(\text{number of visits to parent nodes}\right)}{\text{number of node visits}}}$$

$V + CB = UCB$ value of the node. The global alpha-beta bounds, alpha-minus and beta-plus, are initialized to negative infinity and positive infinity respectively, and each time, depending on whether the node is minimizing or maximizing, either the alpha-minus or beta-plus will be updated if the previous alpha or beta value is less than or greater than the lower or upper bound of the value, respectively. This way, an alpha and beta bound can be maintained even without trying all the possibilities.

In their paper [7], the Gomoku set-up was a 15×15 board with no dynamic boundary box. It started off with an empty board, and implemented the *pie rule*: the first player makes their move, and then the second player is free to swap with the first player's position, or make their own move. The results of the paper's Gomoku experiment indicated that ancestor-based alpha-beta bounds alone do not impact the performance of MCTS significantly, and it was hypothesized that this was due to Gomoku's wide-to-narrow branching factor. Amazons, the other game in the paper that showed little improvement, also shares the same branching factor scaling traits. It was noted that with a higher iteration budget, the improvement is more pronounced, potentially indicating that with a smaller iteration budget and a larger initial action space, the ancestor-based alpha-beta may not have been activated enough to have an impact [7].

To test out this hypothesis, we decided to test out ancestor-based alpha-beta bounds MCTS on various board set-ups. Changing the initial board size affects the initial action space size. If the theory about the alpha-beta bounds not activating is true, then with the same iteration budget, the bot should do better on smaller board sizes and worse on larger board sizes. Introducing a dynamic boundary box to the game flips the branch factor scaling, from *wide-to-narrow* to *narrow-to-wide*. The boundary box increases in size as time goes on, and although it is possible to make moves without increasing the boundary box size, each expansion increases the action space size significantly, until there is no more space on the board to increase the boundary box. If the wide-to-narrow branching factor caused the ineffectiveness in the original experiment, then we should be able to see more effective results with bounding boxes.

We implemented a bot using this method, which will hereafter be referred to as alpha-beta MCTS.

3 Gomoku Settings

3.1 Board Set-Ups

For board size, we tested the following sizes: 11×11, 15×15, 19×19. 19×19 is the traditional board size, shared with the game of Go. 15×15 is the most commonly used board size today, as well as the size used in the ancestor-based alpha-beta bounds paper [7]. 11×11 is a board size we chose in order to provide an even smaller board for comparison.

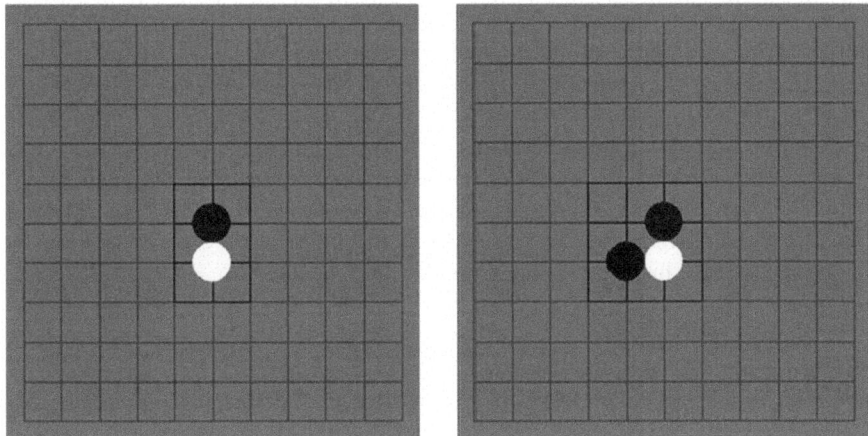

Fig. 3. Dynamic boundary expansion, 11×11 with radius 1.

We test dynamic boundary boxes with radius 1, 2, and infinity. Radius 1 means that the boundary is always 1 grid away from the all the current pieces on the board, unless the boundary is already on the edge of the board. Figure 3 illustrates how this boundary expands dynamically with more pieces placed. Radius 2 means that the boundary is always 2 grids away from the existing pieces on the board. Infinity means that there is no boundary box; this is the usual Gomoku version, where pieces can be placed anywhere on the board from the start.

In total, we have 9 board set-ups: every combination of the 3 board size settings (11×11, 15×15, 19×19) and 3 dynamic boundary settings (radius 1, 2, infinity).

Because we wanted to design the boards to be as simple as possible in order to focus on the results of changing the branching factor and the size of the action space as the game progresses, many of these boards would necessitate a change in the usual strategies used to play. For example, many of the board configurations that are displayed in a 1993 paper analyzing human experts and threat space search require pieces to be played in the corners and edges early on [1], which is not possible with dynamic bounding boxes of a small radius. A smaller 11×11

board may also make it harder to execute some of the strategies by reducing the amount of space that can be inserted between pieces. More research would have to be done on seeing how human players have to adapt to the changes in these rules.

3.2 Action Space Size Calculations

In order to simplify the game, we removed the pie rule and instead ensured fairness when testing both the alpha-beta MCTS and regular MCTS against each other by running an equal number of simulations where each bot goes first as black. For simplicity and to ensure greater variability of games with a limited training budget, we also initialized the board with black having a piece in the center and white having a piece placed randomly within the boundary of the black piece. The first piece in the center was to ensure consistency for the dynamic boundary boxes. Starting with a piece in the corner, for example, would greatly change the size of the bounding box, as two of the box's sides are frozen from the start and unable to expand. By ensuring the boxes are centered at the beginning, we set the game up to allow the boundary boxes to grow as evenly as possible throughout the game.

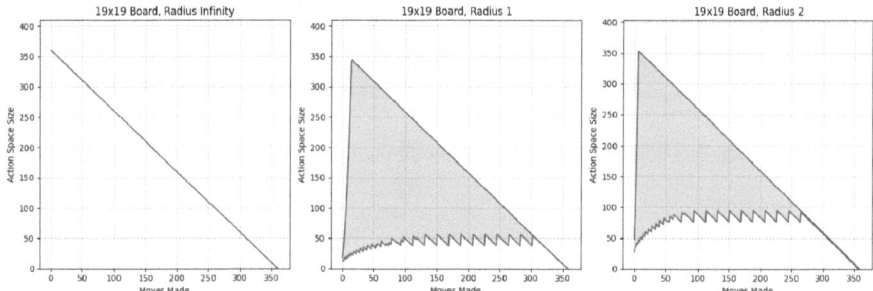

Fig. 4. Action space progression of all 3 boundary radii, for a 19×19 board. Green is the upper bound. Blue is the lower bound. The shaded area covers all the possible action spaces at that move.

The leftmost graph in Fig. 4 shows the action space sizes changing across moves for a 19×19 board with a boundary of radius infinity, or no boundary. In the 11×11 board for example, 2 of the 121 grid intersections are filled, so at move 1 (black), 119 actions are available. By move 2 (white), only 118 actions are available, and so on. The graph displays the upper bound for the action space sizes, not accounting for wins/losses before the whole board is filled. For the 15×15 and 19×19 boards, the only difference is that the starting number of actions is 223 and 359, respectively.

The center graph in Fig. 4 shows the action space sizes changing across moves for a 19×19 board with a boundary of radius 1. The upper bound was calculated through adding a piece in the corner of the box, ensuring maximum expansion per move, until the boundary covers the entire board, after which it shrinks along with the infinity radius graph. The lower bound was calculated taking the minimum at each step of multiple combinations of picking where to expand (side vs corner) and then always filling in the box completely before expanding again. All the possible action space size progressions are contained between these two bounds. The upper bound represents a narrow-to-wide branching factor before becoming wide-to-narrow like the unbounded Gomoku, while the lower bound represents an overall consistent branching factor, albeit with an action space size much smaller than the unbounded Gomoku. However, most games (games within the bounds) generally follow a narrow-to-wide branching factor until the board is fully expanded.

The rightmost graph in Fig. 4 shows the action space sizes for a 19×19 board with a boundary of radius 2. It is very similar to the graph for radius 1, but the entire board is covered in fewer moves, so it hits the radius infinity line sooner. The games of this type can also have a narrow-to-wide branching factor, although the action space sizes will likely be larger than that of radius 1 at most steps.

These graphs were generated for a 19×19 board, but the 11×11 and 15×15 boards follow nearly identical trends, only with the maximum spaces on the board being 121 and 225 respectively, rather than 361.

4   Results

This section presents the results of 9 experiments on game boards of three different sizes.

4.1   Selection Phase

In the selection phase of MCTS, a node is picked for rollout. This node must be a terminal node (leaf node). Starting from the root node, the next node to traverse to is selected either for expansion (expanding the tree with all child nodes for the selected nodes, since it is a new node with no prior visits), with the regular UCT formula for selecting the best child (upper confidence bound applied to trees [6]), or the modified UCT formula that uses alpha-beta bounds to select the best child. We used "AB" to represent the total number of times the alpha-beta UCT was used, with "Final AB" being the subset where the final leaf node that got a rollout was chosen with alpha-beta UCT. "Reg" represented the times regular UCT was used, with "Final Reg" being the subset where the final child was chosen with regular UCT. "Expansions" represents the number of times the node chosen had no prior visits (neither UCT formula could be used).

In every selection iteration, both bots got a rollout budget of 5000. This means that for each move, expansions + final AB + final reg = 5000.

Graphs generally follow the same trend for their radius type. Figure 5 shows that on an 11×11 board with radius 1, alpha-beta UCT does activate a noticeable amount, although only a very small, but non-zero, number of final selected

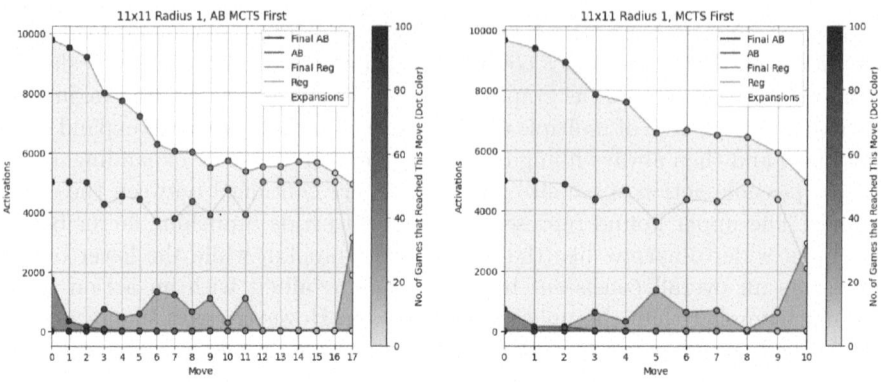

Fig. 5. Activations (separated by type) per move, 11×11 with radius 1.

nodes are chosen with alpha-beta UCT. However, as time passes, a lot of nodes are finally selected by regular UCT, rather than expansions, meaning that enough explorations have been done for the best child formulas to start activating. When MCTS goes first, a lot fewer moves are made by the AB MCTS, indicating it is either losing or winning faster. 15×15 and 19×19 radius 1's graphs also follow a similar trend. Radius 1 follows a narrow-to-wide branching factor that is low initially compared to the unbounded Gomoku games, and these results show that these configurations do allow for the initialization and usage of the alpha-beta bounds, even though it is not used a lot. Figure 6 shows the trend for dynamic boundaries of radius 2. The trend is similar to that of radius 1, but there are visibly less UCT activations of any kind throughout the selection phase as opposed to nodes selected for expansions. Unlike in Fig. 5, where the number of final nodes selected for expansions dropped significantly throughout the game, with radius 2 there was a constant need to pick nodes for expansion

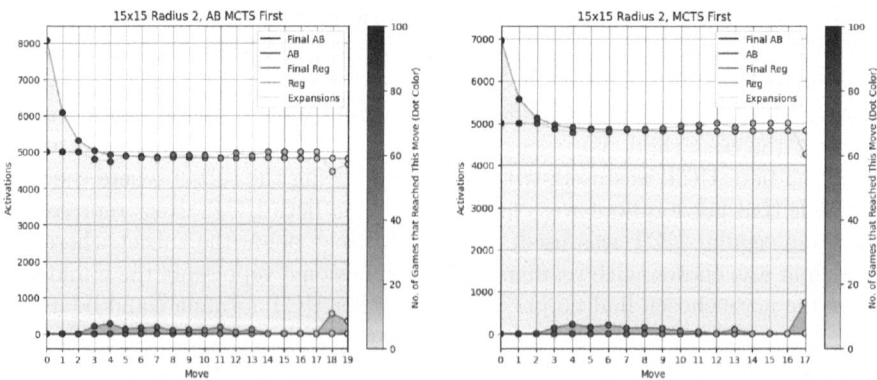

Fig. 6. Activations (separated by type) per move, 15×15 board radius 2.

throughout. This is likely due to the action space becoming too large after a few moves, creating a large influx of unexplored nodes that need to be expanded.

The upper bound of the 15 × 15 radius 2 action space size progression, assuming first two pieces are placed, goes as follows:

- (move 0 black, 23 spaces)
- (move 1 white, 46 spaces)
- (move 2 black, 77 spaces)
- (move 3 white, 116 spaces)
- (move 4 black, 163 spaces)
- (move 5 white, 218 spaces)
- (move 6 black, 217 spaces)
- ...

The sudden jump from 163 spaces to 218 spaces on the 3rd black move coincides with the sudden drop of regular UCT activation around moves 2–3 for AB MCTS, indicating that UCT activation in the selection process does correlate with the action space size. However, as soon as the action space stabilized, a small portion of final selections started to be made with regular UCT, as opposed to being entirely expansions. The ending spike in final regular UCT selections could be an outlier, since only 1–2 games out of 100 reached those last two games. Those games could be due to having a particularly favorable branching factor.

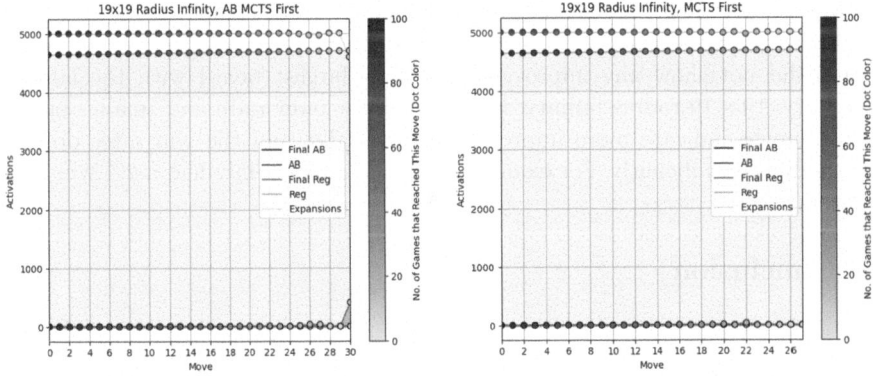

Fig. 7. Activations (separated by type) per move, 19 × 19 radius infinity.

In Fig. 7, the general trend for radius infinity is shown. Although both the radius 2 graphs and radius infinity graphs select expansions for almost all of their rollouts, regular UCT was used a lot less in the first few moves for radius infinity; in fact, regular UCT selection was consistent throughout the game. This is likely because the action space was at its largest throughout, unlike in radius 2, where the first few moves could work with a limited action space.

4.2   Win Rates

Table 1 shows the win rates from running alpha-beta MCTS and MCTS against each other 200 times per board, with 100 games where alpha-beta MCTS goes first and 100 games where regular MCTS goes first.

Table 1. Comparison of alpha-beta MCTS and MCTS win rates as black across all board configurations. 100 games were run per board for each bot. Bold numbers are at least 20% higher.

Configuration	alpha-beta MCTS First	MCTS First
11×11 Radius 1	**74%**	53%
11×11 Radius 2	70%	56%
11×11 Radius ∞	67%	60%
15×15 Radius 1	57%	55%
15×15 Radius 2	68%	68%
15×15 Radius ∞	52%	44%
19×19 Radius 1	66%	48%
19×19 Radius 2	**80%**	60%
19×19 Radius ∞	50%	50%

Alpha-beta MCTS showed a significant improvement compared to MCTS on the board with the smallest board with the tightest boundary, 11×11 radius 1, and did not show any improvement on the largest board with the largest boundary, 19×19 radius infinity. However, while lower radii and smaller board sizes seem to indicate more improvement generally, and the win rates do not always differ significantly. For example, 15×15 radius 1 only had a 2% win rate difference.

5   Conclusion

The alpha-beta MCTS algorithm we tested in this paper did show different results from the paper where the method was first described, which only used the 15×15 radius infinity board. 11×11 radius 1 in particular showed a significant win rate different (21%) with our rollout budget of 5000, showing that the branching factor of action space sizes can impact the results greatly even with a limited budget. Our experiments also give a concrete example of how the action space sizes can directly correlate with different features of MCTS, like the types of activations in the selection phase for alpha-beta MCTS.

The results of this paper help quantify how branching factors in games may affect the efficiency of different types of MCTS algorithms. While this is a useful tool to quantify the improvements made to MCTS, it can also be used help quantify the difficulty of the games in general, therefore helping game designers adjust the game playing experiences for all game players, AI or human.

Acknowledgments. We want to thank Tom Pepels, the author of "Ancestor-Based $\alpha - \beta$ Bounds for Monte-Carlo Tree Search," [7] who provided a lot of feedback on our implementation of his algorithm and insights on how to measure different qualities on its usage. We would also like to thank Dr. Sicun Gao, the professor of CSE 150B (Introduction to Artificial Intelligence: Search and Reasoning) at UC San Diego, for providing the original idea of a bounding box in Gomoku to limit the action space.

Disclosure of Interests. The authors have no competing interests to declare that are relevant to the content of this article.

References

1. Allis, L., Herik, H., Huntjens, M.: Go-Moku and threat-space search. Comput. Intell. **12** (1994)
2. Alus, L., van den Herik, 1, H., Huntjens, M.P.: Go-Moku solved by new search techniques. Comput. Intell. **12**(1), 7–23 (1996)
3. Browne, C.B., et al.: A survey of monte Carlo tree search methods. IEEE Trans. Comput. Intell. AI Games **4**(1), 1–43 (2012)
4. Campbell, M.S., Marsland, T.A.: A comparison of minimax tree search algorithms. Artif. Intell. **20**(4), 347–367 (1983)
5. Coulom, R.: Efficient selectivity and backup operators in Monte-Carlo tree search. In: International Conference on Computers and Games, pp. 72–83. Springer (2006)
6. Kocsis, L., Szepesvári, C.: Bandit based Monte-Carlo planning. In: European Conference on Machine Learning, pp. 282–293. Springer (2006)
7. Pepels, T., Winands, M.H.: Ancestor-based α-β bounds for Monte-Carlo tree search. In: 2024 IEEE Conference on Games (CoG), pp. 1–4. IEEE (2024)
8. Piazzo, L., Scarpiniti, M., Baccarelli, E.: Gomoku: analysis of the game and of the player wine. arXiv preprint: arXiv:2111.01016 (2021)
9. Rimmel, A., Teytaud, O., Lee, C.S., Yen, S.J., Wang, M.H., Tsai, S.R.: Current frontiers in computer go. IEEE Trans. Comput. Intell. AI Games **2**(4), 229–238 (2010)

Games and Artificial Intelligence

Art and Animation: Procedural Content Generation for Sprite Sheet Creation

Pedro M. Fernandes⑩, Carlos Martinho⑩, and Rui Prada(✉)⑩

INESC-ID and Instituto Superior Técnico, Lisbon, Portugal
{pedro.miguel.rocha.fernandes,carlos.martinho,
rui.prada}@tecnico.ulisboa.pt

Abstract. This paper explores the application of Procedural Content Generation (PCG) in the creation of sprite sheets for humanoid character art and animation. PCG offers the possibility of automating the generation of diverse characters and dynamic animations, which is a critical issue in games that use PCG to generate diverse Non-Player Characters (NPC) and enemies. Our approach proposes the use of cellular automata algorithms to create the components of sprites needed for the representation of the character in the game and the core movement animations. The advantages of PCG include significant time savings, enhanced variation and diversity, and scalability for game development. However, challenges related to maintaining artistic control and ensuring quality assurance must be addressed. The paper emphasizes the potential of PCG for humanoid character design in 2D game development, unlocking new creative possibilities supported in the generation of many characters in the game.

Keywords: Procedural Content Generation · Sprite Sheet · Game Art Generation · Character Animation

1 Introduction

Procedural Content Generation (PCG) has the potential of allowing designers to create vast and diverse game worlds with minimal manual effort. While PCG is commonly associated with level design and terrain generation [1,2,9], it has also promising applications in the creation of sprite sheets for game characters, encompassing both art and animation. A sprite sheet is a set of images representing a game object that requires animation, being commonly used in 2D games. A sprite sheet for a game character needs to have images representing the several sides of that character along with images that, when played in succession, generate the illusion of movement. A sprite sheet generated by the algorithm here proposed can be found in Fig. 10, with further sprite sheets found in Fig. 11, 12 and 13.

We will begin by analyzing the use of PCG for sprite sheet creation, highlighting its advantages, challenges, and potential for enhancing game development.

A. Marto et al. (Eds.): VJ 2024, CCIS 2324, pp. 215–227, 2025.
https://doi.org/10.1007/978-3-031-81713-7_15

1.1   The Power of Procedural Content Generation for Sprite Sheets

Sprite sheets are essential elements in 2D game development, comprising a collection of 2D images known as sprites that represent characters, objects, or special effects. These images are presented in sequence by the game engines to create the illusion of movement. Traditionally, artists create these sprite sheets by hand, investing significant time and effort into crafting each frame of animation. However, PCG offers a new approach, leveraging algorithms and rules to generate sprite sheets automatically, saving valuable development time and resources.

Art Generation. By employing PCG, developers can generate sprite sheets with a vast array of unique characters. These algorithms can be programmed to generate sprites with different body types, facial features, colors and movements. With the ability to define rules and parameters, PCG enables the creation of diverse visual assets, opening up new possibilities for game designers.

Animation Generation. PCG is not limited to static art; it can also be used to generate animated sprite sheets. Animations bring characters and objects to life, adding depth and immersion to games. With PCG, developers can define animation rules and parameters such as movement patterns, attack sequences, or idle behaviors. These rules can then be applied to generated sprites, resulting in unique and dynamic animations.

1.2   Advantages of PCG for Sprite Sheet Creation

PCG provides several advantages when it comes to sprite sheet creation:

Time Efficiency. PCG significantly reduces the time required for sprite sheet creation. Instead of manually designing and animating each sprite, developers can rely on procedural algorithms to generate a multitude of sprites and animations quickly. This allows for rapid prototyping, iteration, and experimentation, leading to accelerated game development cycles.

Variation and Diversity. Traditional sprite sheet creation often involves limitations due to the time and effort required to create different variations of characters or objects. With PCG, the possibilities for variation are virtually limitless. By specifying parameters, developers can generate countless unique sprites and animations, enhancing the visual diversity and richness of their games.

Scalability. PCG provides scalability advantages as games grow in size and complexity. By automating the sprite sheet creation process, developers can easily generate additional assets or expand upon existing ones. This flexibility allows for seamless integration of new characters and enemies, ensuring that the game world feels dynamic and immersive.

1.3   The Cellular Automata Approach

In this paper, we will use a cellular automata algorithm to generate the body parts for humanoid pixel art characters that range in size from 16 by 16 pixels to 64 by 64 pixels. The cellular automata algorithm, explained in detail in Sect. 3.1, defines the value of each pixel based on the values of the neighboring pixels that surround it. This makes it a bottom up PCG approach with unpredictable results. To tame this characteristic, we use the algorithm to make body parts, which we then join to create the full sprite. This creation of individual body parts not only allows us to control a bit more of the final appearance of the sprites, making them humanoid like, but also proves crucial to create animations.

2   Related Work

As far as the authors could find, there have been no previously documented approaches of using cellular automata for the creation of game sprite sheets. As such, this related work section will be divided into two categories: the use of cellular automata for the generation of game content; and the automatic generation of game sprites.

Cellular Automata for PCG. Cellular automata algorithms have been used extensively for the creation of game maps and dungeon caves [4,7,8]. Their use in these areas stems from their ability to generate "organic looking" structures, meaning structures without a well-defined symmetrical structure that look similar to what is found in real natural caves and formations. Their lack of intrinsic symmetry when using the variation of the algorithm here used was something that needed to be solved when generating game sprites, given these often require symmetry.

PCG of Game Sprites. The works found that tackled the generation of game sprites focused mostly on the generation of game art. One such example can be found in the work of Horsley et al. [6], which uses convolutional generative adversarial networks (GANs) to generate single sprite images, without any movement or direction considerations.

Also using GANs, Coutinho et al. [3] focus on generating sprites that face different directions based on a single front-facing sprite. This work aligns with the work of this paper in identifying the need to generate more than a single image when generating sprites that will be used in a video game. Notwithstanding, this paper does not tackle the problem of generating animation.

Another approach that generates sprites from training data using neural networks can be found in Gonzalez et al. [5]. This work focuses mostly on generating different variations of a single sprite, like a novel snow-living variant of a monster that normally lives in the desert.

None of the work found that tackles the problem of generating game sprites attempts to generate complete sprite sheets that include animations, focusing only on generating a single image or discrete variations of an image.

3   Generative Algorithm

3.1   Cellular Automata for Body Parts

The art generation done throughout this project will be based on cellular automata. This is a common algorithm used for PCG map generation, being here employed for its ability to create natural/biological-like structures. The main idea behind this algorithm is the creation of a randomly populated 2D matrix of 0s and 1s. The algorithm then goes to each position of the matrix and transforms it (altering its value or letting it be) depending on the values of the surrounding neighboring positions. This is repeated for a set number of cycles. The density of the initially randomly populated matrix, the number of similar neighbors needed to alter a position, the depth of the neighborhood analyzed (if only the 8 surrounding positions or more), and the number of cycles are all parameters that can be tweaked. The sprites shown in this paper were created using a matrix with 20% of its positions populated (set to 1), the neighborhood threshold needed to change a position (number of surrounding 1s) was 3, only the 8 closest neighbors were taken into consideration and the cycle ran 3 times.

Having a matrix of 0s and 1s with an organic shape, we then run through this matrix and transformed all the borders (locations where 0s bordered 1s) into a third value, 2. This was done because the positions with 0s will be discarded from the image, that is, will be transparent. We give a different value to the borders so that these can be colored differently. The nature of the algorithm makes it so that often, a shape has internal borders as well, as can be seen in Fig. 4. This coloring of the borders creates patterns that add to the diversity of the generated body parts.

If we were generating a level or a cave, the 1s would represent walls or rocks and the 0s the floor. The 2s would be the borders between rocks/walls and the floor. Here, the 1s and 2s were made to represent colors. These colors are assigned using a custom made color palette generator, which leverages knowledge in musical theory and applies it to RGB values, creating sets of harmonious color much as one would create harmonious chords. Whenever a new sprite is to be created, a palette of colors is decided upon and all body parts will be colored with colors randomly chosen from that palette.

The front-facing head is further randomly populated with dots of yet another value, 3, which will be assigned a third color. These are added to generate facial features and "eyes".

Finally, some of the body parts, like the front-facing head and torso, need to be mirrored to create symmetry. This is a needed step to generate humanoid sprites as such symmetry is expected in those body parts. More symmetry is added by mirroring the legs and the arms, meaning the left arm is a mirrored reflection of the right one, as is the left leg of the right leg.

An example of the generated components for a sprite can be found in Figs. 1, 2, 3, 4, 5 and 6. Their combination to form a single sprite can be found in Fig. 10.

A sprite that can only face a single direction is not enough for the majority of games where sprites need to move around. As such, we generate sprites facing

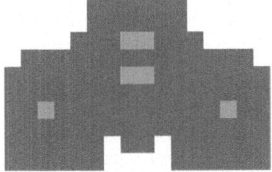

Fig. 1. A generated head.

Fig. 2. A generated back part of the head.

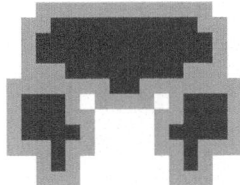

Fig. 3. A generated torso.

four different directions: front; back; left and right. This is the minimum number of directions used in 2D top-down games. The examples previously shown concern a front-facing sprite. We implemented backwards-facing sprites using slight alterations to the components and a different layering and positioning of said components. An example of the backwards-facing sprite corresponding to the front-facing sprite presented in Fig. 7 can be found in Fig. 8.

Finally, we need to generate the side-facing sprites. This is a more complex challenge, as the generated body parts can't be used directly. We also need to ensure that the geometrical properties of the sprite aren't violated. For example, the head needs to have the same height facing forward and sideways: we don't expect a head to shrink suddenly when turning. The generation of the sideways sprites was one of the most complex challenges in this work. We finally decided to use a minor optical illusion in our favor as well as inspiration in the cubism art form.

The sideways-facing components of the torso and the head are in reality simply the original generated shapes without the forced vertical symmetry. We expected that this solution wouldn't work and that it would be too obvious, but in most cases, this solution works well and generates pleasing transitions

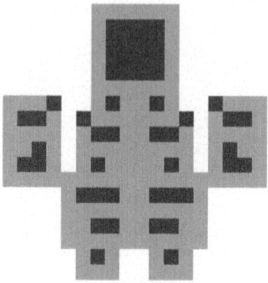

Fig. 4. A generated back part to the torso.

Fig. 5. A generated arm.

Fig. 6. A generated leg.

Fig. 7. Combination of the several components to generate a front-facing sprite art.

between front-facing and side-facing sprites. A left-facing sprite can be found in Fig. 9. Notice how the head of Fig. 1 is the left side of Fig. 9's head mirrored.

Fig. 8. Combination of the several components to generate a back-facing sprite art.

Fig. 9. Combination of the several components to generate a left-facing sprite art.

3.2   Animation

After creating the art for the several components of the sprite, we can use them to create several different iterations of the sprite. These iterations will then be played in sequence to create the animation. We are using slight changes in the positioning of the components along with changes in the layering of the components to give the impression of movement. The final sprite sheet, containing an animated sequence of 3 sprites for each of the 4 directions can be found in Fig. 10.

3.3   Sizes

We used the here described approach to generate sprites of many sizes. The examples given to this point are all for generating a 32×32 pixels sprite. However, we used to approach to also create 16×16 pixels sprites and 64×64 pixels sprites. Three other examples of 32×32 sprites can be found in Fig. 12. Examples of 16×16 sprites can be found in Fig. 11 and of 64×64 sprites in Fig. 13. We believe the approach works very well for 16×16 and 32×32 sprites. However, for

Fig. 10. The final 32 × 32 sprite sheet, containing 4 directions of movement along with 3 different positions for each direction so the illusion of movement can be created.

64 × 64 sprites, the algorithm begins to generate an exceeding amount of detail that makes the sprites less visually pleasing. This is, of course, subjective and the 64 × 64 sprites could still be used to represent more monstrous creatures. These examples can be used to judge the diversity of shapes and colors that the algorithm can generate.

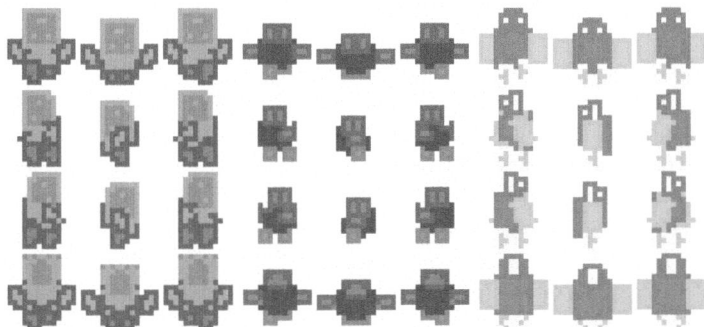

Fig. 11. Three sprite sheets exemplifying the results obtained for 16×16 sprites.

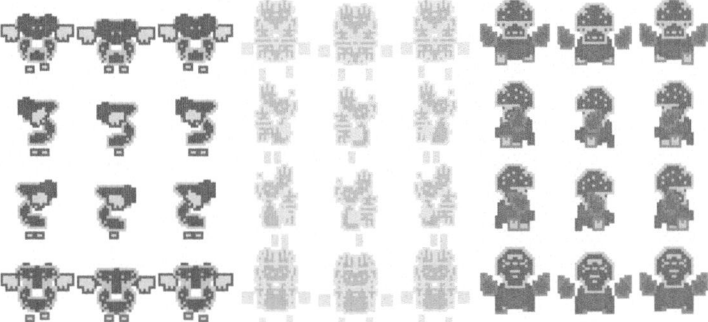

Fig. 12. Three sprite sheets exemplifying the results obtained for 32×32 sprites.

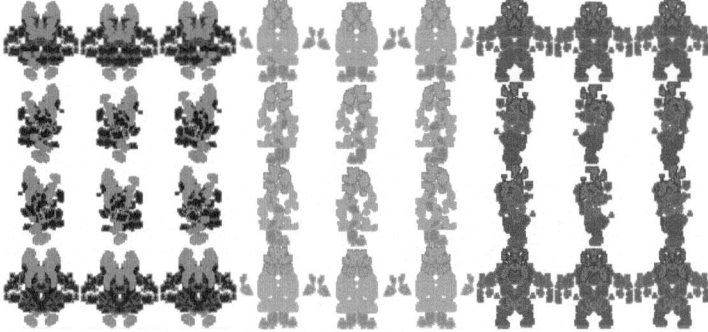

Fig. 13. Three sprite sheets exemplifying the results obtained for 64×64 sprites.

4   Designer GUI

To make this approach usable by game designers and artists, we decided to create a Graphical User Interface (GUI) that allows users to tweak the parameters of the algorithm and create sprites that match their functional and artistic goals. The GUI is currently very simple, with only two sets of sliders: one set that controls the size of the sprite itself and the maximum size of the several components of the sprite (Fig. 14); and a set that controls the animations of the sprite (Fig. 15).

The GUI allows the users to generate novel sprites based on the parameters given by clicking the "Generate!" button. For 16×16 and 32×32 sprites, this generation is instantaneous. For 64×64 sprites, it can sometimes take a couple of seconds. The users can also reset all of the parameters to their predefined values by clicking the "Reset All Parameters" button. They also have control over the speed of the animated sprite in the middle of the screen, using the "GIF Delay" slider, so they can predict how the sprite looks moving at different speeds. Finally, the generated sprite sheet is shown on the right side of the GUI and below it, a "Save!" button allows the users to save the sprite sheet along with an animated GIF of the sprite moving. The users can specify the save path along with the name.

A name is generated for every generated sprite, using a combination of one adjective followed by a noun. The name generated for the sprite in Fig. 14 and Fig. 15 is "Somber Departure". Other examples of generated names are: "Wordy Metal"; "Honest Hall"; "Rectangular Comfortable"; "Tidy Witness"; "Early Visit"; and "Infamous Candy". Although random in nature, these names help grant some character to the generated sprites.

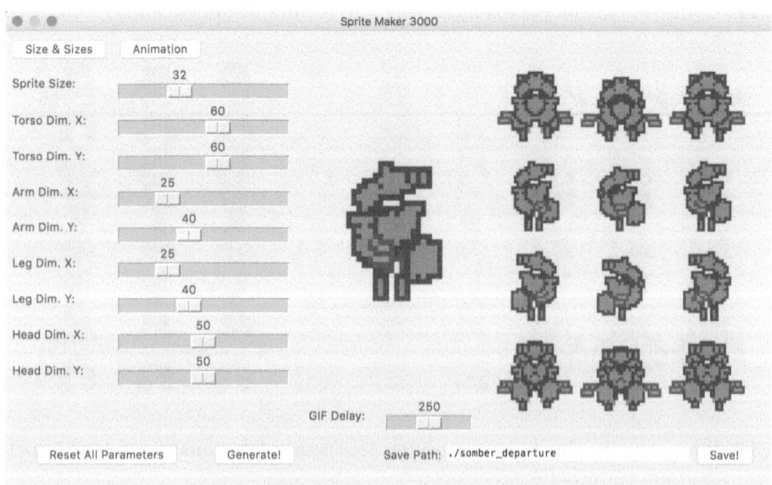

Fig. 14. The GUI showing the available parameters for altering the size of the sprite and the several components of the sprite.

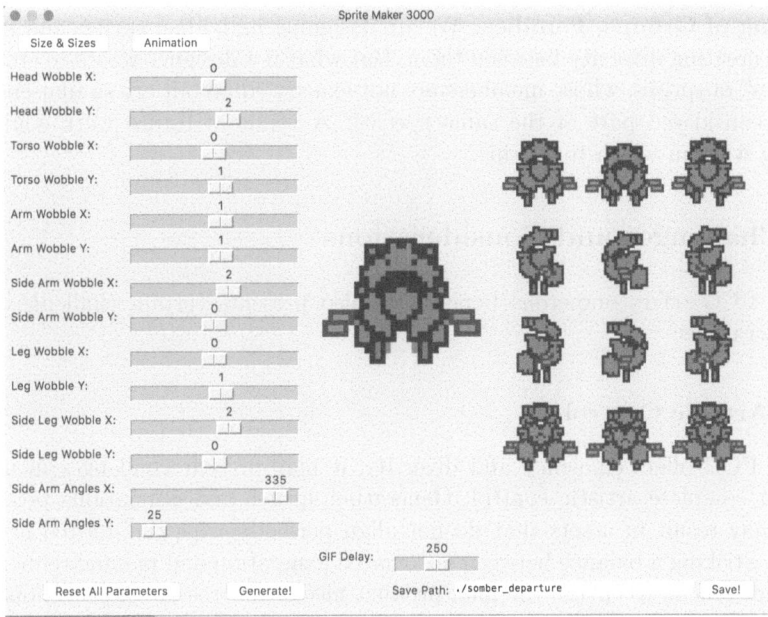

Fig. 15. The GUI showing the parameters that control several characteristics of the animation.

5   Future Work

Consider a Novel Approach for PCG Animation Generation. We are curious to see if we can explore novel methods for PCG animation, taking into consideration more complex movements and shaping the art to create more realistic movements.

Develop a Method to Automatically Filter the Not so Appealing Sprites. The current methodology generates sprite art and animations but has no method of knowing if such creations have any quality. Defining what qualities and characteristics are of interest and which methods could be used to measure them and filter the sprites in terms of quality is a strong venue for future work.

Increase Control over Generated Sprites. The developed GUI already grants developers and artists a certain degree of control over the generated sprites. However, this control could be expanded to more areas of sprite generation. Users should be able to specify the color palette to be used by hand and have further control over the relative location of the several body components of the sprites. The changes in parameters should also reflect directly on the sprite being shown and not only on the next generated sprite, as currently happens.

Making of Groups, Families. We are designing individual sprites and focusing on creating diversity between them. But what if a designer wants to create a "species" of sprites, whose members are not exactly equal but are similar enough to be considered part of the same species? A venue of future work could be finding ways in which to do this.

6   Challenges and Considerations

While PCG offers numerous benefits, it also presents certain challenges and considerations:

6.1   Artistic Control

While PCG offers efficiency and diversity, it may present challenges in maintaining complete artistic control. Generating sprites and animations procedurally may result in assets that do not align perfectly with the desired artistic vision. Striking a balance between automated generation and manual refinement becomes crucial to ensure the final product meets the creative expectations.

6.2   Quality Assurance

With PCG, it is essential to thoroughly test and validate the generated sprite sheets. Quality assurance becomes crucial to identify and rectify any visual or animation anomalies that may occur during the procedural generation process. Iterative refinement is often necessary to achieve the desired level of quality and polish. Automatically generating images also poses the danger of generating images that might be offensive by pure chance, which is a hard thing to filter or evaluate automatically.

7   Conclusion

Procedural Content Generation for sprite sheet creation changes the way developers approach character and object design in 2D game development. The combination of art and animation generation through PCG provides time efficiency, variation, and scalability. We believe we have achieved relative success in creating animated sprites, along with a GUI that grants designers a certain degree of control over the created content. There is, however, much yet that can be improved. Link to code: https://github.com/soulfir/sprite-generator.

Acknowledgement. This work was supported by Fundação para a Ciência e a Tecnologia (FCT) grant UIDB/50021/2020 (https://doi.org/10.54499/UIDB/50021/2020) and grant 2020.05865.BD.

References

1. Adrian, D.F.H., Luisa, S-G.C.A.: An approach to level design using procedural content generation and difficulty curves. In: 2013 IEEE Conference on Computational Inteligence in Games (CIG), pp. 1–8. IEEE (2013)
2. Charity, M., Khalifa, A., Togelius, J.: Baba is y'all: collaborative mixed-initiative level design. In: 2020 IEEE Conference on Games (CoG), pp. 542–549. IEEE (2020)
3. Coutinho, F., Chaimowicz, L.: Generating pixel art character sprites using GANs. In: 2022 21st Brazilian Symposium on Computer Games and Digital Entertainment (SBGames), pp. 1–6. IEEE (2022)
4. Earle, S., Snider, J., Fontaine, M.C., Nikolaidis, S., Togelius, J.: Illuminating diverse neural cellular automata for level generation. In: Proceedings of the Genetic and Evolutionary Computation Conference, pp. 68–76 (2022)
5. Gonzalez, A., Guzdial, M., Ramos, F.: Generating gameplay-relevant art assets with transfer learning. arXiv preprintarXiv:2010.01681 (2020)
6. Horsley, L., Perez-Liebana, D.: Building an automatic sprite generator with deep convolutional generative adversarial networks. In: 2017 IEEE Conference on Computational Intelligence and Games (CIG), pp. 134–141. IEEE (2017)
7. Johnson, L., Yannakakis, G.N., Togelius, J.: Cellular automata for real-time generation of infinite cave levels. In: Proceedings of the 2010 Workshop on Procedural Content Generation in Games, pp. 1–4 (2010)
8. Pech, A., Hingston, P., Masek, M., Lam, C.P.: Evolving cellular automata for maze generation. In: Chalup, S.K., Blair, A.D., Randall, M. (eds.) ACALCI 2015. LNCS (LNAI), vol. 8955, pp. 112–124. Springer, Cham (2015). https://doi.org/10.1007/978-3-319-14803-8_9
9. Viana, B.M.F., dos Santos, S.R.: Procedural dungeon generation: a survey. J. Interact. Syst. **12**(1), 83–101 (2021)

From Images to Stories: Exploring Player-Driven Narratives in Games

Edirlei Soares de Lima[1]([✉])[iD], Margot M. E. Neggers[1][iD],
Marco A. Casanova[2][iD], and Antonio L. Furtado[2][iD]

[1] Academy for AI, Games and Media, Breda University of Applied Sciences,
Breda, The Netherlands
{soaresdelima.e,neggers.m}@buas.nl
[2] Department of Informatics, PUC-Rio, R. Marquês de São Vicente 225,
Rio de Janeiro, Brazil
{casanova,furtado}@inf.puc-rio.br

Abstract. This paper presents a method for generating player-driven
narratives from visual inputs by exploring the visual analysis capabilities of multimodal large language models. By employing Bartle's taxonomy of player types—Achievers, Explorers, Socializers, and Killers—our
method creates stories that are tailored to different player characteristics. We conducted a fourfold experiment using a set of images extracted
from a well-known game, generating distinct narratives for each player
type that are aligned with the visual elements of the input images and
specific player motivations. By adjusting narrative elements to emphasize
achievement for Achievers, exploration for Explorers, social connections
for Socializers, and competition for Killers, our system produced stories
that adhere to established narratology principles while resonating with
the characteristics of each player type. This approach can serve as a helping tool for game designers, offering new insights into how players might
engage with game worlds through personalized image-driven narratives.

Keywords: Story Composition · Image Analysis · Intelligent Agents ·
Player Types · GPT-4o Vision · Text-to-Image

1   Introduction

Images are, first and foremost, *descriptive*. A landscape does more than show
a location—it communicates its atmosphere and scale—just as a portrait does
more than outline a face; it conveys the subject's identity and emotional profile.
But they can also be *narrative*, displaying an event occurring in a given place
and enacted by one or more persons. When a sequence of two or more images
is presented, such as step-by-step successive scenes constituting an event, the
narrative aspect becomes even more evident. The popular adage that "a picture
is worth a thousand words" is not only true in the sense that images clarify
what may be difficult to understand in a textual explanation, but also in view

of the rich variety of widely different stories that come to the imagination when contemplating the images. This is particularly relevant in the context of game design, where visual storytelling plays a crucial role in creating immersive and engaging experiences for players.

In literary works, such as novels or poems, the text is the dominant component, whereas images, when inserted as *illustrations*, play the secondary role of inducing readers to mentally visualize the textual content, while serving to embellish the printed pages. Famous illustrators like Gustave Doré have enriched classics such as Dante's *Divine Comedy* and Perrault's fairy tales. However, images long preceded writing, with *Narrative Art* tracing back to ancient times. A recent *Nature* article [22] describes a cave painting at Leang Karampuang in Sulawesi, Indonesia, dating back at least 51,200 years, making it "the earliest known surviving example of representational art, and visual storytelling, in the world".

In games, images such as concept art, character designs, environment sketches, and even in-game screenshots are not just illustrations but are integral to the narrative fabric that shapes the player's journey. They can help convey complex emotions, hint at backstories, and suggest potential plot developments, all of which enhance player engagement. Recognizing this, we propose a tool named ImageTeller that enables a user to put words to the narrative power of images and image sequences to compose stories inspired by the portrayed scenes.

To further explore the potential of ImageTeller in game design, we present a new approach to narrative generation by incorporating a genre framework based on player archetypes. Drawing from Bartle's taxonomy [1], which identifies distinct player types—Achievers, Explorers, Socializers, and Killers—our proposed method leads to plots that reflect different player motivations and preferences. By using visual inputs such as game concept art or screenshots, designers can utilize ImageTeller to create stories that resonate with the unique characteristics of each player type. This perspective not only allows for the development of more personalized narrative experiences but also offers game designers a way to experiment with storytelling that adapts to diverse player types. The tool can offer game designers new insights into how different players might engage with the game world and assist in redefining game narratives.

The paper is structured as follows. Section 2 reviews related work. Section 3 provides full technical details about the architecture, functionalities, and usage of ImageTeller. Section 4 reports the experiments conducted with ImageTeller. Section 5 presents concluding remarks.

2   Related Work

The concept of generating stories from visual input has been explored in some previous related works. Early research in this area, such as the work by Farhadi et al. [5], focused on generating descriptive sentences from images by comparing estimations of meaning derived from both visual and textual data. Subsequent studies explored the potential of AI models to create more elaborate stories based on images or image sequences. For instance, Huang et al. [8] proposed a system that uses machine learning to generate descriptive captions for images and then constructs short stories from these captions. In a similar work, Smilevski et al. [28] introduced a Sequence-to-Sequence model with separate encoders for visual and narrative components, aiming to generate more human-like stories that go beyond basic image descriptions. Despite these advances, the generated stories tended to be limited in length and narrative richness, often failing to capture intricate plot structures or character arcs. The need for discourse coherence in visual storytelling is explored by Cardona-Rivera and Li [4], who introduced a system that generates stories around user-supplied photographs by considering discourse constraints during fabula generation.

Similarly, researchers have integrated player types into narrative generation to create personalized experiences. Thue et al. [30] introduced a player modeling approach for interactive storytelling, where narratives are dynamically adapted based on the player's inferred style and preferences. By categorizing players according to their behavior, their system generates story content that aligns with individual player motivations. Orji et al. [23] investigated tailoring persuasive strategies in serious games to different player types, utilizing the BrainHex typology [21]. Their research demonstrated that customizing game content and narratives to match player types enhances player engagement and the effectiveness of persuasive messages. de Lima et al. [14] presented a method to interactive storytelling in games where quests and ongoing stories are determined based on individual personality and behavioral attitudes of the player, which are modeled using the Big Five personality traits [7].

While previous works have shown the benefits of using images for automated storytelling and integrating player types and personality traits into narrative generation, they often focus on short, descriptive narratives rather than longer, structurally complex stories. ImageTeller builds upon these previous efforts by not only generating richer and more engaging stories from images using the multi-modal capabilities of GPT-4o and its vision counterpart but also introducing the concept of player-driven narratives. By employing Bartle's taxonomy of player types [1], ImageTeller generates stories that reflect diverse player motivations and behaviors, offering a novel approach to personalized narrative generation that combines visual input analysis with player type modeling.

3   The ImageTeller Prototype

ImageTeller was designed to explore the narrative potential of images by employing AI agents, namely GPT-4o, GPT-4o Vision, and FLUX.1 diffusion model, to analyze visual content and generate stories. Unlike purely text-driven approaches, ImageTeller's integration of visual data aims to produce stories that harness both the imaginative potential of AI and the contextual richness of visual storytelling. The prototype is accessible at: https://narrativelab.org/image teller/.

3.1   User Interface

The user interface of ImageTeller is presented in Fig. 1. To create a story, users can upload images on the system's initial screen (Fig. 1(a)) by clicking on the "+" button (Action 1 in Fig. 1(a)). In the example illustrated in Fig. 1, the user has uploaded three screenshots from the game *The Legend of Zelda: Breath of the Wild* (Nintendo, 2017) without providing additional captions. After adding the images, the user can select a genre for the story. In this case, the user selected the "Explorer" player-driven genre (Action 2 in Fig. 1(a)).

Once the images are added and a genre is selected, the user can click on the "Generate Story" button to initiate the narrative generation process (Action 3 in Fig. 1(a)). The system then transitions to the story composition screen (Fig. 1(b)), where the user can visualize the generated story and its chapters, each accompanied by a title and an illustration.

For each chapter, the user has the option to either regenerate the illustration or regenerate the entire chapter by using the respective buttons available on the interface (Action 4 in Fig. 1(b)). These options allow the user to refine the narrative and its visual representation to better suit their preferences. At the bottom of the screen, the user can click on the "Save Story" button to save the generated story (Action 5 in Fig. 1(b)), which will then be added to the user's personal library (Fig. 1(c)). The full story generated for the example presented in Fig. 1 is available at: https://narrativelab.org/imageteller/#/story/394.

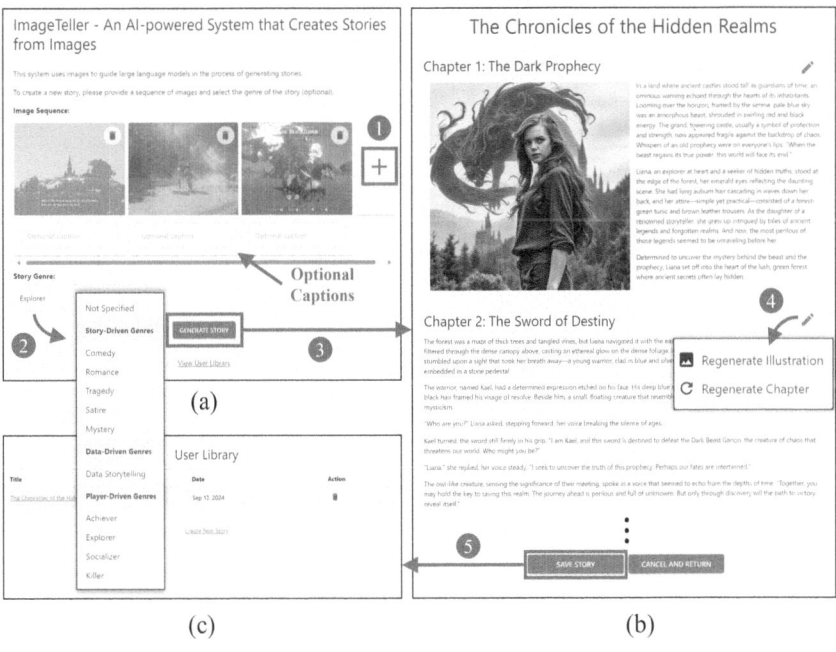

Fig. 1. The user interface of ImageTeller.

3.2   System Architecture

As illustrated in Fig. 2, the architecture of ImageTeller is based on the multi-AI-agent approach for narrative generation using large language models proposed in our previous works [11,13,18,20,26], which we adapted in the present work to support the generation of narratives from visual inputs. This architecture incorporates a set of AI agents, each with a distinct role in the narrative generation process: (1) a Visual Analyzer AI Agent, which uses the GPT-4o Vision to interpret and analyze the content of input images; (2) a Storywriter AI Agent, powered by the GPT-4o model, responsible for generating the textual narrative based on the results of the visual analysis; and (3) an Illustrator AI Agent that uses a text-to-image diffusion model to create visual representations that complement and enhance the story. A Plot Manager coordinates the interactions between agents and the narrative structure, similar to the plot management strategies used in interactive storytelling systems [16].

User interaction is facilitated through a user-friendly interface, allowing users to generate stories by providing images, captions, and selecting a genre. The interface displays the generated narratives in a format resembling illustrated books, with each chapter featuring a title, narrative text, and illustration. Users can provide interact by requesting alternative chapters or illustrations for iterative refinement of the results.

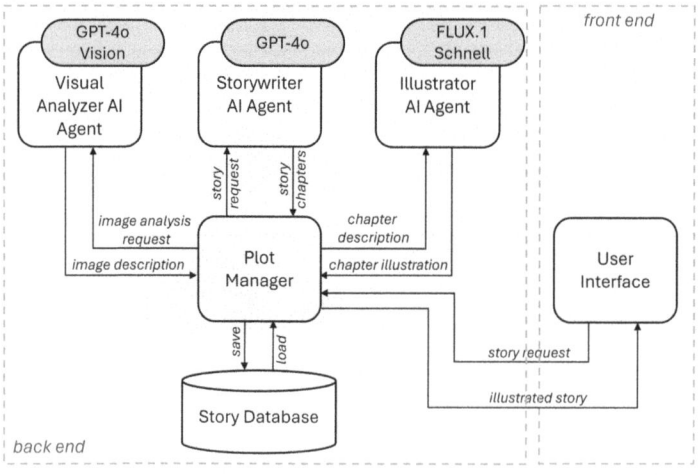

Fig. 2. The multi-AI-agent architecture of ImageTeller.

The AI agents are implemented using a plugin architectural approach. As indicated in Fig. 3, the Visual Analyzer AI Agent utilizes GPT-4o Vision via the OpenAI API for image analysis. The Storywriter AI Agent also utilizes GPT-4o model for text generation, while the Illustrator AI Agent is built upon a distilled version of the FLUX.1 text-to-image diffusion model (FLUX.1 Schnell), hosted on a private server and accessed through a REST API. This modular design enhances the system's adaptability, performance, and scalability.

3.3 Image Analysis

Analyzing and describing visual content is a core function of ImageTeller. This capability is powered by the advanced vision features of the GPT-4o model, which can extract detailed and contextually rich descriptions from each input image. These descriptions are produced using a prompt designed to extract and generate textual descriptions that are suitable for narrative generation.

The prompt used for image analysis (Prompt 1) is crafted to guide the model in identifying and describing key elements of the image, focusing on subjects, their actions, appearance, significant interactions, and environmental details that shape the scene's mood and context. The description is generated as a single cohesive paragraph to ensure seamless integration of all relevant aspects.

Prompt 1. *Analyze the provided image and write a short description of the subjects present in the image, including their actions and appearance. Focus on key elements such as the interactions between subjects and any significant environmental details that contribute to the mood and context of the scene. Write the description as a single paragraph without using headers or lists.*

If a user opts to write a caption, it is directly incorporated into the analysis. The prompt is dynamically adjusted with the instruction: "*When generating the description, take into account that the image has the following caption: $<C_i>$*", where C_i represents the provided caption for image i. This allows the model to align the image analysis with the user's desired interpretation, ensuring that the generated description is consistent with any specific narrative cues suggested by the caption. This feature is especially useful when users want to steer the narrative, specify character names, or highlight certain aspects of the image.

For example, when analyzing the second image used as input in the example illustrated in Fig. 1(a) (also shown in Fig. 3(b)), the model produced the description presented in Output 1, which captures the scene's key elements and atmosphere.

Output 1. *The image depicts a character standing in a lush, mystical forest, with beams of sunlight filtering through the dense canopy above, casting a warm glow on the scene. The character, wearing a blue tunic and armor, is gripping the hilt of a legendary sword embedded in a stone pedestal, suggesting a moment of significant challenge or destiny. The surrounding environment features thick trees, verdant foliage, and the faint presence of mist, creating an atmosphere of ancient magic and mystery. The interplay of light, shadows, and greenery adds a sense of depth and tranquility, while the character's determined stance and focused expression convey resolve and purpose.*

3.4   Story Generation

The story generation process in ImageTeller is designed to transform the visual and contextual data obtained from the image analysis into a cohesive and engaging narrative. This process is powered by the GPT-4o model, which is guided by a structured prompt system that varies depending on the user's input, particularly regarding the type and genre of the narrative.

Three types of narratives are handled by ImageTeller: story-driven, player-driven, and data-driven. A *story-driven narrative* focuses on traditional storytelling and can either follow the conventions of a specific genre or be more general, without strict adherence to genre-specific elements. In our current prototype, we adopted the conventions of five fundamental genres that we have specified in a previous work [19]: Romance, Comedy, Tragedy, Satire, and Mystery. A *player-driven narrative* explores the motivations and preferences of different types of players according to Bartle's taxonomy [1] (Achievers, Explorers, Socializers, and Killers). This type of story is designed to resonate with distinct player motivations by tailoring narrative elements—such as goals, challenges, exploration, social interactions, and conflicts—to align with the interests and behaviors of each player type. Table 1 presents the definitions of the player-driven genres according to Bartle's taxonomy. Finally, a *data-driven narrative* is designed to emphasize the communication of insights and data points, prioritizing clarity and relevance while presenting complex information in an engaging way.

The prompt system used for the narrative generation process is composed of six modular components that are combined to create the final instruction set for the model. These components are:

1. **General Narrative Instruction** ($P_{general}$), which offers core guidance for generating the story, including the structure, title format, and chapter division. This instruction ensures that the story follows a cohesive narrative flow and effectively brings the visual descriptions to life;
2. **Story Instruction** (P_{story}), which focuses on transforming the image descriptions into a rich, character-driven narrative. This instruction guides the development of characters, their interactions, and the overall story dynamics, ensuring the narrative captures the essence of the scenes while maintaining a traditional storytelling style;
3. **Data Instruction** (P_{data}), which shifts the focus of the narrative towards communicating insights and connections in a compelling way, tailoring the story to effectively present data-driven content with clarity and engagement;
4. **Story-Driven Genre Specification** (P_{genre}), which defines the story genre and incorporates genre-specific elements such as setting, tone, and character types, ensuring the story adheres to the conventions of the chosen genre;
5. **Player-Driven Genre Specification** (P_{player}), which adapts the narrative style and content to align with different player types according to Bartle's taxonomy, ensuring the story reflects specific player motivations; and
6. **Image Descriptions** (P_{image}), which contains the sequentially ordered descriptions of the images generated during the image analysis phase.

Table 2 presents the parameterized text for the six components of the prompt system. The final prompt used to guide the GPT-4o model in the narrative generation process is assembled by combining these components based on the user's preferences:

1. For a story-driven narrative: $P_{final} = P_{general} + P_{story} + P_{genre} + P_{image}$
2. For a player-driven narrative: $P_{final} = P_{general} + P_{story} + P_{player} + P_{image}$
3. For a data-driven narrative: $P_{final} = P_{general} + P_{data} + P_{image}$
4. If no genre is specified: $P_{final} = P_{general} + P_{story} + P_{image}$

Table 1. Player-driven genre definitions based on Bartle's taxonomy [1].

Player Type	Prompt Text
Achiever	*An Achiever player focuses on completing challenges, gaining rewards, and demonstrating mastery. Therefore, the narrative must be structured around goal-oriented tasks, measurable progress, milestones, and the desire for recognition. The protagonist is driven by accomplishment and overcoming increasingly difficult challenges*
Explorer	*An Explorer player is focused on discovery, curiosity, and unraveling hidden secrets. Therefore, the narrative must be structured around uncovering mysteries, exploring unknown locations, and the joy of finding something new. The protagonist is driven by a sense of wonder, investigation, and the thrill of the unknown*
Socializer	*A Socializer player focuses on relationships, interactions, and communication. Therefore, the narrative must be structured around friendships, alliances, community building, and the social dynamics between characters. The protagonist is driven by the need for connection, cooperative tasks, and shared experiences*
Killer	*A Killer player focuses on competition, dominance, and asserting power over others. Therefore, the narrative must be structured around conflict, strategic moves, and competition. The protagonist is driven by a desire for victory, control, and exerting influence over other characters*

A key feature of $P_{general}$ is its directive that instructs the model to generate the narrative in the expected output format, with the story divided into chapters and markdown headers used to identify titles. This structured format ensures the narrative is organized with clear distinctions between the story title, chapter titles, and chapter content. By analyzing the markdown headers, the system can parse the generated narrative and process each chapter individually to produce the final formatted version of the narrative to be displayed to users. An example of a complete prompt and the raw narrative content generated by GPT-4o is presented in our complementary open-access paper [12].

Table 2. Components of the prompt system for narrative generation, where $<g_i^{name}>$ identifies the genre name or player type, $<g_i^{description}>$ represents the definition of the genre or player type, and $<I_{description}>$ is an ordered set of image descriptions.

Component	Prompt Text
$P_{general}$	Using the following set of sequentially ordered image descriptions, write a cohesive, engaging, and complete story resembling a book narrative. The story must have a creative title written using markdown header level 1 (#). Follow the sequence of events as represented in the descriptions, maintaining consistency in mood and context. Ensure the narrative has a clear beginning, middle, and end. Divide the story into a natural number of chapters, each with a name and number, using markdown header level 2 (##) for the chapter titles
P_{story}	When writing the story, assign names to the subjects, describe their visual attributes (eye color, hair color, hairstyle) and clothes (type, color, style), and develop their characters based on their described actions and appearances. Ensure the story incorporates all key elements from the image descriptions, including the subjects, their interactions, and significant environmental details. Focus on creating a narrative and dialogue that brings the story to life, rather than merely describing the images. Exclude any details specific to the image format, such as speech balloons, text boxes, or floating text
P_{data}	Highlight key insights, connect the dots between different data points, and present a clear and engaging story. Use metaphors, analogies, and emotional appeal to make the story engaging. Avoid referring directly to the images or describing them. Instead, weave the data into the narrative naturally, as if explaining the insights to someone who has never seen the images. Also suggest potential implications, actions, or reflections based on the insights presented
P_{genre}	Incorporate genre-specific elements such as setting, tone, themes, and character types to clearly reflect the $<g_i^{name}>$ genre. Use the following definition for the $<g_i^{name}>$ genre: $<g_i^{description}>$
P_{player}	The protagonist's journey presented in the narrative must be based on Bartle's definition of the $<g_i^{name}>$ player type. Use the following definition for the $<g_i^{name}>$ player type: $<g_i^{description}>$
P_{image}	Image descriptions: $<I_{description}>$

3.5   Image Generation

In ImageTeller, each story is structured into multiple chapters, with each chapter comprising a title, text, and an illustration. The process for generating illustrations for the chapters takes advantage of the recent advancements in text-to-image machine learning models, such as DALL-E, Midjourney, and Stable Diffusion. In our current implementation, we use the FLUX.1 Schnell model,[1] which

---

[1] https://huggingface.co/black-forest-labs/FLUX.1-schnell.

is a distilled version of the FLUX.1 text-to-image transformer model capable of generating high-quality images in 1 to 4 steps.

The generation of an illustration begins with the GPT-4o model identifying and describing a significant event in the chapter text. This description is then used as input for the image generation model. Prompt 2 shows the prompt used to instruct GPT-4o for this task, where C_i^{text} is the text of a chapter C_i.

Prompt 2. *Identify a single significant event in the provided story chapter and write a concise description (maximum of 80 words) to be used as input for a text-to-image model to generate an illustration. Focus on the characters' actions, interactions, and the environment. When describing the characters, include all details related to their visual attributes (you must describe their age group, hair color, hairstyle, eye color, clothing type, and clothing colors). Write only the description with no formatting elements. Story Chapter: $<C_i^{text}>$.*

While directly providing the generated description to a text-to-image model would suffice for producing an illustrative representation for a chapter, we implemented additional prompt optimizations to improve image quality. First, we incorporated a default illustration style ("*Photography, realistic detailed skin, 4k, highly detailed, diffused soft lighting, shallow depth of field, sharp focus, hyperrealism, cinematic lighting*"), which is combined with the chapter description to define a consistent visual aesthetic for all generated images. Second, we added a general negative prompt to guide the model toward generating more accurate images by explicitly identifying undesirable visual elements in narrative events, as proposed in our previous work [20]. Third, we implemented a technique to maintain character consistency across consecutive images by replacing character names with those of well-known actors familiar to the text-to-image model. This approach ensures consistent character representations, enhancing the visual coherence of the chapter illustrations.

Prompt 3 shows the final prompt generated for the text-to-image model considering Chap. 1 of the story presented in Fig. 1(b) as input. The generated illustration is also shown in Fig. 1(b).

Prompt 3. *(Emma Watson:1.2) stands at the edge of a lush, green forest, her emerald green eyes fixed on the distant horizon. Her long auburn hair cascades in waves down her back, and she wears a forest-green tunic and brown leather trousers. Behind her, an amorphous beast, shrouded in swirling red and black energy, looms ominously over a towering castle, contrasting against the serene, pale blue sky. The atmosphere is tense, filled with a sense of impending danger. Photography, realistic detailed skin, 4k, highly detailed, diffused soft lighting, shallow depth of field, sharp focus, hyperrealism, cinematic lighting.*

4 A Fourfold Experiment on Player-Driven Narratives

To evaluate ImageTeller's ability to generate narratives for each of Bartle's player types [1], we conducted an experiment using the same sequence of three images shown in Fig. 3 as input. These images are screenshots captured from *The Legend of Zelda: Breath of the Wild*, a Nintendo video game known for its rich storytelling [29]. For each player type, we generated a narrative and analyzed its adherence to narratology principles and alignment with the characteristics of that type, drawing parallels with well-known video game protagonists who exemplify each player type.

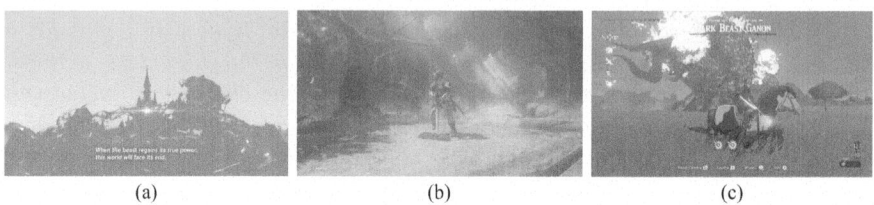

(a)   (b)   (c)

Fig. 3. The screenshots from *The Legend of Zelda: Breath of the Wild* used in the experiment: (a) Calamity Ganon circling around Hyrule Castle; (b) Link drawing the Master Sword from its pedestal; and (c) Link fighting against Dark Beast Ganon.

4.1   Achiever-Oriented Story - The Quest for Glory

– Resulting story: https://narrativelab.org/imageteller/#/story/395.

The generated story is a straightforward instance of Frye's romance genre [6], with an epic plot [25], narrating the quest of Cedric, a protagonist charismatically marked by his unique ability to extract and wield the legendary Sword of Light—aligning with Propp's 17th function [24] and reminiscent of King Arthur's Excalibur episode [10]. Driven by the pursuit of achievement, Cedric faces the challenge of preventing the rise of the Beast of Shadows. His victory not only saves his world but also earns him the recognition he always sought.

The narrative clearly aligns with the achiever player type, focusing on accomplishment, mastery, and tangible rewards. As Lacy discusses [9], the term "achieving", in particular in the expression "achievement of the Grail", had a double sense in the original French romances, since the French word "achever" conveyed both the success in accomplishing the goal of the Grail quest, and the termination of the related adventures. Similarly, the generated story concludes with Cedric's personal triumph and the realization that no challenges remain, satisfying the Achiever's desire for both accomplishment and closure. This resembles the original journey of Link in *The Legend of Zelda* series, where Link undertakes epic quests to defeat evil forces, mastering skills and acquiring powerful items to save Hyrule, ultimately gaining recognition as a legendary hero.

4.2   Explorer-Oriented Story - The Chronicles of the Hidden Realms

– Resulting story: https://narrativelab.org/imageteller/#/story/394.

The generated narrative introduces a female protagonist, Liana, the daughter of a renowned storyteller, leading to an incursion into epistemic plots involving mystery narratives [25]. Determined to uncover the mystery behind the calamitous beast and the prophecy, she embodies the explorer player type by focusing on discovery and unraveling hidden truths. Instead of the fragile victim stance typical of the heroic monomyth pattern [3,31], Liana combines the auxiliary hetaira role—guiding the sword-bearer with her exploratory talents—and the active amazon role—aggressively participating in the successful combat—as in the feminine psyche proposal of one of Carl Jung's collaborators [32].

The narrative aligns with the explorer player type, emphasizing curiosity, exploration, and the thrill of uncovering secrets. The story satisfies the explorer's desire for new experiences and knowledge, mirroring modern heroic feminine renditions such as Lara Croft in the *Tomb Raider* game series, who embodies the spirit of exploration by venturing into unknown territories, uncovering ancient artifacts, and solving mysteries driven by her curiosity and determination.

4.3   Socializer-Oriented Story - The Bonds of Destiny

– Resulting story: https://narrativelab.org/imageteller/#/story/404.

The generated story features a female protagonist, Elara, introduced as a "weaver of connections", suggesting she ranks high in extraversion and agreeableness according to personality trait theories [7]. To fulfill her calling she ought to be particularly skilled in acts of verbal communication, the type of action that characterizes dramatic plots [25]. By allying herself with the sword-bearer, she gently persuades him to recruit a team of helpers, reflecting Propp's 14th function (F9: "various characters place themselves at the disposal of the hero") [24] and Vogler's sixth stage ("Tests, Allies, Enemies") [31]. The inclusion of an old sage, corresponding to Vogler's fourth stage ("Meeting with the Mentor") [31], emphasizes themes of trust and unity, which the protagonist rephrases into an appeal that each one of them should be "the light for each other". As the beast is subdued, Elara's greatest reward is seeing her friends "beaming with relief and triumph", highlighting the importance of social bonds.

The plot directly aligns with the socializer player type by emphasizing relationships, cooperation, and communal success. It satisfies the Socializer's desire for connection and shared experiences, demonstrating how unity and trust lead to overcoming challenges. The story integrates social interaction as the core mechanism for achieving victory, consistent with Bartle's definition of socializer motivations [1]. This focus on building bonds and working together is a core narrative element in the *Mass Effect* game series, where the protagonist, Commander Shepard, forms alliances with a diverse crew, and their collective efforts and relationships are crucial to overcoming the galaxy's threats.

4.4   Killer-Oriented Story - The Conqueror's Path

– Resulting story: https://narrativelab.org/imageteller/#/story/406.

The male protagonist in the generated narrative, Roran, is singularly driven by a competitive desire to dominate, embodying Roger Caillois's concept of the "agon" form of play [2], where a player obsessively seeks to outperform others in qualities like strength, skill, and ingenuity without external assistance, so as to defeat all adversaries. Described as having spent years "honing his skills, driven by an insatiable hunger to dominate and manipulate forces beyond human ken," he exhibits no inclination toward socialization or collaboration. This mirrors the legendary figure of Roland, who refused to summon aid during the Battle of Roncevaux, relying solely on his prowess and his famed Durendal sword [27].

The story directly aligns with the killer player type, emphasizing competition, dominance, and personal power. After his solitary victory, the protagonist feels that "the path of the conqueror was his, and he would let nothing stand in his way", illustrating the Killer's desire for control and unchallenged supremacy. A contemporary parallel can be drawn with the protagonist of the *Doom* game series. Like Roran, Doom Slayer is a lone warrior driven by personal motivations to wage a one-man war against demonic forces.

5   Concluding Remarks

In this paper, we explored the generation of player-driven narratives from visual inputs. By employing AI agents that integrate advanced language models with visual analysis capabilities, we showed how ImageTeller can produce cohesive and engaging stories tailored to different player types, aligning with Bartle's taxonomy [1]. The generated stories can serve as a source of inspiration, offering game designers new insights into how different players might engage with the game world and providing a helpful resource for refining game narratives.

Our initial experiments with ImageTeller, detailed in our complementary open-access paper [12], suggest that analyzing image sequences using the prototype provides a practical and effective approach to generating coherent narratives. The fourfold experiment presented in this paper demonstrated that ImageTeller effectively generates distinct narratives that resonate with the specific characteristics of the targeted player type. The analysis revealed that the generated stories not only adhere to established narratology principles but also resemble well-known narratives, including game franchises, reinforcing the relevance and potential applications of this method in the context of game design.

While the results are promising, there are limitations to consider. Our analysis of the results is primarily technical, focusing on alignment with narratology principles and player types, but comprehensive user studies are needed to assess the approach's effectiveness in practical settings. A more in-depth investigation into the practical applications and usability of this method for assisting game designers will allow us to fully understand its potential benefits and challenges within the game industry. Additionally, the current implementation focuses on a predefined set of player types and genres. Expanding the tool to encompass a broader spectrum of player types taxonomies, such as the BrainHex typology [21], could enhance its versatility and applicability in game design scenarios.

As future work, we intend to conduct extensive user studies to refine the prototype. Such studies will help us better understand user preferences, improve the quality of the generated content, and tailor the system's outputs to meet the needs of both players and designers. Additionally, integrating user modeling techniques presents a promising avenue for future research. By guiding the story generation process according to user preferences for narrative content [15,17] or personality traits [14], we expect to enhance personalization and engagement.

References

1. Bartle, R.: Hearts, clubs, diamonds, spades: players who suit muds. J. MUD Res. **1**(1) (1996)
2. Caillois, R.: Man, Play, and Games. University of Illinois Press, Champaign (2001)
3. Campbell, J.: The hero with a thousand faces. New World Library (2008)
4. Cardona-Rivera, R., Li, B.: Plotshot: generating discourse-constrained stories around photos. In: Proceedings of the AAAI Conference on Artificial Intelligence and Interactive Digital Entertainment, vol. 12, pp. 2–8 (2021). https://doi.org/10.1609/aiide.v12i1.12860
5. Farhadi, A., et al.: Every picture tells a story: generating sentences from images. In: Daniilidis, K., Maragos, P., Paragios, N. (eds.) ECCV 2010. LNCS, vol. 6314, pp. 15–29. Springer, Heidelberg (2010). https://doi.org/10.1007/978-3-642-15561-1_2
6. Frye, N.: Anatomy of Criticism. Princeton University Press, Princeton (2020)
7. Goldberg, L.R.: An alternative "description of personality": the big-five factor structure. J. Pers. Soc. Psychol. **59**(6), 1216–1229 (1990). https://doi.org/10.1037//0022-3514.59.6.1216
8. Huang, T.H.K., et al.: Visual storytelling. In: Knight, K., Nenkova, A., Rambow, O. (eds.) Proceedings of the 2016 Conference of the North American Chapter of the Association for Computational Linguistics: Human Language Technologies, pp. 1233–1239. Association for Computational Linguistics, San Diego, California (2016). https://doi.org/10.18653/v1/N16-1147
9. Lacy, N.J.: The Grail, the Quest, and the World of Arthur. Boydell & Brewer, Martlesham (2008)
10. Lacy, N.J., Wilhelm, J.: The Romance of Arthur. Routledge, Milton Park (2012)
11. de Lima, E.S., Casanova, M.A., Feijó, B., Furtado, A.L.: Semiotic structuring in movie narrative generation. In: Ciancarini, P., Iorio, A.D., Hlavacs, H., Poggi, F. (eds.) Entertainment Computing – ICEC 2023, pp. 161–175. Springer, Singapore (2023). https://doi.org/10.1007/978-981-99-8248-6_13
12. de Lima, E.S., Casanova, M.A., Furtado, A.L.: Imagining from images with an AI storytelling tool. In: ArXiv e-prints, arXiv: 2408.11517 [cs.CL] (2024). https://doi.org/10.48550/arXiv.2408.11517
13. de Lima, E.S., Feijó, B., Casanova, M.A., Furtado, A.L.: Chatgeppetto - an AI-powered storyteller. In: Proceedings of the 22nd Brazilian Symposium on Games and Digital Entertainment, pp. 28–37. ACM (2024). https://doi.org/10.1145/3631085.3631302
14. de Lima, E.S., Feijó, B., Furtado, A.L.: Player behavior and personality modeling for interactive storytelling in games. Entertainment Comput. **28**, 32–48 (2018). https://doi.org/10.1016/j.entcom.2018.08.003

15. de Lima, E.S., Feijó, B., Furtado, A.L.: Adaptive storytelling based on personality and preference modeling. Entertainment Comput. **34**, 100342 (2020). https://doi.org/10.1016/j.entcom.2020.100342

16. de Lima, E.S., Feijó, B., Furtado, A.L.: Managing the plot structure of character-based interactive narratives in games. Entertainment Comput. **47**, 100590 (2023). https://doi.org/10.1016/j.entcom.2023.100590

17. de Lima, E.S., Feijó, B., Furtado, A.L., Gottin, V.M.: Personality and preference modeling for adaptive storytelling. In: 2018 17th Brazilian Symposium on Computer Games and Digital Entertainment (SBGames), pp. 187–18709 (2018). https://doi.org/10.1109/SBGAMES.2018.00030

18. de Lima, E.S., Neggers, M.M.E., Casanova, M.A., Feijó, B., Furtado, A.L.: A pattern-oriented AI-powered approach to story composition. In: Figueroa, P., et al. (eds.) ICEC 2024, pp. 1–16. Springer, Cham (2024). https://doi.org/10.1007/978-3-031-74353-5_11

19. de Lima, E.S., Neggers, M.M.E., Furtado, A.L.: Multigenre AI-powered story composition. In: ArXiv e-prints, arXiv: 2405.06685 [cs.CL] (2024).https://doi.org/10.48550/arXiv.2405.06685

20. de Lima, E.S., Neggers, M.M., Feijó, B., Casanova, M.A., Furtado, A.L.: An AI-powered approach to the semiotic reconstruction of narratives. Entertainment Comput. **52**, 100810 (2025). https://doi.org/10.1016/j.entcom.2024.100810

21. Nacke, L.E., Bateman, C., Mandryk, R.L.: Brainhex: a neurobiological gamer typology survey. Entertainment Comput. **5**(1), 55–62 (2014)

22. Oktaviana, A.A., Joannes-Boyau, R., Hakim, B., Burhan, B., Sardi, R., Adhity-atama, S., et al.: Narrative cave art in Indonesia by 51,200 years ago. Nature **631**(8022), 814–818 (2024). https://doi.org/10.1038/s41586-024-07541-7

23. Orji, R., Vassileva, J., Mandryk, R.L.: Modeling the efficacy of persuasive strategies for different gamer types in serious games for health. User Model. User-Adap. Inter. **24**(5), 453–498 (2014)

24. Propp, V.: Morphology of the Folktale. University of Texas Press, Austin (1968)

25. Ryan, M.-L.: Interactive narrative, plot types, and interpersonal relations. In: Spierling, U., Szilas, N. (eds.) ICIDS 2008. LNCS, vol. 5334, pp. 6–13. Springer, Heidelberg (2008). https://doi.org/10.1007/978-3-540-89454-4_2

26. Schetinger, V., Bartolomeo, S.D., de Lima, E.S., Meinecke, C., Rosa, R.: n walks in the fictional woods. In: ArXiv e-prints, arXiv: 2308.06266 [cs.HC] (2023). https://doi.org/10.48550/arXiv.2308.06266

27. Scott-Moncrieff, C.K.: The Song of Roland. The Project Gutenberg (1996). https://gutenberg.org/cache/epub/391/pg391-images.html

28. Smilevski, M., Lalkovski, I., Madjarov, G.: Stories for images-in-sequence by using visual and narrative components. In: Kalajdziski, S., Ackovska, N. (eds.) ICT 2018. CCIS, vol. 940, pp. 148–159. Springer, Cham (2018). https://doi.org/10.1007/978-3-030-00825-3_13

29. Stark, C.: Breath of the wild is one of the best games of this decade, and perhaps the most impactful of the next decade (2019). https://www.polygon.com/2019/11/11/20955542/legend-of-zelda-breath-of-the-wild-best-games-decade

30. Thue, D., Bulitko, V., Spetch, M., Wasylishen, E.: Interactive storytelling: a player modelling approach. In: Third Artificial Intelligence and Interactive Digital Entertainment Conference, pp. 43–48 (2007). https://doi.org/10.1609/aiide.v3i1.18780

31. Vogler, C.: The Writer's Journey. Michael Wiese Productions (2007)

32. Wolff, T.: Structural Forms of the Feminine Psyche. C. G, Jung Institute (1956)

Deep Learning Anti-cheat System Based on Player Behaviour for Minecraft

Teresa Sousa[1] , Roberto Ribeiro[2]([✉]) , and Gustavo Reis[2]

[1] ESTG, Polytechnic University of Leiria, Leiria, Portugal
[2] CIIC, ESTG, Polytechnic University of Leiria, Leiria, Portugal
`roberto.ribeiro@ipleiria.pt`

Abstract. The evolution of anti-cheat systems has been facing increasing criticism because of the implementation of evasive protection techniques that require kernel-level access, granting the software the highest level of privileges within the player's system. This may pose a security risk for user's devices if the anti-cheat system has flaws or vulnerabilities, as its driver operates in the same virtual memory space as the operating system. If the anti-cheat driver crashes, the entire system can be brought down.

This study proposes an alternative approach using deep learning models, 2D Convolutional Neural Networks (2D-CNN), and Long Short-Term Memory (LSTM) to analyze player behavior through mouse dynamics and find cheating patterns. It will work locally on the player device, being a client-side anti-cheat, and doesn't need kernel access or additional privileges.

Both models were trained with a custom dataset containing data from 13 users playing Minecraft for 40 min. Every player had to record data from normal and cheating gameplay, in which the cheats used were aimbot and automining, developed for the context of the study. A custom data recording tool was also created by developing a mouse listener mod for Minecraft, which captured mouse movements and clicks and added them to a file for future analysis.

The final results show an F-score of 99.68% for 2D-CNN and 99.42% for LSTM, demonstrating the model's high effectiveness in detecting cheaters based on behavioral data. This approach provides an alternative to traditional anti-cheat systems, which tend to evolve reactively in response to new cheating techniques.

Keywords: Anti-Cheat Systems · Cheating Detection · Behavioral Data · Deep Learning · Player Behavior · Video Games

1 Introduction

In recent years, we have witnessed a significant evolution in the video game industry, and the number of players worldwide continues to increase. This growth in popularity and the rise of multiplayer online games resulted in the necessity

A. Marto et al. (Eds.): VJ 2024, CCIS 2324, pp. 243–258, 2025.
https://doi.org/10.1007/978-3-031-81713-7_17

for better security mechanisms to protect clients' data and detect and block players using fraudulent game manipulation techniques, also known as cheats.

However, many traditional anti-cheat systems use aggressive monitoring techniques with the addition of kernel-level access, which can pose a significant risk to user privacy.

As the gaming industry continues to grow, both in terms of the number of players and its monetary value, the need for more effective anti-cheat solutions becomes a priority. The rise of competitive gaming, which led to professional tournaments offering prizes reaching millions of dollars, has further emphasized this need.

At the same time, the presence of cheaters in video games that succeed in acquiring an unfair advantage over other players is still a common reality. For instance, in League of Legends, the presence of one cheater for every 15 games is estimated, and those using scripts to automate mechanics in the game win 80% of the time [1]. Such scenarios compromise the experience of genuine players, which can lead to frustration or even abandonment of the game.

Anti-cheat systems play a crucial role in countering fraudulent behaviors in video games. However, as cheats become more sophisticated, anti-cheats often respond reactively, creating a "cat and mouse" dynamic [2]. This reactivity has led to the adoption of invasive techniques like real time process monitoring, driver blocking, and access to elevated privileges that can boot the anti-cheat alongside the operating system.

These techniques used by traditional anti-cheats contradict the primary goal of these tools, being the protection of genuine players, which demonstrates the importance of finding new alternatives for cheat detection in video games. A promising approach is integrating artificial intelligence, more specifically, deep learning (DL) time series algorithms that can analyze and classify player behavior by detecting suspicious patterns generated by cheats.

This study offers the following contributions:

- Creation of a dataset composed of biometric behavior data from 13 participants, captured while playing Minecraft both with and without cheats.
- Development of a mouse listener tool to capture player interactions through mouse movements and clicks.
- Creation of cheating tools, aimbot, and automining, which will be executed in the background and can interact with the game.
- Proposal of an anti-cheat system using deep learning models, specifically 2D-CNN and LSTM.
- Testing the anti-cheat system in the game Minecraft as a practical case.
- Validation of the anti-cheat system through metrics results.

The structure of this study is the following:

- **Section** 2: Provide some context on some popular types of cheats vastly used in video games, describe the difference between client-side and server-side anti-cheats, and finally it will explain some anti-cheat systems that are already integrating artificial intelligence into their solutions.

- **Section** 3: Explain five works related to the proposition of this study and provide contributions to the area.
- **Section** 4: Describe all the practical processes followed to validate the anti-cheat system proposed.
- **Section** 5: Evaluation of final results from analyzing metrics like f-measure and ROC curve graphs.
- **Section** 6: Final study conclusions and future work.

2   Background

This chapter will provide some fundamental concepts on the most common types of cheats used by the video game community and the difference between client-side and server-side anti-cheats. Furthermore, it will explain how some anti-cheats have already implemented artificial intelligence solutions into their techniques.

2.1   Types of Cheats

Cheat techniques are included in multiple categories. Some of the most popular in the market are the following [3]:

- **Cheating through Cooperation**: There is cooperation between two or more players in order to acquire an advantage over other players. An example of this type of cheating may be Account Sharing, where an experienced player is hired to grow a low level account.
- **AI-Enhanced Cheating**: These cheats use artificial intelligence models to learn how to do tasks in a game automatically. For example, AI Bots learn to identify objects in the game and do tasks like mining ores, cutting wood, and attacking enemies.
- **Client Infrastructure Modification**: The user modifies their device by manipulating software and hardware. Some Wallhacks cheats work by changing the graphics driver configuration.
- **Distributed Denial-of-Service (DDoS) Cheating**: These cheats manipulate network packets to cancel or delay other players connection. For instance, some First Person Shooter (FPS) games send the player position through packets to the enemy. The cheater can delay this packet, which delays the enemy's response time.
- **Bug or Loophole Exploitation**: These are in game flaws that developers may not be aware of, and cheaters can exploit these mistakes instead of reporting them. For example, the price of a virtual item is extremely low in the game market, and players take advantage of that, damaging the economy.

Cheating through artificial intelligence has been advancing with the integration of humanized techniques. This hides the robotic behavior, making the cheat harder to detect. For example, if the mouse needs to automatically move from point A to point B, a normal cheat would make a straight trajectory between the two points. However, with this new technique, the mouse will make the trajectory with extra curved movements to mimic humane behavior [4]. This evolution can be a challenge for the proposed anti-cheat system, which needs to identify cheating behavior.

In Fig. 1 it is represented the normal straight mouse trajectory generated by an automated tool, along with the humanized result after applying the algorithm.

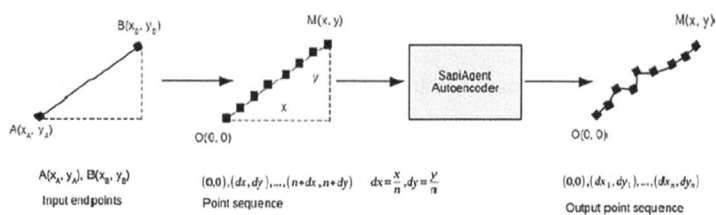

Fig. 1. Humanized mouse movements. Original source [4]

2.2   Types of Anti-cheats

Anti-cheat systems are categorized into client-side and server-side. Client-side systems are installed locally on the user's device and use techniques like [5]:

- **Memory Scanning**: Analyze in game memory was modified through code injections or variable manipulations.
- **File Integrity Checks**: Verifies if game files were modified without authorization by comparing the current state of the files with a secure state where the integrity was not compromised.
- **Code Encryption**: Encrypts game code to avoid reverse engineering and modifications.
- **Kernel-Based Anti-Cheat Drivers**: Installs a driver in the lowest level of the operating system (kernel), achieving the highest privileges, which provides total access to the machine and monitors every background process to detect cheating tools. Figure 2 illustrates the different hierarchy levels in a computer, with the kernel level being the lowest ring, having the highest privileges.

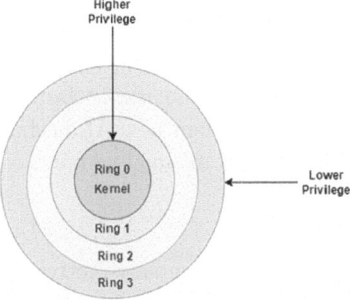

Fig. 2. Protection domains hierarchy

Server-side anti-cheat systems run on the game server, which receives and analyzes game data to detect cheats. Some techniques used by this system may include [5]:

- **Statistical Analysis**: Collection and analysis of game data to identify anomalies. For example, in an FPS game, a player that usually does a norm of 10 kills each game increases that number to 60.
- **Resistant Application Protocol**: Implement secure protocols, UDP and TCP, to guarantee secure communication between the client and the server, ensuring the integrity and confidentiality of data transported through network packets.
- **Obfuscating Network Traffic**: Usage of obfuscation techniques in network traffic in order to prevent packet sniffing attacks and packet manipulation.

The anti-cheat proposed in this study, to detect cheats through player behavior analysis, is a client-side system in which the data capture and analysis tools are running locally on the user's device. The techniques used to detect cheating behavior are exclusively AI techniques, different from the traditional methods used by client-side anti-cheat systems mentioned previously. Figure 3 illustrates the architecture of the proposed anti-cheat system.

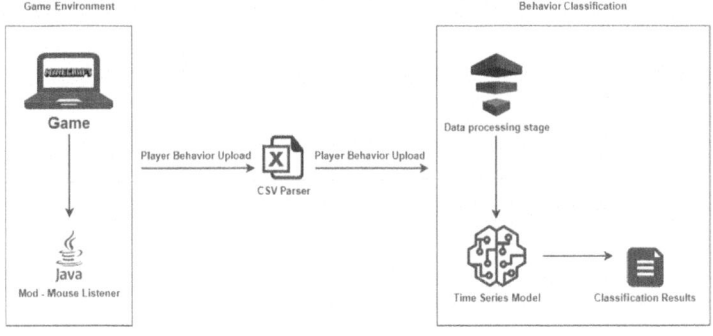

Fig. 3. Proposed anti-cheat system architecture client-side

2.3 AI Anti-cheats

The integration of artificial intelligence in anti-cheat systems has been explored, not only by new anti-cheat systems in the market but also by traditional systems that are starting to study and implement new AI mechanisms into their systems.

This section will explore some promising examples of anti-cheat systems integrating AI into their solutions.

RICOCHET Anti-cheat. By the company Activision, this system has a tool called Replay Investigation Tool, that data during the game, convert them into video to be analyzed by a team. Using an AI model trained to detect cheating behavior, it can analyze 1000 videos per day, prioritize the ones showing suspicious behavior, and send them to an analysis team, optimizing the cheating detection response time [6].

VACnet. A deep learning technology developed for Valve's anti-cheat system, VAC. VACnet was trained with a dataset of results from the technology Overwatch, a trial where user cheating reports are sent in video format to be analyzed by an expert team and classified as cheating behavior or not. The new deep learning model will optimize this system by reporting suspicious behavior to Overwatch instead of only relying on player reporting. It is estimated that from all the reports received by VACnet, 80% to 95% is classified as cheating, compared to 15% to 30% when the report is sent from a player [7].

Anybrain. This new anti-cheat system is a strong example of how deep learning can be successfully used as a primary technique in anti-cheat systems without relying on kernel level access. It uses a Software Development Kit (SDK) to capture game data from mouse and keyboard dynamics and send them to a server for future analysis and classification by the deep learning model, a hybrid client-side and server-side solution. This model can be trained with data from different types of games and multiple cheats, making it an adaptable solution [8].

SARD. This system combines traditional anti-cheat techniques with kernel level access and artificial intelligence models. SARD detects cheating through an SDK responsible for real time monitoring and analysis of player behavior [9].

3 Related Work

Several studies have proposed solutions for integrating machine learning and deep learning models in anti-cheat systems.

The authors of the article [10] address the issue of account sharing, using the game League of Legends as a practical example. They propose a biometric based system that analyzes user behavior through mouse and keystroke dynamics to identify users. Using data from 56 players, they evaluated neural network algorithms, including Multilayer Perceptron (MLP), Bayesian Networks, and Radial Basis Function (RBF). The best results were obtained by the algorithm MLP, achieving an accuracy of 86.27% in user identification.

The study [11] explored a behavior based approach focused on identifying multiple types of aimbots in FPS games, using the game Trojan Battles, created exclusively for the study. They used 18 features from in-game logs, like player position and distance between players, to train machine learning models, Logistic Regression, and Support Vector Machines (SVM). The models were trained with different classification techniques, and the one that achieved the best results was creating one model of each type of cheat.

The authors of the study [12] address the problem of bots in online games through the analysis of in-game logs. Using the game Aion: The Tower of Eternity, it was captured data from 49739 players, of which 7702 were identified as bots. They used time series models, MLP and Correlation based Feature Selection (CFS). The best results were achieved with two layers in the MLP model, with a f-measure of 98%.

The study of the article [13] proposes an AI based cheat detection system by analyzing mouse dynamics, keystroke dynamics, and console logs in the popular game CS:GO. The algorithms evaluated were Decision Tree, SVM, and Naive Bayes. Each model was trained with a type of data each time, and the best results were obtained by the Decision Tree trained with the keystroke dynamics model with a precision of 95.45% in the classification of normal behavior.

The study [14] proposed a deep learning approach using a multivariate time series algorithm to detect cheating behavior through mouse and keystroke dynamics. It captured data from the game CS:GO from genuine players and players using aimbot or triggerbot cheat. The algorithm evaluated was Convolutional Neural Networks (CNN), achieving an accuracy of 99.2% in triggerbot detection and 98.9% in aimbot detection.

4   Methodology

This chapter will explain the methodology used during this study. Section 4.1 will have a brief explanation of the game Minecraft, used to capture player behavior. Then in Sects. 4.2 and 4.3 it will be explained both cheats created for the context of this study, aimbot and automining. After that, in Sect. 4.4, it is described the mouse listener mod created for Minecraft to capture mouse dynamics data. Following up, in Sect. 4.5 is is explained the technics used to prevent both models from overfitting. Then in Sect. 4.6 it is present a sample of the final dataset and a sequence. Finally, in Sects. 4.7 and 4.8, the architecture created for both models is explained.

4.1 Minecraft

Minecraft was chosen as the data recording environment. It is a sandbox game without specific narrative or objectives. It is a singleplayer game that can be played online on community servers.

Since this study is focused on mouse dynamics, it is important to understand how the mouse is used in the game. Mouse movements control the camera, the left click is used to attack, the right click is for placing blocks or using items, and the scroll is to go through items in the inventory. There are multiple events that the player can do in Minecraft, like fighting, mining, fishing, and building. The two events selected for the data recording were fighting and mining, in which the cheats aimbot and automining were tested.

4.2 AI Aimbot

To test cheating behavior during Minecraft combat events, it was developed an aimbot.

An AI Aimbot was developed using YOLOv5, trained with a dataset of Minecraft enemy images in order to detect and target them in real time. The aimbot will be executed in the background and read the game windows. Every time it detects the presence of an enemy, it will move the mouse automatically to the enemy's position and attack it. This will generate mouse movements and clicks without the player's intervention. Figure 4 shows how the aimbot identifies enemies.

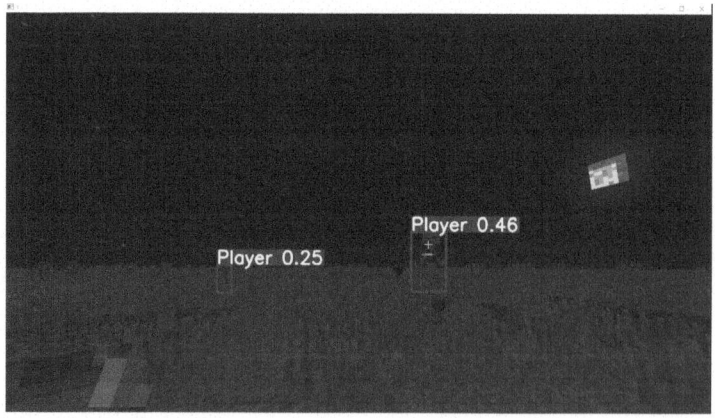

Fig. 4. AI Aimbot cheat in Minecraft

4.3 Automining Script

For the mining event, it was created an automining script using PyAutoGui. The script runs in the background, and without player intervention, it will start mining using a specific pattern. This generates mouse movement, left click, and right click actions. The pattern is represented in Fig. 5.

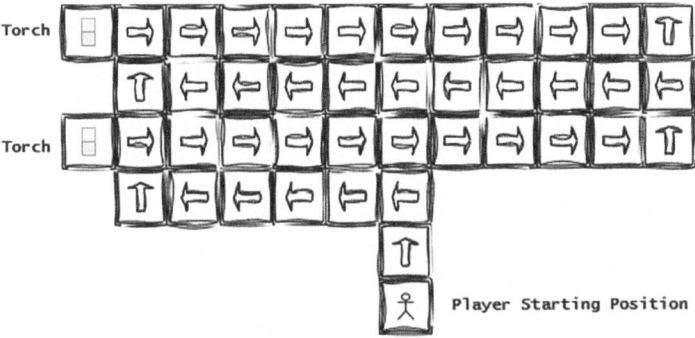

Fig. 5. Automining script cheat in Minecraft

4.4 Mouse Listener Mod

The method used for capturing mouse dynamics data was the creation of a Minecraft mouse listener mod. This mod will add a menu to the game, illustrated in Fig. 6, in which the player can select the type of event that wants to record, being "Normal Fighting" and "Normal Mining" for normal gameplay without cheats enabled, and "Fighting Cheat" and "Mining Cheat" for gameplay with cheats, aimbot or automining, activated.

Fig. 6. Mouse Listener Mod menu in Minecraft

The mouse listener mod captures the following raw metrics: timestamp; cursor x coordinate; cursor y coordinate; movement velocity; movement acceleration; cursor distance; button pressed; action (example: right click); action id game (Fighting or Mining); click press (0 or 1); click release (0 or 1); user; is cheating (0 - normal; 1 - aimbot; 2 - automining).

4.5   Models Optimization

Throughout the development of the proposed system, some overfitting challenges were faced in both 2D-CNN and LSTM models. To solve this issue, some optimization techniques were necessary.

Dropout. It temporarily excludes neurons and the corresponding input and output values in each train iteration. Dropout uses an exclusion rate value for each neuron, which is the probability of this being excluded temporarily, reducing the neural network size [15].

L2 Regularization. Introduce a penalty to the model's loss function, Cross Entropy, to reduce weights, pushing them closer to zero and creating a balance between them. This ensures that no single feature becomes overly dominant [16].

$$\text{Loss}_{\text{total}} = \text{Loss}_{\text{orig}} + \alpha \sum_{i=1}^{n} w_i^2$$

Optuna. It is a framework that tests multiple hyperparameter values for each model and selects those that achieve better results [17].

Table 1 represents the hyperparameters chosen by the Optuna framework for the 2D-CNN and LSTM models.

Table 1. Final hyperparameter for 2D-CNN and LSTM models.

2D-CNN	Hyperparameter	Value
	Learning Rate	0.0002
	Batch Size	64
	Dropout Rate	0.69

LSTM	Hyperparameter	Value
	Learning Rate	0.00045
	L2 Regularization	0.0029
	Hidden Size	53
	Number of Layers	1

Feature Engineering. The raw metrics were used to create new features that provide more information value to the models during training. The new features created were the following: click duration; time difference between actions; session; hour; time between action and session start time; direction of mouse movement; left click frequency; and right click frequency.

4.6   Dataset

After collecting behavior data from 13 players, each playing for 40 min and performing feature engineering, the dataset consisted of 22 columns and 440.545 rows. The models receive sets of 20 consecutive timestamps during training, known as sequences, to find cheating patterns. Figure 7 represents a sequence in the dataset.

Fig. 7. Final features in the dataset

4.7   2D-CNN Model

In order to detect and classify behavior patterns, time series deep learning algorithms can exceed this task. A 2D-CNN model was developed for the proposed anti-cheat system. The model receives the sequence of 20 timestamps and reads them as a grayscale image, where each feature is read as a pixel.

The model needs four inputs to work:

- **Batch Size**: Number of sequences processed in one input.
- **Number of Channels**: Number of input channels, which is 1 for grayscale images.
- **Image Height**: Number of timestamps defines for a time window, being 20 timestamps.
- **Image Width**: Number of features used through training.

The architecture created for the 2D-CNN model is represented in Fig. 8.

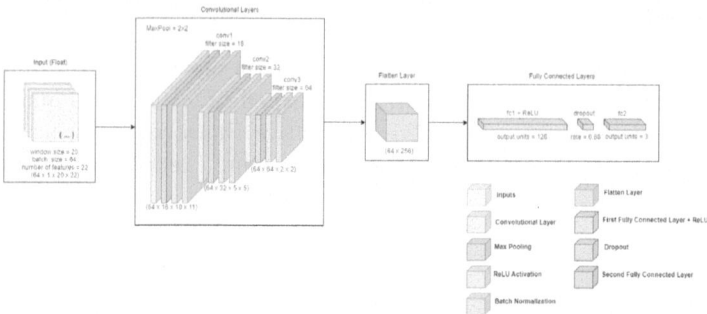

Fig. 8. 2D-CNN architecture

4.8   LSTM Model

Another time series deep learning algorithm chosen for the anti-cheat system is the LSTM network. LSTMs are equipped with internal memory to identify long term dependencies in data sequences. This capability is especially useful for the anti-cheat system, as the model can retain information from previous inputs and associate it with new data, helping to detect patterns and dependencies over time.

To develop this model, it is necessary the input of three dimensions:

– **Batch Size**: Number of sequences processed in one input.
– **Number of Features**: Quantity of features used to train the model.
– **Sequence Length**: Number of timestamps in one sequence of data, which is 20 timestamps.

The architecture developed for the LSTM model is represented in Fig. 9.

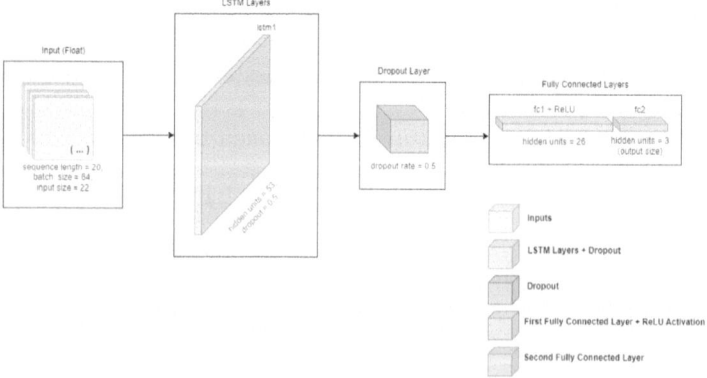

Fig. 9. LSTM architecture

5   Results Analysis

Both models demonstrated excellent results. The primary evaluation metric was the F-score, which provides an average of precision and recall (true positive rate). The 2D-CNN model reached an F-score of 99.68%, and LSTM reached 99.43%. Even though 2D-CNN showed slightly better results, both demonstrated their efficiency in learning how to recognize behavior patterns, making them excellent choices for anti-cheat systems solutions.

In Table 2, it is shown the results of each metric calculated during the evaluation process for both models.

Table 2. General metrics results for 2D-CNN and LSTM models.

2D-CNN	Metric	Value	LSTM	Metric	Value
	Accuracy	0.9969		Accuracy	0.9946
	Precision	0.9960		Precision	0.9965
	Recall	0.9975		Recall	0.9922
	F-Score	0.9968		F-Score	0.9943

Another metric studied was the Receiver Operating Characteristic (ROC) curves of each model. The ROC curve illustrates the relationship between the true positive rate and the false positive rate. A model with a curve closer to the top left corner is considered superior, as it indicates a low false positive rate and a high true positive rate.

From Fig. 10, it is possible to conclude once again that the 2D-CNN model achieved better classification results.

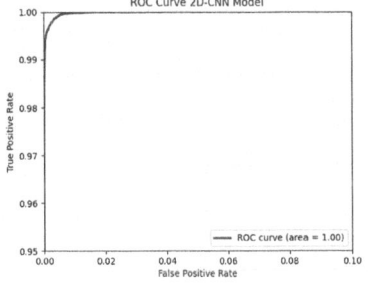 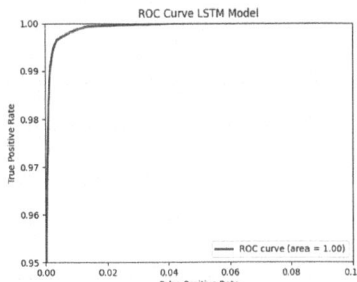

Fig. 10. ROC curves for 2D-CNN and LSTM models

6   Conclusion

Traditional anti-cheat systems have been adopting kernel level solutions to detect cheaters in video games. These methods are intrusive and can compromise a user's security if a vulnerability is found since the anti-cheat driver shares the same virtual memory location as the operating system.

This study addresses a different approach to anti-cheat systems by integrating deep learning models, 2D-CNN and LSTM, to analyze player behavior through mouse dynamics metrics and identify patterns created or modified by cheats. It is categorized as a client-side system and doesn't need kernel access to execute its functionalities.

To test the proposed system it was created two cheats, an aimbot, and an automining script, to run in the background. It was also developed a Minecraft mod, which was a mouse listener that recorded mouse dynamics such as mouse velocity, coordinates, and acceleration and then saved them into a file to be processed and used to create the final dataset. The models were evaluated with the final dataset and achieved excellent F-score results of 99.68% for the 2D-CNN model and 99.43% for the LSTM model. These results validate the power of deep learning models against the continuous challenge of cheaters in video games.

Despite the good results, this approach has certain limitations. The models were specifically trained to recognize behaviors within the game Minecraft and are currently limited to detecting cheats such as aimbots and automining scripts.

Using behavior data, such as mouse dynamics, instead of techniques like memory and process scanning or driver shutdowns at the kernel level, represents a more transparent, ethical, and less intrusive approach, potentially leading to a positive shift in the anti-cheat system evolution.

Future work for this study can include the evaluation of a hybrid model that leverages the 2D-CNN's ability to automatically select the most relevant features for training, combined with the LSTM's capability to detect dependencies over time through internal memory which is essential to detect variations in behavioral patterns. The development of a different mouse listener tool is also important since the mod works exclusively with Minecraft. For this, it would be necessary to create a tool that could be installed in the user's device and executed in the background to listen and capture mouse dynamics in the game when the anti-cheat is activated. It is also essential to expand the training dataset with a larger and more diverse group of players and a wider range of recorded cheats. Given that the proposed system is client-side, it is necessary to create a performance loss test to verify the necessity of creating a server-side integration to analyze the data.

Acknowledgments. This work was supported by national funds through the Portuguese Foundation for Science and Technology (FCT), I.P., under the project UIDB /04524/2020.

References

1. Stubbs, M.: One In 15 League Of Legends' Matches Had A Cheater In It. https://www.forbes.com/sites/mikestubbs/2024/04/11/one-in-15-league-of-legends-matches-had-a-cheater-in-it/. Accessed 25 Sept 2024
2. Durst, D., Taylor, C.: Reversing Anti-Cheat's Detection-Generation Cycle With Configurable Hallucinations. https://www.activision.com/cdn/research/hallucinations. Accessed 25 Sept 2024
3. Yan, J., Randell, B.: A systematic classification of cheating in online games. https://www.researchgate.net/publication/221391501_A_systematic_classification_of_cheating_in_online_games. Accessed 26 Sept 2024
4. Antal, M., Buza, K., Fejer, N.: SapiAgent: a bot based on deep learning to generate human-like mouse trajectories. https://ieeexplore.ieee.org/document/9530664. Accessed 26 Sept 2024
5. Lehtonen, S.: Comparative study of anti-cheat methods in video games. https://helda.helsinki.fi/items/b1141406-eb65-48a5-8922-d1b23d4cfe51. Accessed 26 Sept 2024
6. Blizzard Entertainment: RICOCHET: anti-cheat progress report - launch readiness, machine learning and new features. https://news.blizzard.com/en-us/blizzard/24030414/ricochet-anti-cheat-progress-report-launch-readiness-machine-learning-and-new-features. Accessed 26 Sept 2024
7. GDC: Robocalypse now: using deep learning to combat cheating in counter-strike: global offensive. https://www.youtube.com/watch?v=kTiP0zKF9bc. Accessed 26 Sept 2024
8. Anybrain: Protect your game with AI anti-cheat! https://www.anybrain.gg/product. Accessed 26 Sept 2024
9. SARD: Sard Anti-Cheat. https://www.sard.ac/sard-anti-cheat. Accessed 26 Sept 2024
10. Silva, V., Costa-Abreu, M.: An empirical biometric-based study for user identification with different neural networks in the online game League of Legends. https://ieeexplore.ieee.org/document/8489164. Accessed 26 Sept 2024
11. Alayed, H., Frangoudes, F., Neuman, C.: Behavioral-based cheating detection in online first person shooters using machine learning techniques. https://ieeexplore.ieee.org/document/6633617. Accessed 26 Sept 2024
12. Bernardi, M., Cimitile, M., Martinelli, F., Mercaldo, F.: A time series classification approach to game bot detection. https://www.researchgate.net/publication/318717059_A_time_series_classification_approach_to_game_bot_detection Accessed 26 Sept 2024
13. Spijkerman, R., Ehlers, E.: Cheat detection in a multiplayer first-person shooter using artificial intelligence tools. https://dl.acm.org/doi/10.1145/3440840.3440857. Accessed 26 Sept 2024
14. Pinto, J., Pimenta, A., Novais, P.: Deep learning and multivariate time series for cheat detection in video games. https://link.springer.com/article/10.1007/s10994-021-06055-x. Accessed 27 Sept 2024

15. Srivastava, N., Hinton, G., Krizhevsky, A., Sutskever, I., Salakhutdinov, R.: Dropout: a simple way to prevent neural networks from overfitting. https://www.cs.toronto.edu/~rsalakhu/papers/srivastava14a.pdf. Accessed 27 Sept 2024

16. Google for Developers: Overfitting: L2 regularization. https://developers.google.com/machine-learning/crash-course/overfitting/regularization. Accessed 27 Sept 2024

17. Akiba, T., Sano, S., Yanase, T., Ohta, T., Koyama, M.: Optuna: a next-generation hyperparameter optimization framework. https://arxiv.org/pdf/1907.10902.pdf. Accessed 27 Sept 2024

Enhancing Player Experience Through Generative Artificial Intelligence: Custom Interaction in Game Design

Cláudia Vale Oliveira[1] (ID) and Pedro Neves Rito[2](✉) (ID)

[1] DigiMedia, University of Aveiro, Aveiro, Portugal
claudia.mvoliveira@ua.pt
[2] Higher School of Education of Viseu, Ci&DEI, Polytechnic Institute of Viseu, Viseu, Portugal
rito@esev.ipv.pt

Abstract. The rapid advancement of Generative Artificial Intelligence (GenAI) has significantly impacted game design, offering novel ways to enhance player experiences. GenAI is reshaping game design by introducing new opportunities for human-AI collaboration. This paper presents a review on how GenAI tools can be leveraged to create more meaningful, highly dynamic and personalized interactions within games. We begin by exploring the evolution of GenAI and its current applications in the gaming industry, highlighting its role in facilitating creative and immersive environments.

A main area of investigation is the use of GenAI to enhance player experience, through procedural content generation, and adaptive gameplay mechanics. We examine how GenAI enables players to co-create within game worlds, enhancing their engagement. We delve into the customization of player experiences, analyzing how GenAI-driven systems can adapt to individual preferences and emotional responses to provide a tailored gaming experience.

The paper also addresses the challenges and ethical concerns inherent in deploying GenAI in games, such as ensuring privacy, avoiding emotional manipulation, and providing appropriate content. By synthesizing findings from recent literature, this document offers insights about the integration of GenAI in game design to create enriched and personalized gaming experiences.

Keywords: Game design · Generative Artificial Intelligence · Human-AI collaboration · Player Experience

1 Introduction

Immersive technologies used the potential of virtual reality and augmented reality up until recently, while nowadays games also exploit the potential of artificial intelligence (AI). Machine learning (ML) and AI have transformed society around technology, influencing not only industries but also everyday life [14]. The video games industry has been significantly affected, transforming not only the player experience but also game development. Generative AI (GenAI) has emerged as a transformative tool in game design, offering unprecedented opportunities to create immersive and personalized experiences [19]. The integration of GenAI into game development has recently enabled the creation of more dynamic game environments and promoted deeper interactions between the player and the game world. AI has transformed the player experience, not only in the way they consume content, but also in their influence on virtual worlds through their preferences and actions [21].

This paper aims to provide an analysis of the current context and landscape of GenAI in game design and explore how these tools can be used to enhance immersion as well as develop playability and meaningful interactions with players to enrich the overall gaming experience. A significant application of GenAI lies in its potential to stimulate creativity in the game development process [14, 19]. GenAI promotes the development of video games that adapt to the unique needs of players, providing opportunities to create personalized content [19, 24] and interactive narratives [16, 19] by allowing players to become co-creators in game environments. GenAI fosters the creation of an environment where they can express themselves creatively, get involved in problem-solving and develop new skills through play. However, while GenAI offers interesting perspectives for increasing immersion in games, it also presents challenges and raises ethical questions such as ensuring privacy and protection against emotional manipulation, crucial elements for the integration of GenAI [3, 19].

To understand how AI interferes in the player's experience through personalized interactions and how the integration of these tools is approached by game design, this paper is composed by five sections, this being the first. The second section frames GenAI in the context of video game development and design.

The third section highlights how GenAI allows players to become co-creators, directly influencing the narrative, the development of characters and how emotional involvement changes the dynamic between the system and the player, leading to a significant increase in player engagement and immersion.

The fourth section addresses the challenges and ethical issues related to data privacy, emotional manipulation and bias, highlighting the responsibility of developers to ensure that AI-based systems enhance experiences without exploiting vulnerabilities or compromising the safety and well-being of players.

The last section presents the final remarks of this paper, framing game design, GenAI intelligence and the player's experience within the scope of the initial research being conducted which connects these subjects to the co-creativity process between these systems and children.

2   Generative AI and Game Design

Videogames history can be told from two perspectives: by focusing on their cultural dimension or on technological advances [17]. Meanwhile, the story that is currently being written has two main strands: on the one hand, the evolution of development tools and game engines, and on the other hand, the introduction of GenAI which is changing the way people interact, create and consume multimedia content [5, 17]. Combining these two factors *"the decade might just witness a new era of developing games with generative AI(GAI)-Human collaboration"* [17] revolutionizing the understanding of gaming. AI has become a fundamental tool for game development [5], with the potential to substantially increase efficiency and creativity in the videogame design process [15]. For example, GenAI can automatically generate several views or compositions for a character which is extremely useful for creators.

Developing a video game is a process that requires a development plan and a team of specialists with skills in areas such as programming, design, copyright and other technical and creative fields [7]. However, GenAI is influencing all game development stages, changing the role of each team member and even the player, who is becoming a co-author [11, 22].

Until recently, it was necessary to use third-party tools to create and edit all multimedia content. However, new tools are being developed and integrated into existing software, making it possible to manipulate AI-generated content [17]. Integrating these technologies not only optimizes time and resources, reducing videogames development costs [5, 17], but also makes it possible to include the player in the creative process. Although GenAI advanced capabilities are currently under development, integration's success depends on the technological literacy of those who wish to exploit them, users will need *"to develop a new set of skills to work with (not against) generative variability by learning how to create specifcations that result in artifacts that match their desired intent"* [25].

The need to meet users' preferences has driven research into automated models and algorithms to customize digital content and applications [28]. Generative AI techniques, such as ChatGPT and Stable Diffusion, can help automate some parts of the development, such as narrative and graphics creation [26]. Combining procedural generation and AI enables the creation of personalized content which is essential to unfold the narrative [7] making it possible to create more interesting game levels which enhance the player's experience, making AI to play a crucial role *"in addressing balance issues in game development, particularly by dynamically adjusting game difficulty to accommodate players at different skill levels"* [27]. Features such as the adjustment of difficulty levels, according to the player's profile and journey contribute to maintain the ideal ratio between challenge and competence, ensuring that the player feels interested and challenged, without feeling defeated or overwhelmed by the experience. This balance was traditionally managed by game designers, but now it can be controlled by AI [9]. However, finding a balance remains a challenge, as AI must maintain mystery and unpredictability among the options available [5].

Furthermore, real-time adaptation provides a unique experience [5, 21], allowing players to explore the game repeatedly without receiving the same feedback, this means that *"players can continue to explore and enjoy the game long after its initial release, leading to increased player satisfaction and revenue for the game developer"* [5]. As GenAI continues to evolve, its capacity to generate dynamic content means that the game experience can be endlessly personalized, ensuring that players remain engaged and stimulated.

3   GenAI Custom Interaction and Player Experiences

The customization process has been used before, for example in the study carried out by [13], they verified that when participants recorded their own movements, they were more engaged in the process and felt a stronger personal connection to the outcome. This is an important result, as it highlights the potential for customization through action in video games, where engagement and personal connection are key factors. This finding supports the use of players' own motions to customize both animations and responses.

The *"integration of AI into the gaming experience is revolutionizing player interaction and immersion in multifaceted ways"* [27]. AI was first implemented in games to follow a set of predefined rules, performing functions such as being the player's rival in board games - rule-based AI. It then learned to adapt according to the player's behavioral patterns, creating more challenging and unpredictable experiences - learning-based AI. With the introduction of ML, AI has evolved into what can be considered a simulation of behavior of living organisms - Evolutionary AI - going beyond adaptation based on static learning, absorbing the ability organisms have, to evolve and improve over time. The use of AI *"has become much more important in game development, which allows developers to create intelligent and adaptable opponents or bots, enhancing gameplay and offering unique experiences to average players"* [5]. Investing in systems capable of learning and adapting in real time transforms the gameplay, which becomes more immersive and engaging [5]. Virtual worlds and experiences are no longer constrained but rather responsive to players' actions in real time, unlocking their potential for personalization and creativity [14].

Transforming the main character's appearance to resemble the player, by capturing their voice and the look of their face and body, or changing the game's environments, missions and objects are examples of how customization can be used to enhance engagement, as it becomes difficult to distinguish what is real from what is virtual [27]. Through GenAI it is also possible to create non-player characters (NPC) and characters with different personalities and backgrounds [23] capable of adjusting to player actions [12], such as facial expressions, intonation and physiological signals, generating countless possibilities compared to the predefined behaviors of traditional human-computer interaction mechanics [8].

However, integrating these tools into applications in areas such as health, education and customer service, for example, requires these assistants to acquire an understanding and emotional reaction to adapt their response according to the situation they are being exposed to [8]. With the continuous progress of game technology, *"emotion simulation has become an important means to enhance the gaming experience"* [12]. The introduction of AI technologies, along with affective computing [8], has allowed characters such as NPC to simulate emotions which contribute to enriching the player's emotional experience [12]. Implementing these algorithms, which can analyze the player's emotional state, makes it possible to adjust the NPC response, which creates an empathetic interaction between human and computer [14, 27], in which "*emotional intelligence encompasses the capability to perceive, interpret, and respond to human emotions, which is pivotal in fostering more natural and meaningful interaction"* [8]. Emotional intelligence significantly improves the naturalness and fluidity of communication between robots and people.

The experience is lifted to another level through ML algorithms combined with natural language processing (NLP) [21]. Natural interaction with the player is possible through spontaneous dialogues that are not predefined in a script, through a language that simulates human speech [23] removing the impression of interacting with a robot [12]. This language automatically builds trust and intimacy by providing real and tailored responses that consider context, actions and the player's emotional state [12] allowing them to collaboratively overcome missions and experience dynamic and unique narratives, which provides a more immersive experience adapted to the player's choices [27]. This engaging, reactive and behavior-based interaction with NPC is as fundamental to the player's immersive experience as it is to the quality of video games [12].

Thus, bringing emotion simulation into the game process, which is possible with AI technologies, increases not only immersion and interactivity, but it also influences players' decisions [8, 12]. These decisions depend not only on the game goals and mechanics but are also influenced by the complexity of the emotional connection players feel for the characters and their awareness of the consequences their actions might have. Emotional simulation not only strengthens the bond between characters and player, but also draws them into the game world's immensity, forgetting reality and increasing engagement, focus and empathy for the characters, enriching the plot, the story and the players' satisfaction [12].

Integrating GenAI into game development *"presents challenge in balancing immersive elements with meaningful gameplay"*, where *"even a slight misalignment of the content from the game world can cause disruption of immersion for the player"* [17]. For this reason, game design is fundamental to the cohesion between the elements, the world, the experience and the player's emotions. Finding this balance represents a challenge for game designers [19] who, while developing the game, must choose and implement generative tools and AI assistants capable of evaluating and adapting to the player's expectations. During this process, it's essential to understand the player's intentionality, generate behavioral patterns and analyze a substantial amount of data in a collaborative process with AI, a partnership which involves ethical questions.

4   Challenges and Ethical Considerations

Customize video games requires answering three questions: what are the technical needs required to adapt the content within the game, what new design principles are needed and how to create personalized content based on players' needs [28]. One example of customization can be adapting the game level according to the player's performance and goals. For this purpose, AI must understand the player's intention. However, intentionality is not a variable that can be easily observed and quantified, it is subjective. More development needs to be done to ensure that AI can understand creativity and subjectivity. To minimize misunderstandings, the structure and prompting techniques should differ depending on the expected outcome, e.g., whether the prompt is destined to solve a technical or a creative problem [17].

Even so, there are several models developed to identify player goals that detect intent with reasonable results [28]. One of the ways to accurately determine intent can be to classify the player profile. However, assuming that all the necessary data is available, observing and analyzing a large amount of data is a challenging process as it involves categorizing all the information to distinguish behavioral habits. Even so, this distinction does not ensure that it is possible to predict the player's behavior or that this behavior remains within the expected patterns. Despite the ability to compare a player's behavioral pattern with the behavioral patterns from other players with shared profile features to generate predictions, called collaborative filtering [28], these predictions will be based on mass data and not on player individuality. On the other hand, AI may grant privileges that benefit one player over another in multiplayer mode, for example [5]. Therefore, apart from the experience, empathy towards characters and the immersive effect that being able to choose may generate, the player may opt for rational thinking that brings them closer to the ultimate goal in a game - winning - which doesn't necessarily contribute to the experience or correspond to the predicted pattern [28].

Personal data management and classification can be influenced by biases and stereotypes such as race or gender. Furthermore, recognizing and interpreting human emotions is a process influenced not only by intentionality but also by contextual and cultural circumstances [8]. Developers must ensure that *"their AI systems are designed and implemented with fairness and inclusivity in mind"* preventing algorithms from having biased tendencies [3, 5, 19]. Despite the concerns about the risk of manipulation [3, 8], players consider their interaction with AI to be less complex and with a lower chance of manipulation than interactions with other human players [27]. From a social point of view, this observation demonstrates that giving AI emotional intelligence affects not only the gaming experience, but also social dynamics [14]. Furthermore, if the emotional connection increases when the experience is unpredictable and adapted to the player's behavior, it might lead to potentially addictive experiences, a worrying factor [5]. The player's physical and mental well-being must be the priority for game designers when developing these algorithms, to avoid creating addictive experiences. Addiction not only to the experience, but also to the excessive use of AI.

Although the player experience benefits from emotional and customized interaction, there are structural, material and logistical needs required by the system to support the models' responses. The GTP-3 training, for example, involved approximately 45 terabytes of text, which represents a substantial investment [1]. These models are sustained by the management of a significant amount of data which influences implementation and maintenance costs of the required systems to achieve responses that fulfill player's expectations [1, 8, 12]. This challenge begins with the process of acquiring and submitting emotional data to train AI and continues with the need to optimize the algorithm [12].

Another concern since AI began to be introduced into society is that assistants could replace countless professionals, especially in creative fields. In fact, 70% of video game illustrators in China have seen their work replaced by AI-generated images [1]. These technologies have a significant influence on employees' statutes, as their tasks are being replaced by these tools except, for the time being, in their ability to interpret data and make decisions [14]. AI assistance, using the Large Language Model ChatGPT, can influence the quality and creativity of game design, compared to games designed exclusively by humans, the investigation carried out by [4], three game prototypes of different genres were developed: a minimalist game, one with human-designed features and another with features directly implemented from ChatGPT results. It was noted that although ChatGPT has managed to produce structurally solid and satisfactory games, it does not have the capacity to replace the creative goals provided by a human designer, nor their understanding of the context and what "feels" like a good experience for the player. Nevertheless, the games generated by AI were attractive enough for players to want to explore them fully. In addition, human designers recognized their limitations while designing games and gained a new perspective by exploring ChatGPT's suggestions. This suggests that AI can play an effective role in the creative process of games design without hindering the work of human designers.

Meanwhile, AI-generated content has reached levels of quality that cannot be distinguished from content created by humans. Considering that the content used by the neural network to learn and achieve these indistinguishable results is based on a database of content generated by humans [17], it can be said that the content generated by AI ends up being an appropriation and recreation of human work, raising questions about copyright [3, 25] and privacy and data protection [3, 5, 19, 25]. In fact, gamers express concerns about the use of personal information, especially biometric data, to customize their characters or gaming experience [27].

Game designers and developers must take all these factors into account when choosing and implementing tools, in order to overcome all the challenges involved in game design while creating an immersive experience that transcends creative boundaries without prejudice to the players, since *"the possibilities are endless, and game developers must continue to explore and develop new ways to use AI to create next-generation gaming experiences"* [5].

5   Final Remarks

The videogame industry has grown over at a constant pace the years and is expected to continue to evolve in the future [7]. GenAI *"holds significant promise across various business functions, offering a range of benefits that can streamline processes, enhance creativity, and automate tasks"* [1]. Through its ability to generate content dynamically, GenAI offers developers new ways to create interactive, personalized and adaptable game worlds. GenAI integration can be leveraged not only to engage players, but also to turn them into co-creators, empowering richer and meaningful gaming experiences. By allowing players to become co-creators in game environments, GenAI fosters an environment where they can express themselves creatively, engage in problem-solving and develop new skills through play.

These tools are shaping the narrative and creative process with the support of ML assistants that generate multimedia content [6, 10, 20] through a process of co-authorship and co-creativity between the user and the AI [22] while providing new forms of creative expression and media production [20]. AI becomes the game designer's assistant as a creator of personalized content through a dynamic process between the designer, the system and the player's preferences and actions. It operates within the player's interaction with the game elements, especially through the emotional and empathic bond towards the game characters, enhancing and promoting the experience over the task.

Although there is little review about these models for creative areas, as a way of increasing human creativity, justified by the subjectivity found in evaluating the performance of the models in these areas [17], co-creativity between humans and AI is prominent research topic [18]. Considering AI's impact on industry sectors, which raises concerns about the replacement of workers by computers, especially in mechanical and repetitive roles, it is essential to stimulate children's creative thinking and creativity [2]. While allowing children to make meaningful choices which affect the storyline and visual stimulus, these systems not only provide new forms of creative expression and media production [20] but also increase the engagement of children who get involved in this collaborative experience, empowering the gameplay *"to be designed to be collaborative as collaboration is known to strongly influence and motivate creative thinking"* [2].

This review of the current state of the art highlights how GenAI enables dynamic content generation and custom player interactions as part of the research being conducted for a doctoral thesis. While GenAI has potential to enhance player experience and foster creativity, further research is needed to fully explore its applications. Yet the next critical step involves empirical studies. Future research should focus on collecting, analyzing, and testing AI-driven projects, particularly to understand the processes of human-AI co-creativity and co-authorship. While validating these models, we hope to identify the best practices and insights that will help developers create more engaging, creative, and ethical game experiences, and with a special interest, in the context of children's game design.

Acknowledgments. The author (1) would like to thank the FCT - Foundation for Science and Technology for funding this project under the Doctoral Scholarship with reference 2024.00353.BD and the author (2) is funded by National Funds through the FCT - Foundation for Science and Technology, I.P., within the scope of the project Ref^a UIDB/05507/2020 and DOI identifier https://doi.org/https://doi.org/10.54499/UIDB/05507/2020. Furthermore, I (2) would like to thank the Centre for Studies in Education and Innovation (Ci&DEI) and the Polytechnic of Viseu for their support.

References

1. Akhtar, Z.B.: Unveiling the evolution of generative AI (GAI): a comprehensive and investigative analysis toward LLM models (2021–2024) and beyond. J. Elect. Sys. Info. Technol. **11**(1), 22 (2024). https://doi.org/10.1186/s43067-024-00145-1
2. Ali, S., et al.: A social robot's influence on children's figural creativity during gameplay. Int. J. Child-Comp. Interact. **28**, 100234 (2021). https://doi.org/10.1016/j.ijcci.2020.100234
3. Al-kfairy, M., et al.: Ethical challenges and solutions of generative AI: an interdisciplinary perspective. Informatics **11**(3), 58 (2024). https://doi.org/10.3390/INFORMATICS11030058
4. Anjum, A., et al.: The ink splotch effect: a case study on ChatGPT as a co-creative game designer. ACM International Conference Proceeding Series. (2024). https://doi.org/10.1145/3649921.3650010
5. Antony, S., et al.: Artificial intelligence involvement in graphic game development. In: 2023 Second International Conference on Augmented Intelligence and Sustainable Systems (ICAISS), pp. 82–86. IEEE, Trichy, India (2023). https://doi.org/10.1109/ICAISS58487.2023.10250553

6. Antony, V.N., Huang, C.-M.: ID.8: co-creating visual stories with generative AI. ACM Trans. Interact. Intel. Sys. **14**(3), 1–29 (2023). https://doi.org/10.1145/3672277

7. Auxtero, A.L.S.: Game environment design creator using artificial intelligence procedural generation (2023). http://hdl.handle.net/10362/164247

8. Bengani, V.: AI-Driven Emotional Intelligence for Enhanced Human-Robot Interaction (2024). https://doi.org/10.13140/RG.2.2.27511.43689/1

9. Guo, Z., et al.: Rethinking dynamic difficulty adjustment for video game design. Entertainment Computing **50**, 100663 (2024). https://doi.org/10.1016/J.ENTCOM.2024.100663

10. Han, A., Cai, Z.: Design implications of generative AI systems for visual storytelling for young learners. In: Proceedings of the 22nd Annual ACM Interaction Design and Children Conference, pp. 470–474 Association for Computing Machinery, New York, NY, USA (2023). https://doi.org/10.1145/3585088.3593867

11. Jackson, V., et al.: Creativity, Generative AI, and Software Development: A Research Agenda (2024). https://doi.org/10.48550/arXiv.2406.01966

12. Jiang, Z.: Emotional simulation in game AI and its impact on player experience. Int. J. Edu. Humanit. **13**, 11–13 (2024). https://doi.org/10.54097/pbhsqy34

13. Kleinsmith, A., Gillies, M.: Customizing by doing for responsive video game characters. Int. J. Hum Comput Stud. **71**(7–8), 775–784 (2013). https://doi.org/10.1016/J.IJHCS.2013.03.005

14. Kumar, K., et al.: Game-changing intelligence: Unveiling the societal impact of artificial intelligence in game software. Entertainment Computing **52**, 100862 (2025). https://doi.org/10.1016/J.ENTCOM.2024.100862

15. Lee, J., et al.: Empowering game designers with generative AI. IADIS Int. J Comp. Sci. Info. Sys. **18**(2) (2023). https://doi.org/10.33965/IJCSIS_2023180213

16. de Lima, E.S., et al.: Managing the plot structure of character-based interactive narratives in games. Entertainment Computing **47**, 100590 (2023). https://doi.org/10.1016/J.ENTCOM.2023.100590

17. Linkinen, T.: Generative artificial intelligences: challenges and benefits for game development (2024). https://trepo.tuni.fi/handle/10024/155270

18. Margarido, S., et al.: Boosting Mixed-Initiative Co-Creativity in Game Design: A Tutorial (2024). https://doi.org/10.48550/arXiv.2401.05999

19. Moon, J., et al.: Generative artificial intelligence in educational game design: nuanced challenges, design implications, and future research. Technology, Knowledge and Learning, 1–13 (2024). https://doi.org/10.1007/S10758-024-09756-Z/METRICS

20. Onyejelem, T.E., Aondover, E.M.: Digital generative multimedia tool theory (DGMTT): a theoretical postulation. David Publishing Company **14**(3), 189–204 (2024). https://doi.org/10.17265/2160-6579/2024.03.004

21. Rapaka, A., et al.: Revolutionizing learning – A journey into educational games with immersive and AI technologies. Entertainment Computing **52**, 100809 (2025). https://doi.org/10.1016/J.ENTCOM.2024.100809

22. Serbanescu, A.: Human-AI system co-creativity to build interactive digital narratives. In: Zanella, F., et al. (eds.) Multidisciplinary Aspects of Design, pp. 388–398. Springer Nature Switzerland, Cham (2024). https://doi.org/10.1007/978-3-031-49811-4_37

23. Song, Y., et al.: Developing an immersive game-based learning platform with generative artificial intelligence and virtual reality technologies – "LearningverseVR." Computers & Education: X Reality **4**, 100069 (2024). https://doi.org/10.1016/j.cexr.2024.100069

24. Vijaya, J., et al.: AI-based flappy bird game with dynamic level generation. In: 2024 IEEE International Conference on Interdisciplinary Approaches in Technology and Management for Social Innovation, IATMSI 2024 (2024). https://doi.org/10.1109/IATMSI60426.2024.10503355

25. Weisz, J.D., et al.: Design principles for generative AI applications. In: Proceedings of the CHI Conference on Human Factors in Computing Systems, pp. 1–22. Association for Computing Machinery, New York, NY, USA (2024). https://doi.org/10.1145/3613904.3642466

26. Weisz, J.D., et al.: Toward general design principles for generative AI applications. CEUR Workshop Proceedings **3359**, 130–144 (2023). https://doi.org/10.48550/arXiv.2301.05578

27. Zhai, Q.: AI-driven design and enhanced immersion in open-world games. Applied and Computational Engineering **64**(1), 208–216 (2024). https://doi.org/10.54254/2755-2721/64/20241435

28. Zhu, J., Ontañón, S.: Player-Centered AI for Automatic Game Personalization: Open Problems (2020). https://doi.org/10.1145/3402942.3402951

Author Index

A. Marto et al. (Eds.): VJ 2024, CCIS 2324, pp. 269–270, 2025.
https://doi.org/10.1007/978-3-031-81713-7

The manufacturer's authorised representative in the EU is Springer
Nature Customer Service Centre GmbH, Europaplatz 3, 69115 Heidelberg,
Germany. If you have any concerns regarding our products, please
contact ProductSafety@springernature.com

Printed and bound by CPI Group (UK) Ltd, Croydon, CR0 4YY

29/04/2026

02099544-0003

www.ingramcontent.com/pod-product-compliance
Ingram Content Group UK Ltd.
Pitfield, Milton Keynes, MK11 3LW, UK
UKHW040620240426
470322UK00010B/226

Quantum Groups

Algebra has moved well beyond the topics discussed in standard undergraduate texts on "modern algebra." Those books typically dealt with algebraic structures such as groups, rings and fields: still very important concepts! However, *Quantum Groups: A Path to Current Algebra* is written for the reader at ease with at least one such structure and keen to learn the latest algebraic concepts and techniques.

A key to understanding these new developments is categorical duality. A quantum group is a vector space with structure. Part of the structure is standard: a multiplication making it an "algebra." Another part is not in those standard books at all: a comultiplication, which is dual to multiplication in the precise sense of category theory, making it a "coalgebra." While coalgebras, bialgebras and Hopf algebras have been around for half a century, the term "quantum group," along with revolutionary new examples, was launched by Drinfel'd in 1986.

AUSTRALIAN MATHEMATICAL SOCIETY LECTURE SERIES

Editor-in-chief: Professor Michael Murray, University of Adelaide, SA 5005, Australia

Editors:

Professor P. Broadbridge AMSI, The University of Melbourne,
Victoria 3010, Australia

Professor C. C. Heyde, School of Mathematical Sciences,
Australian National University, Canberra, ACT 0200, Australia

Professor C. E. M. Pearce, Department of Applied Mathematics,
University of Adelaide, SA 5005, Australia

Professor C. Praeger, Department of Mathematics and Statistics,
University of Western Australia, Crawley, WA 6009, Australia

Quantum Groups

A Path to Current Algebra

ROSS STREET

Technical Editor:
ROSS MOORE

CAMBRIDGE
UNIVERSITY PRESS

University Printing House, Cambridge CB2 8BS, United Kingdom

Published in the United States of America by Cambridge University Press, New York

Cambridge University Press is part of the University of Cambridge.

It furthers the University's mission by disseminating knowledge in the pursuit of education, learning and research at the highest international levels of excellence.

www.cambridge.org
Information on this title: www.cambridge.org/9780521695244

© R. Street 2007

First published 2007

A catalogue record for this publication is available from the British Library

ISBN 978-0-521-69524-4 Paperback

To Oscar and Jack

Contents

vii

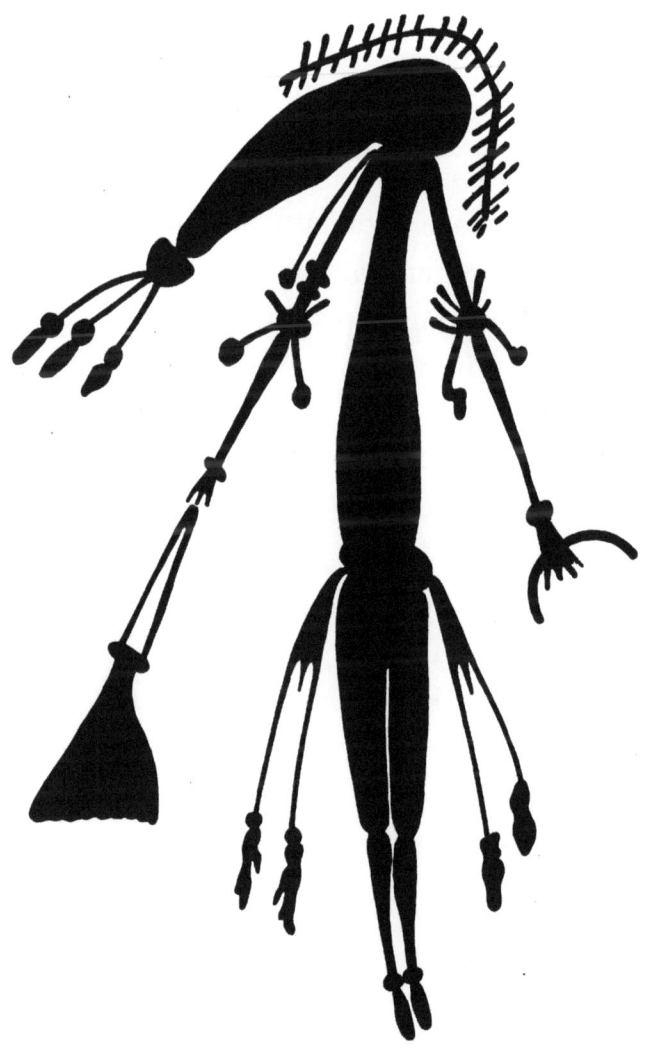

Bradshaw: "Ceremonial Figure", Tassel Bradshaw Group, [Wal94, Plate 20].

Introduction

Algebra has moved well beyond the topics discussed in standard undergraduate texts on "modern algebra". Those books typically dealt with algebraic structures such as groups, rings and fields: still very important concepts! However, *Quantum Groups: A Path to Current Algebra* is written for the reader at ease with at least one such structure and keen to learn the latest algebraic concepts and techniques.

A key to understanding these new developments is categorical duality. A quantum group is a vector space with structure. Part of the structure is standard: a multiplication making it an "algebra". Another part is not in those standard books at all: a comultiplication, which is dual to multiplication in the precise sense of category theory, making it a "coalgebra". While coalgebras, bialgebras and Hopf algebras have been around for half a century, the term "quantum group", along with revolutionary new examples, was unleashed on the mathematical community by Drinfel'd [Dri87] at the International Congress in 1986. Before launching into an explanation of the duality required, I should mention here that an ordinary group gives rise to a quantum group by taking the vector space with the group as basis.

When pushed to provide formal proofs of our claims, mathematicians generally resort to set theory. We build our structures on sets and feel satisfied when we can be explicit about the elements of our constructed objects. Up to the mid twentieth century, algebraic objects were sets with selected operations which assigned elements to lists of elements. Typically, we would have binary operations which might be called addition, multiplication or Lie bracket respectively assigning a sum, product or formal commutator to each pair of elements.

In those days, the importance was recognized of dealing with the homomorphisms between algebraic structures: these were the functions which preserved the operations involved in the kind of structure at hand. The existence of a bijective homomorphism (isomorphism) between two algebraic objects meant that the two objects played the same role. So how could the literal elements be the defining ingredient? The important issue was the way the algebraic object related to others of its own kind by means of homomorphisms into it or out of it. Quite often the elements could be recaptured as homomorphisms from a particular object into the one of interest. For example, the elements of a vector space were in bijection with the linear functions from a selected one-dimensional space.

Homomorphisms into an object might therefore be called "generalized elements" of the object. However, this notion of element of the object will depend on the kind of structure we are studying since that will determine what a homomorphism is (a group homomorphism, a linear function, a ring homomorphism, or whatever).

We quite often wish to add more elements to our sets to improve the properties of the operations: as when we construct the integers from the natural numbers to obtain subtraction; or when we construct the rational numbers from the integers to obtain division; or when we construct the real polynomials from the real numbers to obtain an indeterminate. These constructions can be described explicitly as sets with operations that include the original ones. More importantly, each such construction is unique up to isomorphism with a *universal property*: each homomorphism $X \longrightarrow C$ out of the original structure X, into a set C with the extra structure, extends to a homomorphism $\hat{X} \longrightarrow C$ out of the constructed object $\hat{X}$.

In this way it was realized that knowing the homomorphisms *out of* objects determined the objects just as surely as knowing the homomorphisms *into* them did. It is natural then to call homomorphisms out of an object "generalized co-elements". Once this kind of *duality principle* is acknowledged, interesting facts appear.

Let us take a simple example purely using sets. Consider two sets X and Y. Their *cartesian product* $X \times Y$ is the set whose elements are pairs (x, y) where x lies in X and y lies in Y. We are not studying any structure on these sets except for the property of being a set. So homomorphisms in this case are merely functions. It is clear that functions $f : T \longrightarrow X \times Y$ into $X \times Y$ from a test object T are in bijection with pairs of functions (f_1, f_2) where $f_1 : T \longrightarrow X$ and $f_2 : T \longrightarrow Y$. In other words, T-elements of $X \times Y$ are in bijection with pairs consisting of a T-element of X and a T-element of Y. All that is fairly straightforward.

Now suppose that our sets X and Y have no common elements; if they are not disjoint, replace them by isomorphic sets which are. Write $X + Y$ for the union; we write $X + Y$ rather than $X \cup Y$ to emphasize that it is the disjoint union (if X and Y were finite, the number of elements of $X + Y$ would be the sum of the number of elements in X and the number in Y). A function $f : X + Y \longrightarrow T$ is determined by its restriction to X and its restriction to Y. In other words, the co-T-elements of $X + Y$ are in bijection with pairs consisting of a co-T-element of X and a co-T-element of Y.

We conclude that the constructions $X \times Y$ and $X + Y$ are duals of one another. This is not something that was stressed when we were taught the more abstract multiplication and addition of numbers in infants' school.

If we now look at vector spaces or groups X and Y, the cartesian product $X \times Y$ as sets becomes a vector space or group by means of coordinatewise operations from X and Y; again this has pairs as the generalized

elements $T \longrightarrow X \times Y$. However, to obtain the dual constructions in these cases is quite different from the disjoint union of sets: in the case of vector spaces, we have that $X \times Y$ is self-dual (called direct sum and denoted by $X \oplus Y$); in the case of groups, the dual notion is rather complicated (called the free product by group theorists).

In order to formalize the way in which constructions such as these can be dual, we can use the notion of *category*. I intend to give a definition of this concept in this introduction. Before doing so, I would like to draw an analogy. It was noticed in projective plane geometry that theorems occurred in pairs: one such pair consists of Pascal's Mystic Hexagram Theorem and Brianchion's Theorem; both are about conics. Given one theorem in a pair, the other is obtained by interchanging the role of points and lines, reversing the incidence relation ("lies on" becomes "goes through"). To formally explain this duality, we abstract the notion of *projective plane*.

Here is the essence of the definition. A projective plane $\mathcal{P}$ consists of two sorts of elements: one sort called *points*, the other called *lines*. It also consists of a relation between these elements, called *incidence* (this is a rule telling when a point is incident with a line). There are some axioms which include:

1.   for distinct points P and Q, there is a unique line L such that P and Q are both incident with L; and,

2.   for distinct lines L and M, there is a unique point P such that P is incident with both L and M.

Any system satisfying this is a projective plane! The points do not need to look like points and the lines do not need to look like lines in any sense. Of course, we still draw pictures to help our intuition.

Now we are ready to formalize duality. Given a projective plane $\mathcal{P}$, we obtain a projective plane $\mathcal{P}^{\text{rev}}$ whose points are the lines of $\mathcal{P}$, whose lines are the points of $\mathcal{P}$, and whose incidence relation is the reverse of that of $\mathcal{P}$. Notice that axioms (1) and (2) for $\mathcal{P}^{\text{rev}}$ are respectively axioms (2) and (1) for $\mathcal{P}$. This means that, if we prove a theorem about all projective planes, then the dual theorem is automatically true by applying the original theorem to $\mathcal{P}^{\text{rev}}$.

It turns out that there are not too many interesting theorems assuming only axioms (1) and (2). A further axiom based on a theorem of Pappus can be added and the system remains self-dual. In fact, conics can be defined using an idea of Steiner, and Pascal's Theorem can be proved. Let us now discontinue discussion of this analogy and return to the formalization of the duality at hand.

A *category* $\mathcal{A}$ consists of two sorts of elements: one sort called *objects*, the other called *morphisms* (or *arrows*). It also consists of three functions. The first function assigns to each morphism f a pair (A, B) of objects in

which case A is called the *domain* (or *source*) of f while B is called the *codomain* (or *target*) of f; the notations $f : A \longrightarrow B$ and $A \xrightarrow{f} B$ are used. The second function assigns to each object A a morphism $1_A : A \longrightarrow A$ called the *identity morphism* of A. A pair (f, g) of morphisms is called *composable* when the codomain of f is equal to the domain of g. The third function assigns to each composable pair (f, g) of morphisms, a morphism $g \circ f$, called the *composite* of f and g, whose domain is that of f and whose codomain is that of g. There are two axioms:

1.　　　if (f, g) and (g, h) are composable pairs of morphisms then $(h \circ g) \circ f = h \circ (g \circ f)$; and,

2.　　　if $f : A \longrightarrow B$ is a morphism then $f \circ 1_A = f = 1_B \circ f$.

The standard argument shows that identity morphisms are unique. The notation $\mathcal{A}(A, B)$ (or $\mathrm{Hom}_A(A, B)$) is used for the set of all morphisms in $\mathcal{A}$ from A to B.

There is a category **Set** whose objects are sets, morphisms are functions, and composition is the usual composition of functions. There is a category **Vect**$_\mathbf{k}$ whose objects are vector spaces over a fixed field $\mathbf{k}$ and morphisms are linear functions; composition is as usual. Similarly we have a category whose objects are groups and a category whose objects are rings.

However, there are categories whose objects do not look like sets and whose morphisms do not look like functions. For example, there is a category whose objects are integers, whose morphisms are pairs (m, n) of integers such that the domain of (m, n) is m and the codomain is the product mn; a pair $((m, n), (r, s))$ of morphisms is composable when $mn = r$ and the pair's composite is (m, ns).

Now to duality. Given a category $\mathcal{A}$, there is a category $\mathcal{A}^{\mathrm{op}}$ whose objects are the objects of $\mathcal{A}$, and morphisms are the morphisms of $\mathcal{A}$; however, the domain of a morphism is its codomain in $\mathcal{A}$ while its codomain in $\mathcal{A}^{\mathrm{op}}$ is its domain in $\mathcal{A}$. A pair (g, f) of morphisms is composable in $\mathcal{A}^{\mathrm{op}}$ if and only if (f, g) is composable in $\mathcal{A}$; its composite $f \circ g$ in $\mathcal{A}^{\mathrm{op}}$ is the composite $g \circ f$ in $\mathcal{A}$. We call $\mathcal{A}^{\mathrm{op}}$ the *dual* or *opposite* of the category $\mathcal{A}$.

Perhaps it helps to say that $\mathcal{A}^{\mathrm{op}}$ is the category obtained from $\mathcal{A}$ by reversing arrows: a morphism $f : A \longrightarrow B$ in $\mathcal{A}$ is precisely a morphism $f : B \longrightarrow A$ in $\mathcal{A}^{\mathrm{op}}$. Admittedly, if the objects of $\mathcal{A}$ look like sets (that is, are sets with some structure), the same is true of $\mathcal{A}^{\mathrm{op}}$; but the same cannot be said for morphisms that are functions, since formally reversed functions can scarcely be thought of as functions.

The duality between cartesian product and disjoint union can now be made precise. In a category $\mathcal{A}$, a *product* for two objects A and B consists of an object $A \times B$ and two morphisms $p_1 : A \times B \longrightarrow A$, $p_2 : A \times B \longrightarrow B$ (called *projections*) with the following "universal" property: for all

objects T and morphisms $a : T \longrightarrow A$, $b : T \longrightarrow B$, there exists a unique morphism $T \longrightarrow A \times B$, denoted by (a, b), such that $p_1 \circ (a, b) = a$ and $p_2 \circ (a, b) = b$. This means that T-elements of $A \times B$ are in bijection with pairs consisting of a T-element of A and a T-element of B.

A morphism $h : C \longrightarrow D$ in a category $\mathcal{A}$ is called a *right inverse* for a morphism $k : D \longrightarrow C$ when $k \circ h = 1_C$; we also say that k is a *left inverse* for h. A morphism h is *invertible* (or an *isomorphism*) when it has both a left and right inverse; in this case, a familiar argument shows that the left and right inverse agree and this common morphism is unique, being called the *inverse* of h and denoted by h^{-1}. If there exists an invertible morphism $C \longrightarrow D$ then we say C and D are *isomorphic* and write $C \cong D$. In a category, we think of isomorphic objects as being essentially the same. Any two products of two objects A and B can be proved, by an easy argument, to be isomorphic.

Now we have our duality between cartesian product and disjoint union of sets: cartesian product is the product in the category **Set** while disjoint union is the product in the category $\textbf{Set}^{\text{op}}$.

We can give an even simpler example. An object K of a category $\mathcal{A}$ is called *terminal* when, for all objects A of $\mathcal{A}$, there is precisely one morphism $A \longrightarrow K$. The singleton set 1 is terminal in the category **Set** while the empty set $\varnothing$ is terminal in $\textbf{Set}^{\text{op}}$.

Any concept defined for all categories $\mathcal{A}$ has a dual concept which is the same concept translated to $\mathcal{A}^{\text{op}}$: the prefix "co-" is used. So a product in $\mathcal{A}^{\text{op}}$ is called a *coproduct* in $\mathcal{A}$. A terminal object in $\mathcal{A}^{\text{op}}$, under this system, would be called a *coterminal object* in $\mathcal{A}$; but it is also called an *initial object* of $\mathcal{A}$.

In the spirit of category theory itself, we should consider appropriate morphisms of categories. These are called *functors*. A functor $F : \mathcal{A} \longrightarrow \mathcal{X}$ between categories $\mathcal{A}$ and $\mathcal{X}$ consists of two functions. The first assigns to each object A of $\mathcal{A}$ an object FA of $\mathcal{X}$. The second function assigns to each morphism $f : A \longrightarrow B$ of $\mathcal{A}$ a morphism $Ff : FA \longrightarrow FB$ of $\mathcal{X}$. There are two axioms:

1. $F1_A = 1_{FA}$ for all objects A of $\mathcal{A}$; and,

2. $F(g \circ f) = Fg \circ Ff$ for all composable pairs (f, g) in $\mathcal{A}$.

It is easy to see that functors preserve invertibility of morphisms: in fact they take inverses to inverses.

Let us look at a couple of examples of functors.

- Each object T of a category $\mathcal{A}$ determines a functor $R_T = \mathcal{A}(T, \_) : \mathcal{A} \longrightarrow \textbf{Set}$ called the functor *represented by* T; the elements of $R_T A = \mathcal{A}(T, A)$ are the morphisms $a : T \longrightarrow A$ in $\mathcal{A}$ (that is, the T-elements of A), while the function $R_T f : R_T A \longrightarrow R_T B$ takes such an a to $f \circ a$.

- Suppose K is an object of $\mathcal{A}$ for which a product $K \times A$ exists (and is chosen) for all objects A. There is a functor $F = K \times \_ : \mathcal{A} \longrightarrow \mathcal{A}$ defined on objects by $FA = K \times A$ and on morphisms by $Ff = K \times f$ where $K \times f = (p_1, f \circ p_2) : K \times A \longrightarrow K \times B$.

Categories were invented not only to formalize duality but to formalize the concept of "naturality" in mathematics. The idea was that a natural transformation should be one that involves no ad hoc choices. For example, if V is a vector space and V^* is the vector space of linear functions from V into the base field $\mathbf{k}$, there is a natural linear function $V \longrightarrow V^{**}$ which takes $v \in V$ to the linear function $e_v : V^* \longrightarrow \mathbf{k}$ defined by evaluation at v. However, any linear function $V^{**} \longrightarrow V$ that depends on a choice of basis for V should not be natural.

Suppose $F : \mathcal{A} \longrightarrow \mathcal{X}$ and $G : \mathcal{A} \longrightarrow \mathcal{X}$ are functors between the same categories. A *natural transformation* $\theta : F \longrightarrow G$ is a function. The function assigns to each object A of $\mathcal{A}$ a morphism $\theta_A : FA \longrightarrow GA$ of $\mathcal{X}$. There is a single axiom: for each morphism $f : A \longrightarrow B$,

$$Gf \circ \theta_A = \theta_B \circ Ff.$$

There is an obvious componentwise composition of natural transformations. This defines a category $[\mathcal{A}, \mathcal{X}]$, called a *functor category*, where the objects are functors $F : \mathcal{A} \longrightarrow \mathcal{X}$ and the morphisms are natural transformations. A *natural isomorphism* is an invertible morphism in the functor category. A functor $F : \mathcal{A} \longrightarrow \mathbf{Set}$ is called *representable* when it is isomorphic to R_T for some object T; such a T is called a *representing object* for F. For example, the functor $U : \mathbf{Vect}_\mathbf{k} \longrightarrow \mathbf{Set}$, which takes each vector space to its underlying set and each linear function to that function, is representable: we have $U \cong R_\mathbf{k}$ since the linear functions from the field $\mathbf{k}$ to a vector space V are in natural bijection with elements of V. Many constructions in mathematics are designed to provide representing objects for interesting functors.

Let us look at a couple of examples of natural transformations:

- Suppose $F : \mathcal{A} \longrightarrow \mathbf{Set}$ is a functor and T is an object of $\mathcal{A}$. Each element x of FT determines a natural transformation $\hat{x} : R_T \longrightarrow F$ defined by $\hat{x}_A(a) = (Fa)(x)$. The *Yoneda Lemma* states that this defines a bijection $FT \cong [\mathcal{A}, \mathbf{Set}](R_T, F)$. The inverse bijection is even easier: it takes the natural transformation $\theta : R_T \longrightarrow F$ to the element $\theta_T(1_T)$ of FT.

- Suppose $h : K \longrightarrow L$ is a morphism of a category $\mathcal{A}$ in which products of pairs of objects exist. Then we obtain a natural transformation $f \times \_ : K \times \_ \longrightarrow L \times \_$ whose value at the object A is the morphism $f \times A = (f \circ p_1, p_2) : K \times A \longrightarrow L \times A$.

Modern algebra in the sense of the first half of the twentieth century dealt with sets equipped with operations. Soon after, the idea of co-operations crept into mathematics. The notion of coalgebra is dual to algebra. This is the main concept in this book.

Now I turn to the book's contents. Chapter 1 gives precise definitions of monoids and groups; the axioms are expressed in terms of diagrams ready to be imported to a general category. This importation is carried out in Chapter 2 where we provide the important example of 2×2 matrices in readiness for the quantum version. A duality between geometry and algebra is explained. In Chapter 3, we describe the quantum general linear group of 2×2 matrices as a coalgebra. This comes from lectures by Manin in Montréal. Chapter 4 is about modules over rings; we find it natural to take a 2-sided point of view so that our basic module M has a left action by a ring R and a right action by a ring S which compatibly interact. Chapter 5 concerns finitely generated, projective modules under the mysterious name of "Cauchy modules". It turns out that F. W. Lawvere noticed a concept in enriched category theory which has Cauchy sequences as an example; when interpreted for additive categories, it leads to modules that are finitely generated and projective. Chapter 6 discusses algebras, Lie algebras and the Poincaré–Birkhoff–Witt Theorem. Chapter 7 is about coalgebras and bialgebras. A coalgebra is a vector space with a comultiplication. A bialgebra is an algebra which is also a coalgebra subject to a compatibility condition. The dual vector space of a coalgebra is an algebra, however, the usual dual of an algebra need not naturally be a coalgebra. In Chapter 8 we describe Sweedler's modification (see [Swe69] and [Abe80]) of the dual of an algebra which is a coalgebra.

In Chapter 9, we look at Hopf algebras. These should be thought of as generalized groups. An important part of group theory is the theory of their representations: these are modules over the group ring. In Chapter 10 we look at modules over bialgebras. Then, in Chapter 11, we move to use categories more seriously. We discuss categories equipped with an abstract tensor product: monoidal or tensor categories. We discuss examples involving braids. A deep example, not treated here, is the subject of the paper [JS95]. An important property of the tensor product $U \otimes V$ of vector spaces is that it represents the vector space $[V, W]$ of linear functions from V to W:

$$\mathbf{Vect}_k(U \otimes V, W) \cong \mathbf{Vect}_k(U, [V, W]).$$

In Chapter 12, this idea is lifted to arbitrary tensor categories. Examples from knot theory are provided.

The Yang–Baxter equation, from the branch of physics called statistical mechanics, had a major influence on the new examples of Hopf algebras called quantum groups. In Chapter 13, an algebraic concept of Yang–Baxter operator, which makes sense in any tensor category, is explained. A family of examples from linear algebra is provided in Chapter 14.

In Chapter 15, the notion of monoid is lifted to the level of generality at which "algebras" and "coalgebras" become precise categorical duals. For the first time, I believe, in a text at this level, emphasis is placed on "2-cells" between monoid morphisms, providing the student with a gentle introduction to higher-dimensional category theory.

Each bialgebra has a tensor category of representations. This correspondence is a modern formulation of Tannaka–Krein duality. The treatment of this topic in Chapter 16 makes use of the 2-dimensional structure of monoids from Chapter 15. There is by now a vast literature on Tannaka duality. We satisfy ourselves with a sketch in Chapter 17 of an application to construct universally a Hopf algebra from a bialgebra.

Finally, in Chapter 18, the example of Chapter 3 is revisited in the light of what has been learned. There are exercises at the end of several chapters. Solutions to most of these are provided in Chapter 19.

Acknowledgements ...

Many ideas presented here are my version of joint research with André Joyal. I consider myself very fortunate to have experienced such exciting collaboration.

I would like to thank the students and staff who attended the original course in the first half of 1990 at Macquarie University. I am very grateful to Paddy McCrudden (as a Vacation Scholar under my supervision in January–February 1995) for his careful reading of my handwritten notes, and particularly for his systematic checking of the exercises.

It is a pleasure to acknowledge the support of grants from the Australian Research Council during the preparation of this work.

... the typing process

Many of the chapters were carefully typed by Elaine Vaughan. Technology moved ahead incredibly from that word-processor available in our Department in 1990. Typesetting with TEX and LATEX was begun by Ross Moore and continued by several post-doc and graduate students; namely Sam Williams, Ross Talent (now deceased) and Simon Byrne. Each added further exercises and solutions, as these were encountered in lecture courses.

Most important in this was the use of XY-pic[1] to produce the commutative diagrams that appear throughout this book, and which play such an integral part in the visualization and understanding of much of the material. None of these diagrams has been imported as an image built using other software. All, including the braids, tangles, and 2-cell diagrams, are

---

[1] Originally written by Kristoffer Rose; extended and enhanced by Rose and Moore for mathematical applications and higher quality output. The XY-pic package and documentation is now included with all TEX and LATEX distributions.

specified within the LaTeX source using the Xy-pic package syntax. Indeed the syntax and coding to handle curves and 2-cells was written in 1993–94 by Ross Moore, specifically for use with this book. Since then the Xy-pic package has become a useful tool for presenting diagrammatic material in Category Theory and other branches of mathematics, computer science and linguistics.

As an application of Ross Moore's work with the LaTeX2HTML translation software, an earlier version of the present manuscript was made available via the "world-wide web", now known as the internet. In that form it was used as a source of lecture notes for courses at Macquarie.

A great deal of credit is also due to Simon Byrne (as a Vacation Scholar in January–February 2005) for finishing off the typing of exercises and for assembling the manuscript into a form ready to submit as a proposal for the Australian Mathematical Society Lecture Series. With this go-ahead, the final version of the manuscript, complete with up-to-date Bibliography, Index, front-matter and filler images was prepared by Ross Moore, who is acknowledged here as being the Technical Editor for this monograph.

... the illustrations

The illustrations appearing at the end of some chapters are reproduced from Grahame Walsh, *Bradshaws* [Wal94]. I am very grateful to the Bradshaw Foundation and Edition Limitée for consenting to their inclusion. The original coloured rock paintings, which the silhouettes trace, are the work of Australian people living as much as 50 millennia before our time.

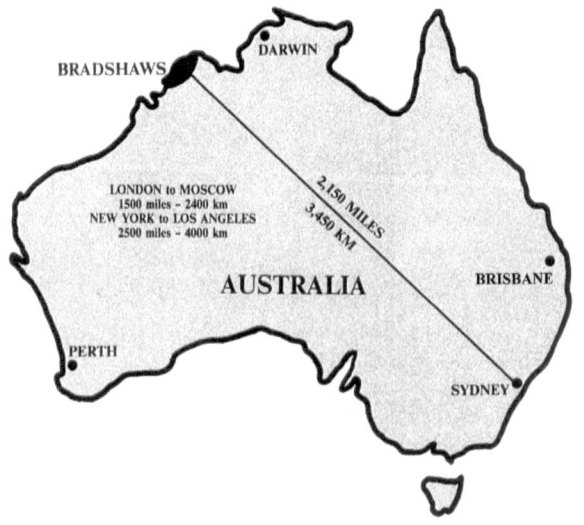

These paintings have been mentioned already in the mathematico-scientific literature, in connection with knots and braids; *viz.*

> How old are knots? It has been suggested that the Stone Age should be called the Age of String. The extraordinary tasselled figures photographed and described by G. L. Walsh in *Bradshaws: Ancient Rock Paintings of Western Australia* (Edition Limitée, 1994) have been suggested to be 50,000 years old. Knots have been intimately linked with the development of humans, through weapons, fishing, hunting, clothing, housing, boating and a myriad of other ways.
>
> The metaphor of knots is found throughout literature, and knots and interlacing are featured in many forms of art.

Ronald Brown, review of: "The Knot Book:
An Elementary Introduction to the Mathematical Theory
of Knots" by Colin C. Adams (W. H. Freeman 1994),
appeared in *Nature*, Vol. 371 (13 October 1994).

Suggested Further Reading

[JS91b] André Joyal and Ross Street. An introduction to Tannaka duality and quantum groups. In *Category Theory (Como, 1990)*, volume 1488 of *Lecture Notes in Mathematics*, pages 413–492. Springer, Berlin, 1991. **MR1173027**

[Kas95] Christian Kassel. *Quantum Groups*, volume 155 of *Graduate Texts in Mathematics*. Springer, New York, 1995. **MR1321145**

[Maj95] Shahn Majid. *Foundations of Quantum Group Theory*. Cambridge University Press, Cambridge, 1995, (paperback, 2000). **MR1381692**

[SS93] Steven Shnider and Shlomo Sternberg. *Quantum Groups: From CoAlgebas to Drinfel'd Algebras*. Graduate Texts in Mathematical Physics, II. International Press, Cambridge, MA, 1993. **MR1287162**

[Yet01] David N. Yetter. *Functorial Knot Theory*, volume 26 of *Series on Knots and Everything*. World Scientific Publishing Co. Inc., River Edge, NJ, 2001. **MR1834675**

1

Revision of basic structures

The *cartesian product* of n sets $X_1, \ldots, X_n$ is the set

$$X_1 \times \cdots \times X_n = \{(x_1, \ldots, x_n) \mid x_i \in X_i\}.$$

There is a *canonical bijection*

$$(X_1 \times \cdots \times X_m) \times (X_{m+1} \times \cdots \times X_n) \cong X_1 \times \cdots \times X_n$$

given by deleting the inside brackets. The *diagonal* function

$$\delta : X \longrightarrow X \times \cdots \times X$$

is given by $\delta(x) = (x, \ldots, x)$.

The cartesian product of no sets is the special set 1, with precisely one element, which should technically be denoted by empty parentheses ( ). Particular cases of the canonical bijections are

$$X \times 1 \cong X \cong 1 \times X.$$

The diagonal $X \longrightarrow 1$ will be denoted by ε rather than δ; it is the *only* function $X \longrightarrow 1$. Functions $f_1 : X_1 \longrightarrow Y_1, \ldots, f_n : X_n \longrightarrow Y_n$ induce a function

$$f_1 \times \cdots \times f_n : X_1 \times \cdots \times X_n \longrightarrow Y_1 \times \cdots \times Y_n$$

given by $(f_1 \times \cdots \times f_n)(x_1, \ldots, x_n) = (f_1(x_1), \ldots, f_n(x_n))$.

The *identity* function $1_X : X \longrightarrow X$ on a set X is given by $1_X(x) = x$.

We noted that $\varepsilon : X \longrightarrow 1$ is uniquely determined. Similarly the diagonal $\delta : X \longrightarrow X \times X$ is unique, determined by commutativity of the diagram

(Identity)

$$1 \times X \xleftarrow[\varepsilon \times 1_X]{} X \times X \xrightarrow[1_X \times \varepsilon]{} X \times 1.$$

with X at top, arrows $\cong$ down to $1 \times X$ and $X \times 1$, and δ down to $X \times X$.

1

Furthermore, the following diagram commutes

(Associativity) $$X \xrightarrow{\;\delta\;} X \times X \underset{1_X \times \delta}{\overset{\delta \times 1_X}{\rightrightarrows}} X \times X \times X .$$

The function $X \longrightarrow X \times X \times X$ so determined is none other than the ternary diagonal.

A *monoid* is a set M together with special purpose functions $\eta : 1 \longrightarrow M$, $\mu : M \times M \longrightarrow M$ such that the following diagrams commute.

(Id)

$$1 \times M \xrightarrow{\;\eta \times 1_M\;} M \times M \xleftarrow{\;1_M \times \eta\;} M \times 1$$

with M at the apex, $\cong$ maps from $1 \times M$ and $M \times 1$, and μ from $M \times M$ to M.

(Assoc) $$M \xleftarrow{\;\mu\;} M \times M \underset{1_M \times \mu}{\overset{\mu \times 1_M}{\leftleftarrows}} M \times M \times M$$

If we write 1 for the value of η at the only element of 1 and we write $x\,y$ for $\mu(x,y)$ then the above diagrams translate to the equations

$$\begin{aligned} 1\,x &= x = x\,1 \\ (x\,y)\,z &= x\,(y\,z) \end{aligned} \qquad \text{for all} \quad x, y, z \in M .$$

This time functions η and μ are not uniquely determined by the set M. However given μ, condition (Id) uniquely determines η while the condition (Assoc) gives an unambiguous ternary operation $\mu : M \times M \times M \longrightarrow M$ which we write as $\mu(x,y,z) = x\,y\,z$. Generally there is an unambiguous multiple product function $\mu : M \times \cdots \times M \longrightarrow M$ determined by the binary μ.

An element $x \in M$ is called *invertible* when there exist $y, z \in M$ such that $y\,x = 1$ and $x\,z = 1$. Notice that

$$y = y\,1 = y(x\,z) = (y\,x)z = 1\,z = z$$

so each invertible element x determines uniquely an element, denoted x^{-1}, satisfying $x^{-1}x = 1 = x\,x^{-1}$.

A *group* is a monoid in which each element is invertible. Then we have a function $\iota : M \longrightarrow M$ such that this next diagram commutes.

(Invertibility)

$$M \xrightarrow{\;\delta\;} M \times M \underset{1_M \times \iota}{\overset{\iota \times 1_M}{\rightrightarrows}} M \times M \xrightarrow{\;\mu\;} M$$

with ε and η maps to I forming the commuting triangle.

Note carefully the dependence of this axiom on the diagonal structure of cartesian product.

For a set X, the n-fold cartesian product $X \times \cdots \times X$ is denoted by X^n. Each permutation ξ on $\{1, \dots, n\}$ induces a bijection

$$\sigma_\xi : X^n \longrightarrow X^n$$

given by $\sigma_\xi(x_1, \dots, x_n) = (x_{\xi(1)}, \dots, x_{\xi(n)})$. In particular, we have the *switch* coming from the non-identity permutation of $\{1, 2\}$:

$$\sigma : X \times X \longrightarrow X \times X \quad , \quad \sigma(x, y) = (y, x) \, .$$

Each σ_ξ is a composite of bijections of the form $1_X \times \cdots \times \sigma \times \cdots \times 1_X$. Notice that the following diagram commutes.

(Commutativity)

$$
\begin{array}{ccc}
 & X & \\
{}^{\delta}\nearrow & & \searrow{}^{\delta} \\
X \times X & \xrightarrow{\ \ \sigma\ \ } & X \times X
\end{array}
$$

A monoid (M, η, μ) is called *commutative* when the following diagram commutes.

$$
\begin{array}{ccc}
 & M & \\
{}^{\mu}\nearrow & & \nwarrow{}^{\mu} \\
M \times M & \xleftarrow{\ \ \sigma\ \ } & M \times M
\end{array}
$$

It follows that the composite $M^n \xrightarrow{\ \sigma_\xi\ } M^n \xrightarrow{\ \mu\ } M$ is independent of the permutation ξ.

Suppose M and N are monoids. A *monoid morphism* (or *homomorphism of monoids*) is a function $f : M \longrightarrow N$ such that the following diagrams commute.

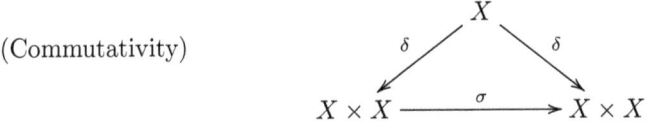

Expressed in terms of elements, these diagrams merely say that $f(1) = 1$ and $f(xy) = f(x)f(y)$. If N has left-cancellation (i.e., $ab = ac$ implies $b = c$; e.g., if N is a group) then $f(1) = 1$ is redundant.

Monoid morphisms preserve invertibility: if $x \in M$ is invertible, $f(x^{-1}) = f(x)^{-1}$. So for groups M and N we have commutativity of the square

$$
\begin{array}{ccc}
M & \xrightarrow{\ f\ } & N \\
\downarrow{\scriptstyle \iota} & & \downarrow{\scriptstyle \iota} \\
M & \xrightarrow{\ f\ } & N\ .
\end{array}
$$

A *rig* is a set R enriched with two monoid structures, a commutative one written additively and the other written multiplicatively, such that the following equations hold:

$$a\,0 = 0 = 0\,a$$

(Distributive) $\qquad a(b+c) = a\,b + a\,c\ ,\quad (a+b)c = a\,c + b\,c\ .$

The natural numbers $\mathbb{N} = \{0,1,2,\dots\}$ provide an example of a rig.

A *ring* is a rig for which the additive monoid is a group. The integers $\mathbb{Z}$ provide an example.

A rig is *commutative* when the multiplicative monoid is commutative. A *field* is a commutative ring for which each element is either 0 or has a multiplicative inverse.

For rigs R and S a *rig morphism* $f : R \longrightarrow S$ is a function which is a monoid morphism for both the additive and multiplicative structures.

Let $\mathbf{k}$ denote a field. A $\mathbf{k}$-*algebra* is a ring A together with a ring morphism $\eta : \mathbf{k} \longrightarrow A$. Notice that either A is trivial (i.e., $1 = 0$), or η is injective $[\,\kappa \neq \kappa' \Rightarrow \kappa - \kappa' \neq 0 \Rightarrow \kappa - \kappa'$ is invertible $\Rightarrow \eta(\kappa - \kappa')$ is invertible $\xRightarrow{1 \neq 0} \eta(\kappa - \kappa') \neq 0 \Rightarrow \eta(\kappa) \neq \eta(\kappa')\,]$. We can define *scalar multiplication* $\mathbf{k} \times A \longrightarrow A$ by $\kappa\, a = \eta(\kappa)\, a$.

For $\mathbf{k}$-algebras A and B, a $\mathbf{k}$-*algebra morphism* $f : A \longrightarrow B$ is a ring morphism such that the next diagram commutes;

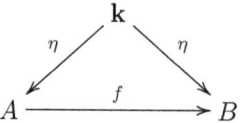

that is, $f(\kappa\, a) = \kappa\, f(a)$. We write $\mathbf{Alg}_{\mathbf{k}}(A, B)$ for the set of $\mathbf{k}$-algebra morphisms $f : A \longrightarrow B$.

An *isomorphism* is a bijective morphism; automatically its inverse function is also a morphism.

2

Duality between geometry and algebra

The purpose of this section is to convince you that commutative algebras are really *spaces* seen from the other side of your brain.

For a compact hausdorff space X, we have the algebra $C(X)$ of continuous, complex-valued functions $a : X \longrightarrow \mathbb{C}$. The addition and multiplication are obtained pointwise from $\mathbb{C}$.

A continuous function $f : X \longrightarrow Y$ gives rise to an algebra morphism $C(f) : C(Y) \longrightarrow C(X)$ (note the reversal of direction!) via $C(f)(b) = a$, where $a(x) = b(f(x))$. In particular, the unique $X \longrightarrow 1$ gives the algebra morphism $\eta : \mathbb{C} = C(1) \longrightarrow C(X)$, while each *point* $x : 1 \longrightarrow X$ of the space gives an algebra morphism $C(X) \longrightarrow \mathbb{C}$.

Actually $C(X)$ is more than just a $\mathbb{C}$-algebra; it is what is called a *commutative C^*-algebra* (there is a norm and an involution $(\_)^*$ coming from conjugation). With this extra structure the duality becomes precise:

> *Each commutative C^*-algebra A is isomorphic to $C(X)$ for some compact hausdorff space X; each C^*-algebra morphism $C(Y) \longrightarrow C(X)$ has the form $C(f)$ for a unique continuous function $f : X \longrightarrow Y$.*

This result is commonly referred to as *Gelfand duality*.

Algebraic geometry is the study of spaces called *varieties*: the solutions to polynomial equations in several variables. In studying the variety given by $x^2 + 2y^3 = z^4$ over the field $\mathbf{k}$, we pass to the $\mathbf{k}$-algebra

$$A = \mathbf{k}[x, y, z] / (x^2 + 2y^3 = z^4).$$

By $\mathbf{k}[x, y, z]$ we mean the $\mathbf{k}$-algebra of polynomials in three commuting indeterminates x, y, z; the elements are expressions

$$\sum_{i,j,k} \alpha_{ijk}\, x^i\, y^j\, z^k$$

where $\alpha_{ijk} \in \mathbf{k}$ and (i,j,k) runs over a finite subset of $\mathbb{N}^3$. The quotient algebra A is obtained from $\mathbf{k}[x,y,z]$ by identifying elements when they may be transformed one into another by means of the equation $x^2 + 2y^3 = z^4$ and the algebra axioms.

Given a $\mathbf{k}$-algebra B, a $\mathbf{k}$-*algebra morphism* $f : \mathbf{k}[x,y,z] \longrightarrow B$ is uniquely determined by its values on x,y,z. In fact we have a bijection

$$\mathbf{Alg}_\mathbf{k}\big(\mathbf{k}[x,y,z],B\big) \cong B^3 .$$

Similarly, we have a bijection

$$\mathbf{Alg}_\mathbf{k}(A,B) \cong \big\{ (u,v,w) \in B^3 \mid u^2 + 2v^3 = w^4 \big\}$$

where A is as above. Again we see that a $\mathbf{k}$-algebra morphism $A \longrightarrow B$ corresponds to a map of varieties in the reverse direction.

For general $\mathbf{k}$-algebras A and B, it is suggestive to call a morphism $f : A \longrightarrow B$ a *B-point* of A. A *point* of (the space corresponding to) A is a $\mathbf{k}$-point, not to be confused with an element of the algebra A itself.

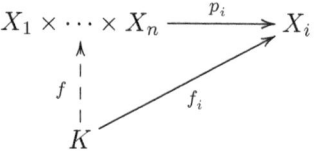

Let $\mathcal{X}$ denote a category. I am thinking of the *objects* of $\mathcal{X}$ as spaces X and Y say, and the *arrows* $X \longrightarrow Y$ as the maps appropriate to that kind of space. Write $\mathcal{X}(X,Y)$ for the set of arrows from X to Y in $\mathcal{X}$.

Let $X_1, \ldots, X_n$ be arbitrary objects of $\mathcal{X}$. A *product* for this list of objects consists of an object $X_1 \times \cdots \times X_n$ together with arrows

$$p_i : X_1 \times \cdots \times X_n \longrightarrow X_i \qquad \text{for } i = 1, \ldots, n$$

such that, given any other object K and arrows

$$f : K \longrightarrow X_i \qquad \text{for } i = 1, \ldots, n$$

there exists a unique arrow $f : K \longrightarrow X_1 \times \cdots \times X_n$ with $p_i \circ f = f_i$.

$$X_1 \times \cdots \times X_n \xrightarrow{\;\;p_i\;\;} X_i$$

$$f \Big\uparrow \qquad \nearrow f_i$$

$$K$$

This means we have a bijection

$$\mathcal{X}\big(K, X_1 \times \cdots \times X_n\big) \cong \mathcal{X}(K,X_1) \times \cdots \times \mathcal{X}(K,X_n) .$$

In particular, the empty product is called a *terminal object*, denoted by 1. We have

$$\mathcal{X}(K,1) \cong 1.$$

Products are unique up to isomorphism (if they exist).

The *diagonal* $\delta : X \longrightarrow X \times \cdots \times X$ is defined by $p_i \circ \delta = 1_X$ for all i. The canonical isomorphisms $f_1 \times \cdots \times f_n$ and isomorphisms σ_ξ can be defined as for sets.

The diagrammatic definition of *monoid* and *group* can be carried into the category $\mathcal{X}$ (provided the products exist; 1 and $M \times M$ are enough). If M is a monoid (group) in $\mathcal{X}$ then each $\mathcal{X}(K, M)$ becomes a monoid (group) using the multiplication $*$ given by

$$f * g = \mu \circ (f \times g) \circ \delta$$

$$K \xrightarrow{\;\delta\;} K \times K \xrightarrow{\;f \times g\;} M \times M \xrightarrow{\;\mu\;} M.$$

A group in the category of topological spaces and continuous maps is called a *topological group*. A group in the category of smooth manifolds and smooth maps is called a *Lie group*.

We are more interested here in groups in the category $(\mathbf{Comm\,Alg_k})^{\mathrm{op}}$ of commutative $\mathbf{k}$-algebras and reversed morphisms; these are called *affine groups* over $\mathbf{k}$. This is the variety point of view. On the algebraic side they are called *commutative Hopf algebras* over $\mathbf{k}$. Product of varieties becomes tensor product $A \otimes_{\mathbf{k}} B$ of $\mathbf{k}$-algebras (more on this later). A commutative Hopf algebra H thus has structure given by the $\mathbf{k}$-algebra morphisms

$$\varepsilon : H \longrightarrow \mathbf{k}, \qquad \delta : H \longrightarrow H \otimes_{\mathbf{k}} H, \qquad \nu : H \longrightarrow H$$

called *counit, comultiplication, antipode* (corresponding respectively to the unit, multiplication, inversion for the group). Now for each commutative $\mathbf{k}$-algebra A, we obtain a group $\mathbf{Alg_k}(H, A)$ of A-points in H.

It will also be necessary to consider the algebraic version of *affine monoids* over $\mathbf{k}$. These are called *commutative bialgebras* over $\mathbf{k}$. They have a counit and comultiplication, but generally no antipode.

Example 2.1 *Let* $M(2)$ *denote* $\mathbf{k}[\,a,b,c,d\,]$ *as a commutative* $\mathbf{k}$*-algebra. A counit* $\varepsilon : M(2) \longrightarrow \mathbf{k}$ *is defined by* $\varepsilon(a) = \varepsilon(d) = 1$, $\varepsilon(b) = \varepsilon(c) = 0$. *Clearly*

$$\mathbf{k}[\,a,b,c,d\,] \otimes_{\mathbf{k}} \mathbf{k}[\,a,b,c,d\,] \cong \mathbf{k}[\,a,b,c,d,a',b',c',d'\,]$$

with the coprojections

$$\mathbf{k}[\,a\,,b\,,c\,,d\,] \longrightarrow \mathbf{k}[\,a\,,b\,,c\,,d\,,a'\,,b'\,,c'\,,d'\,] \longleftarrow \mathbf{k}[\,a\,,b\,,c\,,d\,]$$
$$a\,,b\,,c\,,d \longmapsto a\,,b\,,c\,,d \quad and \quad a'\,,b'\,,c'\,,d' \longleftarrow\!\shortmid\ a\,,b\,,c\,,d\ .$$

The comultiplication $\delta : \mathrm{M}(2) \longrightarrow \mathrm{M}(2) \otimes_{\mathbf{k}} \mathrm{M}(2)$ *is given by*

$$a\,,b\,,c\,,d \longmapsto a\,a' + b\,c'\,,\ a\,b' + b\,d'\,,\ c\,a' + d\,c'\,,\ c\,b' + d\,d'\ .$$

This makes $\mathrm{M}(2)$ *into a commutative* $\mathbf{k}$-*bialgebra. Notice that we have a monoid isomorphism*

$$\mathbf{Alg}_{\mathbf{k}}\big(\mathrm{M}(2)\,,A\big)\ \cong\ \mathbf{Mat}\big(2\,,A\big)$$

where on the right we have the multiplicative monoid of 2×2 *matrices with entries in* A. *Thus* $\mathrm{M}(2)$ *is the* coordinate $\mathbf{k}$-algebra *of the variety of* 2×2 *matrices.*

To obtain the coordinate $\mathbf{k}$-*algebra of the* general linear group, *we take*

$$\mathrm{GL}(2)\ =\ \mathbf{k}[\,a\,,b\,,c\,,d\,,x\,]\,/\,(x\,a\,d - x\,b\,c = 1)\ .$$

There is an epimorphic $\mathbf{k}$-*algebra morphism* $\mathrm{M}(2) \longrightarrow \mathrm{GL}(2)$ *which induces a bialgebra structure on* $\mathrm{GL}(2)$ *from that on* $\mathrm{M}(2)$. *The antipode*

$$\nu\ :\ \mathrm{GL}(2)\ \longrightarrow\ \mathrm{GL}(2)$$
$$a\,,b\,,c\,,d\,,x \longmapsto x\,d\,,\,-x\,b\,,\,-x\,c\,,x\,a\,,a\,d - b\,c$$

makes $\mathrm{GL}(2)$ *into a commutative Hopf algebra.*

Bradshaw: Tassel Bradshaw Group, [Wal94, Plate 22].

3

The quantum general linear group

The passage from quantum to classical mechanics is quite well defined by taking the limit as Planck's constant $\hbar$ tends to 0. The passage in the other direction is not so clear cut, and may not be uniquely determined. On the algebraic side, "quantization" involves deforming commutative algebras to non-commutative ones:

$$\text{e.g.,} \qquad x\,y \;=\; y\,x \qquad \text{becomes} \qquad x\,y \;=\; e^{\hbar}\,y\,x\,.$$

Usually we deal with $q = e^{\hbar}$ rather than $\hbar$, so classical results correspond to the case $q = 1$. Quantum spaces correspond to more general $\mathbf{k}$-algebras, not necessarily commutative.

Let $\mathbf{k}$ be a fixed field and fix $q \in \mathbf{k}$ with $q \neq 0$. Write $\mathbf{k}\langle x_1 , \dots , x_n \rangle$ for the $\mathbf{k}$-algebra of polynomials in *non-commuting* indeterminates $x_1 , \dots , x_n$. As a vector space over $\mathbf{k}$, a basis is given by those elements

$$x_{\xi(1)}^{m_1}\, x_{\xi(2)}^{m_2}\, \cdots\, x_{\xi(r)}^{m_r}$$

for which $r \in \mathbb{N}$ and $m_1, \dots, m_r \in \mathbb{Z}^+$ and $\xi : \{1, \dots, r\} \longrightarrow \{1, \dots, n\}$ is any function. Notice that

$$\mathbf{k}[\,x\,,y\,] = \mathbf{k}\langle x\,,y\,\rangle / (\,x\,y = y\,x\,)\,.$$

The coordinate algebra of the space of *quantum 2×2 matrices* is defined by

$$\mathrm{M}_q(2) \;=\; \mathbf{k}\langle a\,,b\,,c\,,d\,\rangle / R$$

where R is the system of equations

$$ab \;=\; q^{-1}ba \;,\quad ac \;=\; q^{-1}ca \;,\quad cd \;=\; q^{-1}dc \;,\quad bd \;=\; q^{-1}db$$
$$bc \;=\; cb \;,\quad ad - da \;=\; (q^{-1} - q)\,bc\,.$$

(mnemonic)

The monomials $a^{m_1} b^{m_2} c^{m_3} d^{m_4}$ form a basis for the algebra, as a vector space over $\mathbf{k}$.

$$\mathbf{Alg_k}\left(M_q(2),A\right) \cong \left\{ \begin{pmatrix} a & b \\ c & d \end{pmatrix} \in \mathbf{Mat}\,(2,A) \mid R \text{ holds} \right\}.$$

Theorem 3.1 *Let* $\begin{pmatrix} a & b \\ c & d \end{pmatrix}$ *and* $\begin{pmatrix} a' & b' \\ c' & d' \end{pmatrix}$ *be two A-points of* $M_q(2)$ *such that each entry of the first commutes with each entry of the second.*

(i) *The product* $\begin{pmatrix} a & b \\ c & d \end{pmatrix}\begin{pmatrix} a' & b' \\ c' & d' \end{pmatrix}$ *(as matrices) is an A-point of* $M_q(2)$.

(ii) *The "q-determinant"* $\det_q\begin{pmatrix} a & b \\ c & d \end{pmatrix} = (ad - q^{-1}bc)$ *commutes with each of* a,b,c,d *and*

$$\det_q\left(\begin{pmatrix} a & b \\ c & d \end{pmatrix}\begin{pmatrix} a' & b' \\ c' & d' \end{pmatrix}\right) = \det_q\begin{pmatrix} a & b \\ c & d \end{pmatrix} \times \det_q\begin{pmatrix} a' & b' \\ c' & d' \end{pmatrix}.$$

(iii) *If* $\det_q\begin{pmatrix} a & b \\ c & d \end{pmatrix}$ *is invertible in A then*

$$\begin{pmatrix} a & b \\ c & d \end{pmatrix}^{-1} = \left(\det_q\begin{pmatrix} a & b \\ c & d \end{pmatrix}\right)^{-1}\begin{pmatrix} d & -qb \\ -q^{-1}c & a \end{pmatrix}$$

is an A-point of $M_{q^{-1}}(2)$.

The above result can be proved by direct calculation, but this gives little insight into the special nature of the relations R. Examples such as this arose in work of L. D. Fadde'ev [FRT88] and his school on the *quantum inverse scattering transform* (QIST) method. The version I present here comes from some lectures of Yu. Manin [Man88] given at Université de Montréal in June 1988. The following "explanation" of Theorem 3.1 is due to Yu. Kobyzev (Moscow, winter 1986–87).

Introduce the *quantum plane*, as defined by the $\mathbf{k}$-algebra

$$\mathbb{A}_q^{2|0} = \mathbf{k}\langle x,y \rangle/(xy = q^{-1}\,yx).$$

The monomials $x^m y^n$ with $m,n \in \mathbb{N}$ form a basis for this as a vector space. We also need to consider a quantized version of a Grassmannian algebra in two variables:

$$\mathbb{A}_q^{0|2} = \mathbf{k}\langle \xi,\eta \rangle/(\xi^2 = \eta^2 = 0 = \xi\eta + q\,\eta\xi).$$

The monomials $\xi^m \eta^n$ with $m,n \in \{0,1\}$ form a basis for this algebra. The reason for the funny superscripts 2|0 and 0|2 comes from "supergeometry" where dimensions are represented by pairs $d\,|\,d'$ of numbers. This $\mathbb{A}_q^{0|2}$ is a *quantum superplane*.

An A-point of B is called *generic* when the algebra morphism $B \twoheadrightarrow A$ is injective.

Theorem 3.2 *Suppose* (x, y) *is a generic A-point of* $\mathbb{A}_q^{2|0}$ *and* (ξ, η) *is a generic A-point of* $\mathbb{A}_q^{0|2}$. *Suppose* $a, b, c, d \in A$ *all commute with* x, y, ξ, η. *Put*

$$\begin{pmatrix} x' \\ y' \end{pmatrix} = \begin{pmatrix} a & b \\ c & d \end{pmatrix} \begin{pmatrix} x \\ y \end{pmatrix} \ , \quad \begin{pmatrix} x'' \\ y'' \end{pmatrix} = \begin{pmatrix} a & c \\ b & d \end{pmatrix} \begin{pmatrix} x \\ y \end{pmatrix} \ , \quad \begin{pmatrix} \xi' \\ \eta' \end{pmatrix} = \begin{pmatrix} a & b \\ c & d \end{pmatrix} \begin{pmatrix} \xi \\ \eta \end{pmatrix} .$$

If $q^2 \neq -1$, *the following conditions are equivalent* :
(i) (x', y') *and* (x'', y'') *are points of* $\mathbb{A}_q^{2|0}$;
(ii) (x', y') *is a point of* $\mathbb{A}_q^{2|0}$ *and* (ξ', η') *is a point of* $\mathbb{A}_q^{0|2}$;

(iii) $\begin{pmatrix} a & c \\ b & d \end{pmatrix}$ *is a point of* $\mathrm{M}_q(2)$.
[For $q^2 = -1$ *we only have* (iii) $\Rightarrow$ (i) & (ii).*]*

Proof. (i) $\Leftrightarrow$ (iii). (x', y') is a point of $\mathbb{A}_q^{2|0}$ iff $x'y' = q^{-1}y'x'$; that is, iff $(a\, x + b\, y)(c\, x + d\, y) = q^{-1}(c\, x + d\, y)(a\, x + b\, y)$. Multiply out the products using the fact that a, b, c, d each commute with x and y; since (x, y) is generic, we can equate coefficients of $x^2, y^2, x\, y$. So the single equation is in fact equivalent to the following set of three equations:

$$(*) \qquad ac = q^{-1}ca \ , \quad bd = q^{-1}db \ , \quad ad - da = q^{-1}cb - q\, bc .$$

Interchanging b and c we see that (x'', y'') is a point of $\mathbb{A}_q^{2|0}$ iff

$$(**) \qquad ab = q^{-1}ba \ , \quad cd = q^{-1}dc \ , \quad ad - da = q^{-1}bc - q\, cb .$$

Taking the last equations in $(*)$ & $(**)$ we get $q^{-1}cb - q\, bc = q^{-1}bc - q\, cb$; that is, $(q + q^{-1})(bc - cb) = 0$ hence $bc = cb$, provided $q^2 \neq -1$.
So (iii) $\Leftrightarrow$ $(*)$ & $(**)$, which together are equivalent to (i).
(ii) $\Leftrightarrow$ (iii). (ξ', η') is a point of $\mathbb{A}_q^{0|2}$ iff $0 = (a\xi + b\eta)^2 = (c\xi + d\eta)^2 = (a\xi + b\eta)(c\xi + d\eta) + q\, (c\xi + d\eta)(a\xi + b\eta)$. Using $\xi^2 = \eta^2 = 0$ these become $ab\, \xi\eta + ba\, \eta\xi = 0$ and $cd\, \xi\eta + dc\, \eta\xi = 0$ and $ab\, \xi\eta + bc\, \eta\xi + q\, (cb\, \xi\eta + da\, \eta\xi) = 0$. Using $\xi\eta = -q\, \eta\xi$ and the linear independence of η and ξ in A, we get that $-q\, ab + ba = 0$ and that $-q\, cd + dc = 0$ and also $-q\, (ad + q\, cb) + bc + q\, da = 0$. These are equivalent to $(**)$. So (ii) $\Leftrightarrow$ $(*)$ and $(**)$ $\Leftrightarrow$ (i). $\square$

In other words, the relations R are precisely what is needed for $\begin{pmatrix} a & b \\ c & d \end{pmatrix}$ and its transpose to both transform the quantum plane into itself; or for $\begin{pmatrix} a & b \\ c & d \end{pmatrix}$ to transform both the plane and superplane into themselves.

Proof of Theorem 3.1. (i) Let B be the free **k**-algebra containing the indeterminates $a, b, c, d, a', b', c', d', x, y$ subject to the relations on these variables in the hypotheses of Theorems 3.1 and 3.2. Then (x, y) is generic;

$\begin{pmatrix} a & b \\ c & d \end{pmatrix}$ and $\begin{pmatrix} a' & b' \\ c' & d' \end{pmatrix}$ are B-points of $\mathrm{M}_q(2)$. By Theorem 3.2, we have

that $\begin{pmatrix} a & c \\ b & d \end{pmatrix}\begin{pmatrix} x \\ y \end{pmatrix}$ and $\begin{pmatrix} a' & b' \\ c' & d' \end{pmatrix}\begin{pmatrix} x \\ y \end{pmatrix}$ are B-points of $\mathbb{A}_q^{2|0}$. Each coordinate

in the first of these commutes with all of a', b', c', d' while coordinates in the

second commute with a, b, c, d. Also $\begin{pmatrix} a & c \\ b & d \end{pmatrix}\begin{pmatrix} x \\ y \end{pmatrix}$ is generic since when

composed with $B \longrightarrow \mathbb{A}_q^{2|0}$ for which $(a,b,c,d,x,y) \longmapsto (1,0,0,1,x,y)$

we get (x,y), which is generic. Similarly $\begin{pmatrix} a' & b' \\ c' & d' \end{pmatrix}\begin{pmatrix} x \\ y \end{pmatrix}$ is generic. So by

Theorem 3.2 we have $\begin{pmatrix} a' & c' \\ b' & d' \end{pmatrix}\begin{pmatrix} a & c \\ b & d \end{pmatrix}\begin{pmatrix} x \\ y \end{pmatrix}$ and $\begin{pmatrix} a & b \\ c & d \end{pmatrix}\begin{pmatrix} a' & b' \\ c' & d' \end{pmatrix}\begin{pmatrix} x \\ y \end{pmatrix}$

both being B-points of $\mathbb{A}_q^{2|0}$. Again by Theorem 3.2, $\begin{pmatrix} a & b \\ c & d \end{pmatrix}\begin{pmatrix} a' & b' \\ c' & d' \end{pmatrix}$ is a

B-point of $\mathrm{M}_q(2)$.

To obtain the result for the given A apply the morphism $B \longrightarrow A$ for which
$(a,b,\ldots,d',x,y) \longmapsto (a,b,\ldots,d',0,0)$.

(ii) We now get a natural definition of the quantum determinant which immediately yields its multiplicativity: in the notation of Theorem 3.2,

$$\xi'\eta' = (a\xi + b\eta)(c\xi + d\eta) = \det{}_q\begin{pmatrix} a & b \\ c & d \end{pmatrix}\xi\eta\,.$$

(iii) This is left as an exercise for the reader. $\qquad\qquad\qquad\qquad\square$

The *quantum general linear group* is defined from 2×2 matrices by inverting the determinant:

$$\mathrm{GL}_q(2) = \mathrm{M}_q(2)[t]/(t\,a = a\,t\,,\ t\,b = b\,t\,,\ t\,c = c\,t\,,\ d\,t = t\,d\,,\ t\det{}_q = 1)\,.$$

Similarly, the *quantum special linear group* is defined by requiring that the determinant be equal to 1:

$$\mathrm{SL}_q(2) = \mathrm{M}_q(2)/(\det{}_q = 1)\,.$$

Theorem 3.2 describes the representations of these "groups" on quantum spaces $\mathbb{A}_q^{2|0}$ and $\mathbb{A}_q^{0|2}$.

Exercise 3.1 *Give a direct proof of Theorem 3.1 applied to quantum 2×2 matrices.*

4

Modules and tensor products

Let R be a ring (not necessarily commutative). We write R^{op} for the ring with opposite multiplication

$$R \times R \xrightarrow{\ \sigma\ } R \times R \xrightarrow{\ \mu\ } R .$$

(To say R is commutative is to say $R^{\mathrm{op}} = R$.)

A *left R-module* is an abelian group M (written additively) together with a function

$$R \times M \longrightarrow M \quad \text{whereby} \quad (r,m) \longmapsto rm$$

called *scalar multiplication*, such that

$$1\,m = m \quad , \quad (r\,s)\,m = r\,(s\,m)$$
$$(r+r')\,m = r\,m + r'\,m \quad , \quad r\,(m+m') = r\,m + r\,m' .$$

A *right R-module* is defined similarly, with multiplication $M \times R \longrightarrow M$.

A left R^{op}-module structure on an abelian group M "is the same" as a right R-module structure. More precisely, $\mu : R \times M \longrightarrow M$ is a scalar multiplication for a left R^{op}-module iff $M \times R \xrightarrow{\ \sigma\ } R \times M \xrightarrow{\ \mu\ } M$ is one for a right R-module. In this way, we can deal only with left R-modules and omit "left", unless we explicitly stipulate otherwise.

If R is commutative, $R = R^{\mathrm{op}}$ and there is no need to distinguish left and right modules. If R is a field, an R-module is precisely a *vector space* over R. Furthermore, $\mathbb{Z}$-modules are precisely abelian groups since each abelian group A admits a unique $\mathbb{Z}$-scalar multiplication given by $n\,a = a + \cdots + a$ (n terms) for $n \geq 0$ and $n\,a = -\big((-n)a\big)$ for $n < 0$.

A subset X of an R-module M is said to *generate* M (or *span* M) when, for each $m \in M$, there exist $r_1, \ldots, r_n \in R$ and $x_1, \ldots, x_n \in X$ such that

$$(*) \qquad\qquad m = r_1 x_1 + \cdots + r_n x_n .$$

Call M *finitely generated* when it is generated by some finite subset.

A (not necessarily finite) subset X of M is *linearly independent* when for $x_1, \ldots, x_n \in X$ distinct elements, having a relation of the form $r_1 x_1 + \cdots + r_n x_n = 0$ with $r_1, \ldots, r_n \in R$ implies that $r_1 = \cdots = r_n = 0$. Then each expression $(*)$ is unique up to order of factors (with $x_1, \ldots, x_n$ distinct).

An R-module F is said to be *free* when it is generated by some linearly independent subset. Every vector space is free, but this is peculiar to R being a field. It is easy to see that $\mathbb{Z}/(2)$ is not a free abelian group.

Each set X determines an R-module

$$\mathcal{F}_R(X) = \{r_1 x_1 + \cdots + r_n x_n \mid r_i \in R, \, x_i \in X, \, n \in \mathbb{N}\}$$

with addition and scalar multiplication defined in the obvious way. We can identify $x \in X$ with $1\, x \in \mathcal{F}_R(X)$ and see easily that X is linearly independent and generates $\mathcal{F}_R(X)$. So $\mathcal{F}_R(X)$ is free.

For R-modules M and N, a function $f : M \longrightarrow N$ is *(left)R-linear* (or an *R-module morphism*) when $f(m+m') = f(m)+f(m')$ and $f(r\,m) = r\,f(m)$ for all $m, m' \in M$ and $r \in R$. Write $\mathrm{Hom}_R(M, N)$ for the abelian group of R-linear functions $f : M \longrightarrow N$; the addition is given by $(f + g)(m) = f(m) + g(m)$.

Warning: You may think $\mathrm{Hom}_R(M, N)$ becomes an R-module by defining $(rf)(m) = r\,f(m)$. But this rf does not preserve scalar multiplication when R is non-commutative.

For sets X and Y, write Y^X for the set of all functions $f : X \longrightarrow Y$. An R-linear function $f : \mathcal{F}_R(X) \longrightarrow M$ is uniquely determined by its restriction to X. Indeed, this gives an isomorphism of abelian groups

$$\mathrm{Hom}_R(\mathcal{F}_R(X), M) \cong M^X$$

where the addition on M^X is pointwise.

A *submodule* H of an R-module M is a subset which is closed under addition and scalar multiplication. This gives an equivalence relation $\equiv_H$ on M whereby

$$m \equiv_H m' \qquad \text{if and only if} \qquad m - m' \in H.$$

The equivalence class containing $m \in M$ is $m + H = \{m + h \mid h \in H\}$, called the *H-coset* containing m. The set M/H of H-cosets becomes an R-module via

$$(m + H) + (n + H) = (m + n) + H \quad, \quad r(m + H) = r\,m + H.$$

We have a surjective R-linear function $\rho : M \longrightarrow M/H$ for which $\rho(m) = m + H$. For each R-linear $g : M \longrightarrow N$ with $g(m) = 0$ for all $m \in H$,

there exists a unique R-linear $\hat{g} : M/H \longrightarrow N$ with $\hat{g} \circ \rho = g$. The kernel $\ker f = \{m \in M \mid f(m) = 0\}$ of any R-linear $f : M \longrightarrow N$ is a submodule of M; we have a commutative diagram

$$
\begin{array}{ccc}
M & \xrightarrow{\ f\ } & N \\
{\scriptstyle \rho}\big\downarrow & & \big\uparrow \\
M/\ker f & \xrightarrow{\ \cong\ } & \operatorname{im} f
\end{array}
$$

of R-modules, where $\operatorname{im} f = \{f(m) \mid m \in M\}$ is the image of f, the bottom arrow is an R-linear isomorphism, and the right arrow is an inclusion of a submodule.

The submodule (X) *generated* by a subset X of an R-module M is the smallest submodule of M which contains X. As such it is the image of the R-linear function $\mathcal{F}_R(X) \longrightarrow M$ whose restriction to X is the inclusion $X \hookrightarrow M$. Of course (X) is generated by X, but in general not freely.

Suppose that M is a right R-module and N is a left R-module. A function $f : M \times N \longrightarrow A$ into an abelian group A is R-*bilinear* when it satisfies

$$
\begin{aligned}
f(m, n + n') &= f(m, n) + f(m, n') \\
f(m + m', n) &= f(m, n) + f(m', n) \\
f(m r, n) &= f(m, r n) .
\end{aligned}
$$

Write $\mathbf{Bil}_R(M, N; A)$ for the abelian group, which is a subgroup of $A^{M \times N}$, of R-bilinear functions $f : M \times N \longrightarrow A$. Our main goal is to construct a "universal" bilinear function $\lambda : M \times N \longrightarrow M \otimes_R N$.

Let B denote the subset of the abelian group $\mathcal{F}_{\mathbb{Z}}(M \times N)$ consisting of all elements of the form

$$
\begin{aligned}
&(m + m', n) - (m, n) - (m', n), \\
&(m, n + n') - (m, n) - (m, n'), \\
&(m r, n) - (m, r n)
\end{aligned}
$$

for $m, m' \in M$ with $n, n' \in N$ and $r \in R$. Put

$$
M \otimes_R N = \mathcal{F}_{\mathbb{Z}}(M \times N)/(B) .
$$

Then we have abelian group isomorphisms

$$
\begin{aligned}
\operatorname{Hom}_{\mathbb{Z}}(M \otimes_R N, A) &= \operatorname{Hom}_{\mathbb{Z}}\big(\mathcal{F}_{\mathbb{Z}}(M \times N)/(B), A\big) \\
&\cong \{g \in \operatorname{Hom}_{\mathbb{Z}}\big(\mathcal{F}_{\mathbb{Z}}(M \times N), A\big) \mid g \text{ is zero on } B\} \\
&\cong \{f \in A^{M \times N} \mid f \text{ is } R\text{-bilinear}\} \\
&= \mathbf{Bil}_R(M, N; A) .
\end{aligned}
$$

In particular by taking $A = M \otimes_R N$ we get the identity morphism $A \longrightarrow A$ corresponding, under the composite of the above string of isomorphisms, to a bilinear morphism $\lambda : M \times N \longrightarrow M \otimes_R N$. Then we easily see that each R-bilinear $f : M \times N \longrightarrow A$ uniquely determines an abelian group morphism $g : M \otimes_R N \longrightarrow A$ with $g \circ \lambda = f$.

For $(m, n) \in M \times N$, we put $m \otimes n = \lambda(m, n)$. A typical element of $M \otimes_R N$ then has the form

$$\sum_{i=1}^{k} m_i \otimes n_i$$

where $m_1, \ldots, m_k \in M$ and $n_1, \ldots, n_k \in N$. These elements satisfy

$$
\begin{aligned}
(m + m') \otimes n &= m \otimes n + m' \otimes n \\
m \otimes (n + n') &= m \otimes n + m \otimes n' \\
m r \otimes n &= m \otimes r n \,.
\end{aligned}
$$

With R and S rings, a *module M from R to S*, written $M : R \rightarrowtail S$, is an abelian group M enriched with a left R-module structure and a right S-module structure related by

$$(r m)s = r(m s)$$

for all $r \in R$, $m \in M$ and $s \in S$. (In the literature this structure is also known as a *left R-/right S-bimodule*.) In this notation, tensor product can be viewed as a kind of "composition of modules".

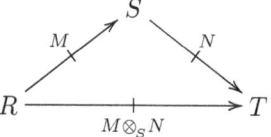

For M and N as above, $M \otimes_S N$ becomes a module from R to T by defining

$$r(m \otimes n)t = (r m) \otimes (n t) \,.$$

This composition of modules is not strictly associative, but is associative up to canonical isomorphisms much like cartesian product of sets. This can be seen by defining a *multiple tensor product* as we now proceed to do.

For rings R and S and any set X, there is a *free module from R to S generated by X*. It is denoted by $\mathcal{F}_R^S(X)$ and its elements have the form

$$r_1 x_1 s_1 + \cdots + r_n x_n s_n \qquad \text{for} \quad r_i \in R, \ s_i \in S, \ x_i \in X, \ n \in \mathbb{N} \,.$$

For each module $M : R \rightarrowtail S$ we have

$$\mathrm{Hom}_R^S\left(\mathcal{F}_R^S(X),M\right) \cong M^X$$

where $\mathrm{Hom}_R^S(N,M)$ is the abelian group which has as elements the left R-/right S-module morphisms $N \longrightarrow M$.

Given rings and modules as in the diagram

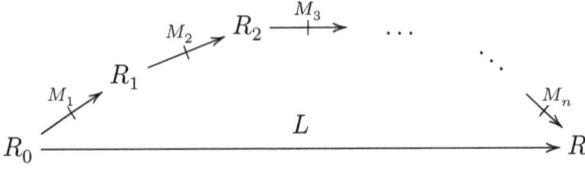

a function $f : M_1 \times \cdots \times M_n \longrightarrow L$ is called *multilinear* when it satisfies the equations

$$
\begin{aligned}
f(m_1,\ldots,m_i + m_i',\ldots,m_n) &= f(m_1,\ldots,m_i,\ldots,m_n) \\
&\quad + f(m_1,\ldots,m_i',\ldots,m_n) \\
r_0\, f(m_1,\ldots,m_n) &= f(r_0\, m_1, m_2,\ldots,m_n) \\
f(m_1,\ldots,m_i\, r_i,\, m_{i+1},\ldots,m_n) &= f(m_1,\ldots,m_i,\, r_i\, m_{i+1},\ldots,m_n) \\
f(m_1,\ldots,m_n)\, r_n &= f(m_1,\ldots,m_{n-1},\, m_n\, r_n)
\end{aligned}
$$

for $r_i \in R_i$ and $m_i, m_i' \in M_i$. Write

$$\mathbf{Mult}\,(M_1,\ldots,M_n;L)$$

for the abelian group of such functions f. It should now be clear how to construct a module

$$M_1 \otimes_{R_1} M_2 \otimes_{R_2} \cdots \otimes_{R_{n-1}} M_n : R_0 \rightarrowtail R_n$$

and multilinear function

$$\lambda : M_1 \times \cdots \times M_n \longrightarrow M_1 \otimes_{R_1} \cdots \otimes_{R_{n-1}} M_n$$

having the universal property that, for each multilinear function $f : M_1 \times \cdots \times M_n \longrightarrow L$, there exists a unique left R_0-/right R_n-module morphism $g : M_1 \otimes_{R_1} \cdots \otimes_{R_{n-1}} M_n \longrightarrow L$ for which $g \circ \lambda = f$. This describes an abelian group isomorphism

$$\mathbf{Mult}\,(M_1,\ldots,M_n;L) \cong \mathrm{Hom}_{R_0}^{R_n}(M_1 \otimes_{R_1} \cdots \otimes_{R_{n-1}} M_n, L)$$

(where $\mathrm{Hom}_R^S(M,N) = \mathbf{Mult}\,(M,N)$ is the abelian group of left R-/right S-module morphisms $M \longrightarrow N$). When there is no ambiguity about the

rings, we write $M_1 \otimes \cdots \otimes M_n$ instead of $M_1 \otimes_{R_1} \cdots \otimes_{R_{n-1}} M_n$. As with cartesian product we have canonical isomorphisms

$$(M_1 \otimes \cdots \otimes M_k) \otimes (M_{k+1} \otimes \cdots \otimes M_n) \cong M_1 \otimes \cdots \otimes M_n .$$

However, the diagonal $M \longrightarrow M \otimes M$ in which $m \longmapsto m \otimes m$, does not preserve addition. The empty tensor product $M_1 \otimes \cdots \otimes M_n$ for $n = 0$ is just R_0 as a module $R_0 \rightarrowtail R_0$, using multiplication in R as scalar multiplication on both sides. We have canonical isomorphisms

$$R \otimes_R M \cong M \cong M \otimes_S S .$$

Given $M, M' : R \rightarrowtail S$, we write

$$f : M \Rightarrow M' : R \rightarrowtail S \qquad \text{or} \qquad R \underset{M'}{\overset{M}{\Downarrow f}} S$$

to mean $f : M \longrightarrow M'$ is a left R- and right S-module morphism. Given the data

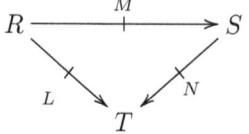

we obtain $f_1 \otimes \cdots \otimes f_n : M_1 \otimes_{R_1} \cdots \otimes_{R_{n-1}} M_n \Rightarrow M_1' \otimes_{R_1} \cdots \otimes_{R_{n-1}} M_n' :$ $R_0 \rightarrowtail R_n$ given by $(f_1 \otimes \cdots \otimes f_n) \circ \lambda = \lambda \circ (f_1 \times \cdots \times f_n)$.

We have seen that tensor products allow us to represent bilinear functions as module morphisms. Another way of doing this uses Hom instead of tensor. Given a triangle of modules

$$R \xrightarrow{\quad M \quad} S$$

we can enrich the abelian group $\mathrm{Hom}_R(M, L)$ (resp. $\mathrm{Hom}^T(N, L)$) of left R- (resp. right T-) module morphisms with a module structure

$$\mathrm{Hom}_R(M, L) : S \rightarrowtail T$$
$$(\text{resp.} \quad \mathrm{Hom}^T(N, L) : R \rightarrowtail S)$$

using the scalar multiplications

$$(s\,f\,t)(m) = f(ms)\,t \qquad (\text{resp. } (r\,g\,s)(n) = r\,g(sn)\,)\,.$$

We then have abelian group isomorphisms

$$\begin{aligned} \mathrm{Hom}_S^T(N, \mathrm{Hom}_R(M, L)) \; &\cong \; \mathbf{Mult}\,(M, N\,;L) \\ &\cong \; \mathrm{Hom}_R^S(M, \mathrm{Hom}^T(N, L)) \end{aligned}$$

induced by the canonical bijections

$$\left(L^M\right)^N \; \cong \; L^{M \times N} \; \cong \; \left(L^N\right)^M\,.$$

Combining these with the earlier results, we have

$$\begin{aligned} \mathrm{Hom}_S^T(N, \mathrm{Hom}_R(M, L)) \; &\cong \; \mathrm{Hom}_R^T(M \otimes_S N, L) \\ &\cong \; \mathrm{Hom}_R^S(M, \mathrm{Hom}^T(N, L))\,. \end{aligned}$$

These isomorphisms are determined by the *evaluation morphisms*

$$\begin{aligned} \mathrm{ev}_M &: M \otimes_S \mathrm{Hom}_R(M, L) \longrightarrow L\,, & m \otimes f &\longmapsto f(m) \\ \mathrm{ev}^N &: \mathrm{Hom}^T(N, L) \otimes_S N \longrightarrow L\,, & g \otimes n &\longmapsto g(n)\,. \end{aligned}$$

Explicitly, the first isomorphism takes any $u : N \longrightarrow \mathrm{Hom}_R(M, L)$ to the composite

$$M \otimes_S N \xrightarrow{\;1_M \otimes u\;} M \otimes_S \mathrm{Hom}_R(M, L) \xrightarrow{\;\mathrm{ev}_M\;} L\,.$$

Exercise 4.1 *For rings R, S, T and any sets X, Y prove that*

$$\begin{aligned} \mathcal{F}_R^T(X \times Y) \; &\cong \; \mathcal{F}_R^S(X) \otimes_S \mathcal{F}_S^T(Y) \\ (x\,,y\,) \; &\longmapsto \; x \otimes y\,. \end{aligned}$$

Hint: *Look at left R-/right T-module morphisms into $M : R \nrightarrow T$.*

Exercise 4.2 *Describe $\mathbb{Z}/(2) \otimes_{\mathbb{Z}} \mathbb{Z}/(5)$.*

Exercise 4.3 (a) *If R, S are rings, describe a canonical ring structure on $R \otimes_{\mathbb{Z}} S$.*

(b) *Is the function from R to $R \otimes_{\mathbb{Z}} S$ taking R to $r \otimes 1$ a ring morphism? Why?*

(c) *Show that $R \otimes_{\mathbb{Z}} S$ is the coproduct of R, S in the category of commutative rings.*

Exercise 4.4 *Show that a module M from R to S amounts to the same thing as a left $R \otimes_{\mathbb{Z}} S^{\mathrm{op}}$-module.*

Exercise 4.5 *Describe explicitly the construction of $M \otimes_S N \otimes_T L$.*

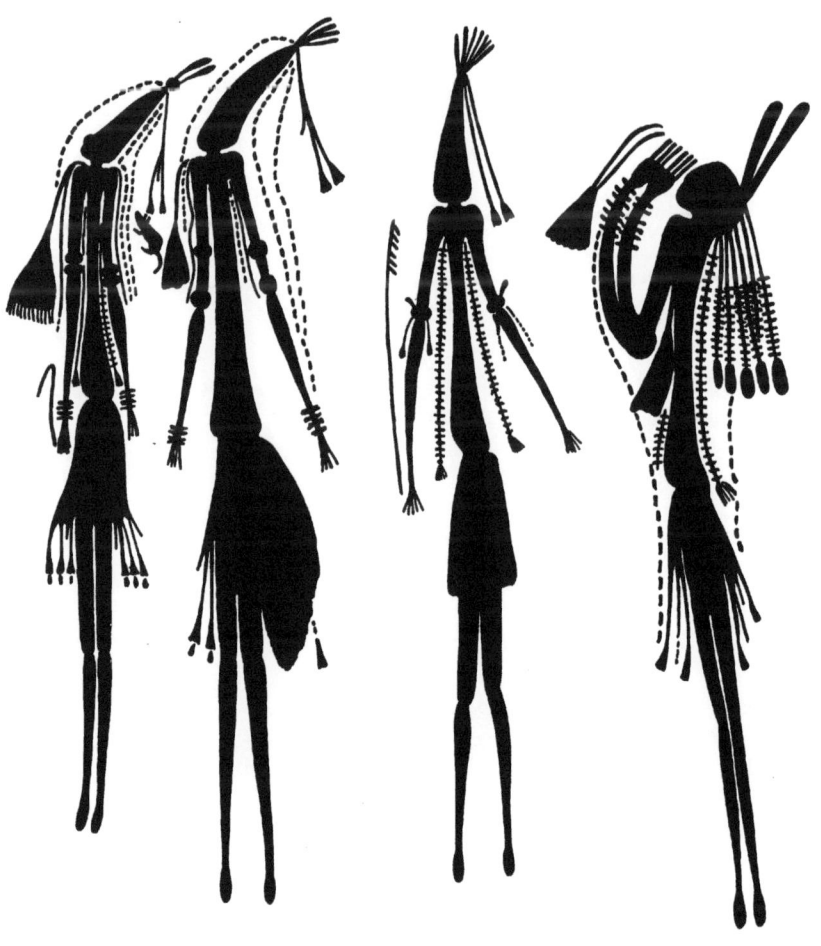

Bradshaw: Tassel Bradshaw Group, [Wal94, Plate 23].

5

Cauchy modules

A module $M : R \rightarrowtail S$ gives rise to a module $M^* = \mathrm{Hom}_R(M, R) : S \rightarrowtail R$ called the *left dual* of M. There is a canonical module morphism

$$\rho_L^M : M^* \otimes_R L \longrightarrow \mathrm{Hom}_R(M, L)$$

given by $\rho_L^M(u \otimes l)(m) = u(m)l$, for each left R-module L.

Call an $M : R \rightarrowtail S$ *Cauchy* when ρ_L^M is an isomorphism for all left R-modules L. Our goal in this section is to characterize Cauchy modules more intrinsically.

A module P is called *projective* when, for all surjective module morphisms $e : L \longrightarrow L'$ and all module morphisms $f : P \longrightarrow L'$, there exists some module morphism $g : P \longrightarrow L$ with $f = e \circ g$.

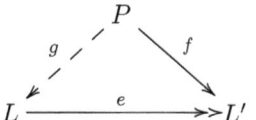

A morphism $r : M \longrightarrow N$ is said to be a *retraction* when there exists a morphism $i : N \longrightarrow M$ with $r \circ i = 1_N$. When a retraction exists from M to N, we call N a *retract* of M.

Proposition 5.1 *A module P is projective iff P is a retract of some free module F.*

Proof. (1) *A retract Q of a projective P is projective.* To see this take $i : Q \longrightarrow P$ and $r : P \longrightarrow Q$ with $r \circ i = 1_Q$. Suppose $e : L \longrightarrow L'$ is a surjective morphism and $f : Q \longrightarrow L'$. Then $f \circ r : P \longrightarrow L'$, and since P is projective, there is a morphism $h : P \longrightarrow L$ with $e \circ h = f \circ r$. But then $e \circ (h \circ i) = (e \circ h) \circ i = f \circ r \circ i = f \circ 1_Q = f$, so $g = h \circ i$ has $e \circ g = f$.

(2) *Free modules $\mathcal{F}(X)$ are projective.* Take $e : L \longrightarrow L'$ surjective and any $f : \mathcal{F}(X) \longrightarrow L'$. Then we can choose (using the axiom of choice)

an element $g(x) \in L$ for each $x \in X$ such that $e(g(x)) = f(x)$. Since $\mathcal{F}(X)$ is free, we can extend g uniquely to a morphism $g : \mathcal{F}(X) \longrightarrow L$; and furthermore $e \circ g = f$ since they agree on X.

(3) *For each module M there is a free module F and a surjective morphism* $e : F \longrightarrow M$. Just take F to be the free module $\mathcal{F}(M)$ on the underlying set of M. To give a morphism $e : F \longrightarrow M$ we only have to give it on M, so we take $e(m) = m$. Clearly this e is surjective.

(4) *If $e : F \longrightarrow P$ is surjective and P projective then e is a retraction.* For we have i as in:

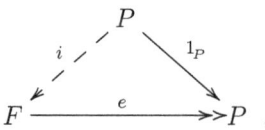

$\square$

This brings us to the *fundamental theorem of "Morita theory"*.

Theorem 5.2 *The following conditions on a module $M : R \longrightarrow S$ are equivalent.*
(i) *M is Cauchy.*
(ii) *There exists a morphism $d : S \Rightarrow M^* \otimes_R M : S \longrightarrow S$ such that both the following two composites are identity morphisms*

$$M \cong M \otimes_S S \xrightarrow{\ 1_M \otimes d\ } M \otimes_S M^* \otimes_R M \xrightarrow{\ \mathrm{ev}_M \otimes 1_M\ } R \otimes_R M \cong M$$
$$M^* \cong S \otimes_S M^* \xrightarrow{\ d \otimes 1_{M^*}\ } M^* \otimes_R M \otimes_S M^* \xrightarrow{\ 1_{M^*} \otimes \mathrm{ev}_M\ } M^* \otimes_R R \cong M^* .$$

(iii) *There exists a module $N : S \longrightarrow R$ and morphisms*

$$e : M \otimes_S N \longrightarrow R \quad , \quad d : S \longrightarrow N \otimes_R M$$

such that the following composite is the identity morphism

$$M \cong M \otimes_S S \xrightarrow{\ 1_M \otimes d\ } M \otimes_S N \otimes_R M \xrightarrow{\ e \otimes 1_M\ } R \otimes_R M \cong M .$$

(iv) *M is a finitely generated projective left R-module.*

Proof. (i) $\Rightarrow$ (ii). Since ρ_M^M is an isomorphism, there is an element of $M^* \otimes_R M$ taken by ρ_M^M to $1_M : M \longrightarrow M$. This element of $M^* \otimes_R M$ now determines a unique morphism $d : S \longrightarrow M^* \otimes_R M$ whose value at $1 \in S$ is the element. Write $d(1) = \sum_i u_i \otimes m_i$. The condition $\rho_M^M(d(1))(m) = m$ becomes $\sum_i u_i(m)\, m_i = m$ for all $m \in M$. This immediately gives that the

first composite of (ii) takes m to m. To see that the second takes $u \in M^*$ to itself we use $u(m) = u(\sum u_i(m)\, m_i) = \sum u_i(m)\, u(m_i)$.

(ii) $\Rightarrow$ (iii). Just take $N = M^*$, $e = \mathrm{ev}_M$ and d as in (ii).

(iii) $\Rightarrow$ (iv). Just put $d(1) = \sum_{i=1}^{k} n_i \otimes m_i \in N \otimes_R M$. From the fact that the composite in (iii) is the identity, we have $\sum_i e(m \otimes n_i)\, m_i = m$ for all $m \in M$. So M is generated by $m_1, \ldots, m_k$. It remains to see that M is projective. Take $s : L \longrightarrow L'$ surjective and $f : M \longrightarrow L'$. Then we can choose $l_1, \ldots, l_k \in L$ with $s(l_i) = f(m_i)$. Define $g : M \longrightarrow L$ by $g(m) = \sum_i e(m \otimes n_i)\, l_i$ and we get $s(g(m)) = \sum_i e(m \otimes n_i)\, s(l_i) = \sum_i e(m \otimes n_i)\, f(m_i) = f(\sum_i e(m \otimes n_i) m_i) = f(m)$, as required.

(iv) $\Rightarrow$ (i). It is easy to see that a retract of a Cauchy module is Cauchy (Exercise 5.3). So it suffices to show that $M = \mathcal{F}_R(X)$ is Cauchy for X a finite set $\{x_1, \ldots, x_k\}$. But then $M^* = \mathrm{Hom}_R(\mathcal{F}(X), R) \cong R^k$ and $\mathrm{Hom}_R(M, L) = \mathrm{Hom}_R(\mathcal{F}(X), L) \cong L^k$. Under these isomorphisms ρ_L^M carries across to the morphism $R^k \otimes_R L \longrightarrow L^k$ with $(r_1, \ldots, r_k) \otimes l \longmapsto (r_1\, l, \ldots, r_k\, l)$ which has inverse $(l_1, \ldots, l_k) \longmapsto \sum_{i=1}^{k} u_i \otimes l_i$, in which $u_i \in R^k$ projects to 0 in all components except the i-th where it projects to 1. So ρ_L^M is an isomorphism. $\square$

Given rings R and S, from any ring morphism $f : S \longrightarrow R$ we obtain two modules ${}_f R : S \longrightarrow R$ and $R_f : R \longrightarrow S$, which have R as underlying abelian group. They have scalar multiplicatons

$$S \times {}_f R \longrightarrow {}_f R \quad , \quad {}_f R \times R \longrightarrow {}_f R$$
$$R_f \times S \longrightarrow R_f \quad , \quad R \times R_f \longrightarrow R_f$$

given by, respectively

$$(s, r) \longmapsto f(s)\, r \quad , \quad (r, r') \longmapsto r\, r'$$
$$(r, s) \longmapsto r\, f(s) \quad , \quad (r', r) \longmapsto r'\, r.$$

For any module $L : R \longrightarrow T$ we have canonical isomorphisms

$$\begin{array}{ccccc} {}_f R \otimes_R L & \cong & L & \cong & \mathrm{Hom}_R(R_f, L) \\ r \otimes l & \longmapsto & l & & \\ & & u(1) & \longleftarrow\!\shortmid & u. \end{array}$$

It follows easily from this that

$$(R_f)^* \cong {}_f R$$

and that R_f is Cauchy.

A module $M : R \longrightarrow S$ is called *convergent* when there exists a ring morphism $f : S \longrightarrow R$ and a module isomorphism $M \cong R_f$.

The *product* $\prod_{i \in I} M_i : R \longrightarrow S$ of a family of modules $M_i : R \longrightarrow S$ with $i \in I$, has as elements the families $\mathbf{m} = (m_i)_{i \in I}$ with $m_i \in M_i$; addition and scalar multiplication are given by

$$\mathbf{m} + \mathbf{m}' = (m_i + m'_i)_{i \in I} \quad , \quad r \, \mathbf{m} \, s = (r \, m_i \, s)_{i \in I} .$$

There are *projections*

$$\mathrm{pr}_j : \prod_{i \in I} M_i \longrightarrow M_j \qquad \text{for each } j \in I$$

given by $\mathrm{pr}_j(\mathbf{m}) = m_j$. There are also injective module morphisms

$$\mathrm{in}_j : M_j \longrightarrow \prod_{i \in I} M_i \qquad \text{for each } j \in I$$

given by $\mathrm{in}_j(m) = \mathbf{m}$ where $m_j = m$ and $m_i = 0$ for all $i \neq j$; we can use these to identify each M_j with the submodule of $\prod_{i \in I} M_i$ consisting of those $\mathbf{m}$ with $m_i = 0$ for all $i \neq j$.

The *direct sum* $\sum_{i \in I} M_i : R \longrightarrow S$ is the submodule of $\prod_{i \in I} M_i$ which consists of those $\mathbf{m}$ for which $m_i = 0$ for all but finitely many $i \in I$. This is the submodule generated by the union $\cup_{i \in I} M_i$, hence we can write $\sum_{i \in I} m_i$ instead of $\mathbf{m} \in \sum_{i \in I} M_i$. Of course the injections in_j actually land in $\sum_{i \in I} M_i$.

Proposition 5.3 *There are module isomorphisms:*

(a)
$$\mathrm{Hom}_R\left(\sum_{i \in I} M_i, L \right) \;\cong\; \prod_{i \in I} \mathrm{Hom}_R(M_i, L)$$
$$f \;\longleftrightarrow\; (f \circ \mathrm{in}_i)_{i \in I} ,$$

(b)
$$\left(\sum_{i \in I} M_i \right) \otimes_S N \;\cong\; \sum_{i \in I} M_i \otimes_S N$$
$$\left(\sum_i m_i \right) \otimes n \;\longleftrightarrow\; \sum_i (m_i \otimes n) .$$

Proof. (a) *Injectivity.* If $f \circ \mathrm{in}_i = 0$ for all $i \in I$ then f is zero on each M_i and hence on $\sum M_i$.

Surjectivity. Given $f_i : M_i \longrightarrow L$ for all $i \in I$, define $f : \sum M_i \longrightarrow L$ by $f(\sum m_i) = \sum f_i(m_i)$.

(b) $\qquad \operatorname{Hom}_R\left(\left(\sum_i M_i\right) \otimes_S N, L\right) \;\cong\; \operatorname{Hom}^S\left(\sum_i M_i, \operatorname{Hom}^R(N, L)\right)$

$$\cong \prod_i \operatorname{Hom}^S(M_i, \operatorname{Hom}^R(N, L))$$

$$\cong \prod_i \operatorname{Hom}_R(M_i \otimes_S N, L)$$

$$\cong \operatorname{Hom}_R\left(\sum_i (M_i \otimes_S N), L\right)$$

and the composite isomorphism is induced by the given map in (b). This proves it. (Why?) $\qquad\qquad\qquad\qquad\qquad\qquad\qquad\qquad\qquad\qquad$ □

When I is finite, notice that $\sum_{i \in I} M_i = \prod_{i \in I} M_i$. This is also frequently written $\oplus_{i \in I} M_i$. So $M \oplus N = M \times N = M + N$.

Exercise 5.1 *Show that a module P is finitely generated and projective if and only if P is a retract of a free module on a finite set.*
Hint: *In part (3) of the proof of Proposition 5.1 we did not need $\mathcal{F}(M)$; only $\mathcal{F}(X)$ for any X generating M.*

Exercise 5.2 *Suppose M is a finitely generated projective module over a commutative ring R. Show that M^* is a finitely generated projective module and that the canonical morphism $M \to M^{**}$ is bijective.*

Exercise 5.3 *Prove directly from the definition of "Cauchy module" that a retract of a Cauchy module is Cauchy.*

Exercise 5.4 *Re-examine the proof of Theorem 5.2 to show that a module $M : R \longrightarrow S$ is Cauchy if and only if ρ_M^M is surjective.*

Bradshaw: Tassel Bradshaw Group, [Wal94, Plate 30].

6

Algebras

Let R be any ring. An *algebra over R* (or *R-algebra*) is a module $A : R \rightarrowtail R$ together with module morphisms

$$\mu : A \otimes_R A \longrightarrow A \quad , \quad \eta : R \longrightarrow A$$

such that

(Associativity)
$$A \otimes_R A \otimes_R A \overset{\mu \otimes 1_A}{\underset{1_A \otimes \mu}{\rightrightarrows}} A \otimes_R A \overset{\mu}{\longrightarrow} A$$

(Identity)
$$A \overset{\eta \otimes 1_A}{\underset{1_A \otimes \eta}{\rightrightarrows}} A \otimes_R A \overset{\mu}{\longrightarrow} A \ .$$
$$\underset{1_A}{}$$

Notice that A becomes a ring with multiplication $a\,b = \mu(a \otimes b)$ and identity $1 = \eta(1)$.

For R-algebras $A, B : R \rightarrowtail R$ an *algebra morphism* $f : A \longrightarrow B$ is a module morphism satisfying

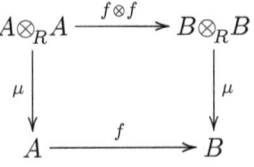   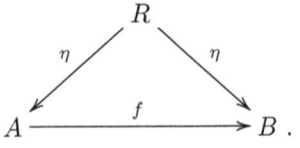

We write $\mathbf{Alg}_R(A, B)$ for the set of algebra morphisms from A to B.

Example 6.1 *For any module* $M : R \rightarrowtail S$, *the* endomorphism *algebras, over S and R respectively, are given by*

$$\begin{aligned}
\operatorname{End}_R(M) &= \operatorname{Hom}_R(M, M) : S \rightarrowtail S \\
\operatorname{End}^S(M) &= \operatorname{Hom}^S(M, M) : R \rightarrowtail R .
\end{aligned}$$

In each case the multiplication is given by composition.

A module morphism

$$\hat{\mu} : A \Rightarrow \operatorname{End}^S(M) : R \rightarrowtail R$$

corresponds to a module morphism

$$\mu : A \otimes_R M \Rightarrow M : R \rightarrowtail S .$$

To say that $\hat{\mu}$ is an algebra morphism is precisely to say that μ is a scalar multiplication enriching M with the structure of left A-module.

Example 6.2 *For any module* $M : R \rightarrowtail R$, *write*

$$M^{\otimes n} = M \otimes_R \cdots \otimes_R M \qquad (n \text{ terms}) .$$

The tensor algebra *on M is defined by the "geometric series"*

$$T(M) = \sum_{n=0}^{\infty} M^{\otimes n}$$

with multiplication $\mu : T(M) \otimes_R T(M) \longrightarrow T(M)$ induced by the canonical isomorphisms

$$M^{\otimes p} \otimes_R M^{\otimes q} \xrightarrow{\ \cong\ } M^{\otimes (p+q)}$$

and unit $\eta : R \longrightarrow T(M)$ equal to the injection

$$\operatorname{in}_0 : R = M^{\otimes 0} \longrightarrow \sum_{n=0}^{\infty} M^{\otimes n} .$$

Composition with the injection $\operatorname{in}_1 : M \longrightarrow T(M)$ gives a bijection

$$\mathbf{Alg}_R(T(M), A) \cong \operatorname{Hom}_R^R(M, A)$$

for any algebra A. The inverse takes $f : M \longrightarrow A$ to $g : T(M) \longrightarrow A$ given by $g(m_1 \otimes \cdots \otimes m_r) = f(m_1) \cdots f(m_r)$. In particular, if we take $M = A$ and $f = 1_A$, we obtain an algebra morphism

$$\mu : T(A) \longrightarrow A \quad \text{with} \quad \mu(a_1 \otimes \cdots \otimes a_r) = a_1 \cdots a_r .$$

Example 6.3 *Let G be any monoid. There is an R-algebra $R(G)$ which is just the free module $\mathcal{F}_R^R(G)$ on the underlying set of G together with the multiplication μ which extends that of G in the sense that*

$$
\begin{array}{ccc}
G \times G & \longhookrightarrow \mathcal{F}_R^R(G \times G) \;\cong\; \mathcal{F}_R^R(G) \otimes_R \mathcal{F}_R^R(G) \\
\mu \downarrow & \qquad\qquad\qquad \downarrow \mu \\
G & \longhookrightarrow \mathcal{F}_R^R(G) \;.
\end{array}
$$

This $R(G)$ is called the monoid R-algebra of G; *or when G is a group, the* group R-algebra of G. *Each monoid morphism $G \longrightarrow A$ into the multiplicative monoid of A extends uniquely to an R-algebra morphism $R(G) \longrightarrow A$.*

A representation of G on M is an $R(G)$-module. Scalar multiplication $R(G) \otimes_R M \longrightarrow M$ can be viewed as a monoid morphism

$$ G \longrightarrow \operatorname{End}_R(M) \;. $$

The subset of M given by $\{gm - m \mid g \in G, m \in M\}$ generates a submodule $(gm - m \mid g \in G, m \in M)$ and we write M/G for the quotient module $M/(gm - m \mid g \in G, m \in M)$.

An *ideal* in an algebra A is a submodule I such that $a\,x\,b \in I$ for all $x \in I$ and $a, b \in A$. There is a unique structure of algebra on the quotient module A/I for which the canonical $\rho : A \longrightarrow A/I$ is an algebra morphism. The kernel of any algebra morphism $f : A \longrightarrow B$ is an ideal in A.

If X is a subset of an algebra A, we write (X) for the smallest ideal of A containing X. This should not cause confusion with the module notation; the ideal (X) is precisely the submodule $(A\,X\,A)$ generated by the subset $A\,X\,A = \{a\,x\,b \mid a, b \in A, x \in X\}$ of A. Now given any algebra morphism $g : A \longrightarrow B$ satisfying $g(x) = 0$ for all $x \in X$, then an algebra morphism $f : A/(X) \longrightarrow B$ is uniquely determined via the equation $f \circ \rho = g$.

Now suppose that R is a commutative ring. Then left R-modules are "the same thing" as right R-modules. Moreover, each left R-module M can be naturally regarded as a module $M : R \longmapsto R$ by defining

$$ r\,m\,s = (r\,s)\,m \qquad \text{for all} \quad r, s \in R \quad \text{and} \quad m \in M \;. $$

In dealing with modules over a commutative ring, we happily regard left modules as two-sided via this process. Thus for R-modules $M_1, \ldots, M_n$ we have a tensor product R-module

$$ M_1 \otimes_R \cdots \otimes_R M_n \;. $$

Furthermore every permutation ξ on the set $\{1, \ldots, n\}$ induces a canonical module isomorphism

$$
\begin{array}{ccc}
\sigma_\xi : M_1 \otimes_R \cdots \otimes_R M_n & \xrightarrow{\ \cong\ } & M_{\xi(1)} \otimes_R \cdots \otimes_R M_{\xi(n)} \\
m_1 \otimes \cdots \otimes m_n & \longmapsto & m_{\xi(1)} \otimes \cdots \otimes m_{\xi(n)} .
\end{array}
$$

Given an algebra A over R with multiplication μ and unit η, we obtain an *opposite algebra* A^{op} on the same module A, with multiplication

$$
\mu^{\mathrm{op}} : A \otimes_R A \xrightarrow{\ \sigma\ } A \otimes_R A \xrightarrow{\ \mu\ } A
$$

and with the same unit $\eta : R \longrightarrow A$. Call A *commutative* when $A^{\mathrm{op}} = A$ as algebras. It follows that the composite

$$
A \otimes_R \cdots \otimes_R A \xrightarrow{\ \sigma_\xi\ } A \otimes_R \cdots \otimes_R A \xrightarrow{\ \mu\ } A
$$

is independent of the permutation ξ.

Example 6.4 *For any set X, the set R^X of all functions from X into the commutative ring R becomes a commutative R-algebra after defining addition, scalar multiplication and multiplication as acting pointwise. The unit $\eta : R \longrightarrow R^X$ is given by $\eta(r)(x) = r$ for all $r \in R$ and $x \in X$.*

Example 6.5 *Let M be any module over the commutative ring R. There is a natural representation of the symmetric group $\mathcal{S}_n$ on $M^{\otimes n}$ given by*
$$\sigma_- : \mathcal{S}_n \longrightarrow \mathrm{End}_R(M^{\otimes n}) ; \text{ that is, } \xi \cdot (m_1 \otimes \cdots \otimes m_n) = m_{\xi(1)} \otimes \cdots \otimes m_{\xi(n)} .$$
The symmetric R-algebra on M is given by the "exponential series"

$$
\mathcal{S}(M) = \sum_{n=0}^{\infty} M^{\otimes n} / \mathcal{S}_n .
$$

Another way of constructing this is as follows. For any R-algebra A we can form a commutative R-algebra by taking the quotient of A by the ideal $(ab - ba \mid a, b \in A)$. Applying this construction to the tensor algebra $T(M)$ gives $\mathcal{S}(M)$.

For every commutative R-algebra A, we have that

$$
\mathbf{Alg}_R(\mathcal{S}(M), A) \cong \mathrm{Hom}_R(M, A) .
$$

In particular, corresponding to the identity map $1_A : A \longrightarrow A$ there is an R-algebra morphism $\mu : \mathcal{S}(A) \longrightarrow A$.

The following diagram of "forgetful" and "free" constructions summarizes some of the above.

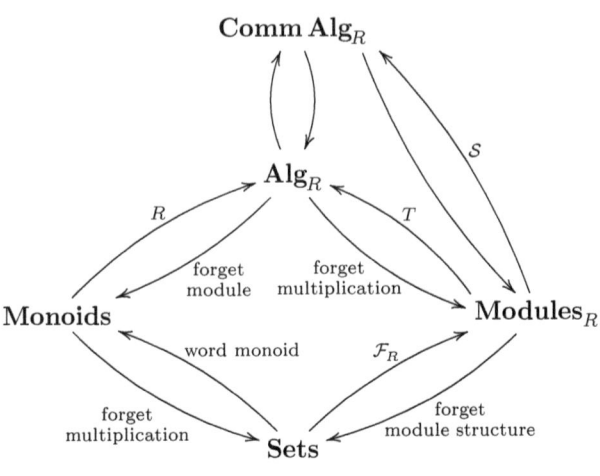

Skew commutativity $ab + ba = 0$ for an R-algebra is too strong as a requirement for *all* $a, b \in A$. For example taking $b = 1$, it would give $(1 + 1)a = 0$. Hence if R is a field of characteristic other than 2 (meaning $1 + 1 \neq 0$ in R), we would get $a = 0$, and so $A = \{0\}$.

An R-algebra A is said to be *skew commutative* when for all $a \in A$ either $a^2 = 0$ or $a \in \eta(R)$. Then, provided none of a, b and $a + b$ is in the image of $\eta : R \longrightarrow A$, we have

$$ab + ba = (a + b)^2 - a^2 - b^2 = 0.$$

Example 6.6 *For any R-algebra A we can form the quotient by the ideal $(a^2 \mid a \notin \eta(R))$ to obtain a skew commutative algebra. If we do this to the tensor algebra $T(M)$ we obtain the* exterior algebra $\Lambda(M)$. *Alternatively, let $\Lambda_n(M)$ be the quotient module of $M^{\otimes n}$ by the submodule generated by the elements $m_1 \otimes \cdots \otimes m_n$ with $m_i = m_j$ for some $i \neq j$ (this submodule is $\{0\}$ when $n = 0$ or 1); then*

$$\Lambda(M) = \sum_{n=0}^{\infty} \Lambda_n(M).$$

We write $m_1 \wedge \cdots \wedge m_n$ for the image of $m_1 \otimes \cdots \otimes m_n$ in $\Lambda(M)$. For all $x, y, z \in M$ we have

$$
\begin{aligned}
x \wedge x &= 0 \quad \text{therefore} \quad x \wedge y = -y \wedge x \\
(rx + sy) \wedge z &= r(x \wedge z) + s(y \wedge z).
\end{aligned}
$$

If $M = \mathcal{F}_R\{x_1, \ldots, x_k\}$ is a free module on a k-element set then $\Lambda_n(M)$ is a free module on a $\binom{k}{n}$-element set; so $\Lambda(M)$ is a free module on a set with 2^k-elements. In particular $\Lambda_k(M)$ is free on the singleton set $\{x_1 \wedge \cdots \wedge x_k\}$, so if

$$y_i = \sum_{j=1}^{k} r_{ij}\, x_j$$

then $y_1 \wedge \cdots \wedge y_k$ must be a unique scalar multiple

$$y_1 \wedge \cdots \wedge y_k = \det(r_{ij})\, x_1 \wedge \cdots \wedge x_k$$

of $x_1 \wedge \cdots \wedge x_k$. This can be taken as a definition of the determinant of $(r_{ij}) \in \mathbf{Mat}\,(k, R)$.

If A is a skew commutative algebra then we have a bijection

$$\mathbf{Alg}_R(\Lambda(M), A) \;\cong\; \mathrm{Hom}_R(M, A)\,.$$

An *R-Lie algebra* is an R-module L together with a module morphism $\beta : L \otimes_R L \twoheadrightarrow L$ satisfying the conditions

$$\beta(x, x) = 0$$

(Jacobi identity) $\quad \beta\big(\beta(x, y), z\big) + \beta\big(\beta(z, x), y\big) + \beta\big(\beta(y, z), x\big) = 0\,.$

Call such a β a *Lie bracket* on the module L.

Example 6.7 *For any R-algebra A the* commutator $[a, b] = a\,b - b\,a$ *defines a Lie bracket on the underlying R-module of A:*

$$[[a, b], c] + [[c, a], b] + [[b, c], a]$$
$$= [a, b]\,c - c\,[a, b] + [c, a]\,b - b\,[c, a] + [b, c]\,a - a\,[b, c]$$
$$= (a\,b\,c - b\,a\,c) - (c\,a\,b - c\,b\,a) + (c\,a\,b - a\,c\,b) - (b\,c\,a - b\,a\,c)$$
$$\quad + (b\,c\,a - c\,b\,a) - (a\,b\,c - a\,c\,b) = 0\,.$$

So A becomes a Lie algebra, denoted by A_L. It turns out (at least when R is a field) that every Lie algebra is a submodule, closed under commutator, of such an example.

Example 6.8 *Let A be an R-algebra and $M : A \twoheadrightarrow A$ a module. Then a* derivation $D : A \longrightarrow M$ *is an R-module morphism satisfying*

(Leibniz rule) $\qquad D(a\,b) = D(a)\,b + a\,D(b)\,.$

Notice that $a = b = 1$ gives $D(1) = 2\,D(1)$, so $D(1) = 0$. Let $\mathbf{Der}_R(A, M)$ denote the submodule of $\mathrm{Hom}_R(A, M)$ consisting of the derivations. We write $\mathbf{Der}_R(A)$ for $\mathbf{Der}_R(A, A)$. It is easy to check that $\mathbf{Der}_R(A)$ is closed under commutator in the algebra $\mathrm{End}_R(A)$; that is, if $D_1, D_2 : A \longrightarrow A$ are derivations then so is $[\,D_1, D_2\,] = D_1 \circ D_2 - D_2 \circ D_1$.

Example 6.9 *The tangent space at the identity of each Lie group is a Lie algebra. The pioneering work of Sophus Lie and Eli Cartan showed how much information about the Lie group is obtainable from the Lie algebra (especially in the compact case).*

The Lie groups $\mathrm{GL}\,(n\,,\mathbb{R})$ and $\mathrm{SL}\,(n\,,\mathbb{R})$ and $\mathrm{O}(n\,,\mathbb{R})$ consist of those matrices $\mathbf{x} \in \mathbf{Mat}(n\,,\mathbb{R})$ for which respectively $\mathbf{x}$ is invertible, $\det \mathbf{x} = 1$ and $\mathbf{x}\mathbf{x}^t = 1$. They have associated Lie algebras

$$
\begin{aligned}
\mathfrak{gl}(n\,,\mathbb{R}) &= \mathbf{Mat}(n\,,\mathbb{R}) \\
\mathfrak{sl}(n\,,\mathbb{R}) &= \left\{\mathbf{x} \in \mathfrak{gl}(n\,,\mathbb{R}) \mid \mathrm{trace}(\mathbf{x}) = 0\right\} \\
\mathfrak{o}(n\,,\mathbb{R}) &= \left\{\mathbf{x} \in \mathfrak{gl}(n\,,\mathbb{R}) \mid \mathbf{x}^t = -\mathbf{x}\right\}.
\end{aligned}
$$

(We shall not stop to prove this here.) The Lie bracket is $[\,\mathbf{x}\,,\mathbf{y}\,] = \mathbf{x}\mathbf{y} - \mathbf{y}\mathbf{x}$ in each case. As an exercise the reader should check that $\mathfrak{sl}(n\,,\mathbb{R})$ and $\mathfrak{o}(n\,,\mathbb{R})$ are closed under commutator.

Suppose L and L' are R-Lie algebras and $f : L \longrightarrow L'$ is an R-module morphism. Then f is a *Lie algebra morphism* when it satisfies

$$
f\big(\beta(x\,,y)\big) \;=\; \beta\big(f(x)\,,f(y)\big)\,.
$$

Write $\mathbf{Lie}_R(L\,,L')$ for the set of Lie algebra morphisms $f : L \longrightarrow L'$.

We saw in Example 6.7 above that each R-algebra A gives rise to an R-Lie algebra A_L using the commutator. We shall describe an "adjoint" for this process: for each R-Lie algebra L we obtain an R-algebra $\mathcal{U}(L)$, called the *universal enveloping algebra* of L, such that there is a natural bijection

$$
\mathbf{Alg}_R\big(\mathcal{U}(L)\,,A\big) \;\cong\; \mathbf{Lie}(L\,,A_L)\,. \tag{$*$}
$$

For this we use the tensor algebra $T(L)$ on the underlying R-module of L, and take the quotient by the appropriate ideal:

$$
\mathcal{U}(L) \;=\; T(L)/\big(x{\otimes}y - y{\otimes}x - \beta(x\,,y) \mid x\,,y \in L\big)\,.
$$

We have a Lie algebra morphism i as in:

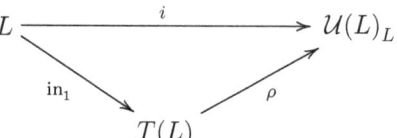

and it is composition with i that induces the bijection $(*)$.

The *direct sum* $L_1 \oplus L_2$ of Lie algebras L_1, L_2 is their direct sum as modules together with the Lie bracket

$$\beta\big((x_1, x_2), (y_1, y_2)\big) \;=\; \big(\beta(x_1, y_1),\, \beta(x_2, y_2)\big).$$

Proposition 6.10 *There is an algebra isomorphism*

$$\mathcal{U}(L_1 \oplus L_2) \;\cong\; \mathcal{U}(L_1) \otimes_R \mathcal{U}(L_2)$$

whose composite with $i : L_1 \oplus L_2 \longrightarrow \mathcal{U}(L_1 \oplus L_2)$ *takes the pair* (x_1, x_2) *to* $x_1 \otimes 1 + 1 \otimes x_2$.

Proof. It is left to the reader to check that

$$
\begin{array}{lll}
L_1 \oplus L_2 \longrightarrow \big(\mathcal{U}(L_1) \otimes_R \mathcal{U}(L_2)\big)_L & (x_1, x_2) \longmapsto x_1 \otimes 1 + 1 \otimes x_2 \\
L_1 \longrightarrow \mathcal{U}(L_1 \oplus L_2)_L & x_1 \longmapsto (x_1, 0) \\
L_2 \longrightarrow \mathcal{U}(L_1 \oplus L_2)_L & x_2 \longmapsto (0, x_2)
\end{array}
$$

are Lie algebra morphisms. These three must therefore be composites with i of algebra morphisms

$$
\begin{array}{lll}
\phi & : \mathcal{U}(L_1 \oplus L_2) & \longrightarrow \mathcal{U}(L_1) \otimes_R \mathcal{U}(L_2) \\
\psi_1 & : \mathcal{U}(L_1) & \longrightarrow \mathcal{U}(L_1 \oplus L_2) \\
\psi_2 & : \mathcal{U}(L_2) & \longrightarrow \mathcal{U}(L_1 \oplus L_2).
\end{array}
$$

Define $\psi : \mathcal{U}(L_1) \otimes_R \mathcal{U}(L_2) \longrightarrow \mathcal{U}(L_1 \oplus L_2)$ by $\psi(a \otimes b) = \psi_1(a)\,\psi_2(b)$. Then we have that

$$
\begin{array}{lll}
\psi\big(\phi(x_1, x_2)\big) & = & \psi(x_1 \otimes 1 + 1 \otimes x_2) = (x_1, 0) + (0, x_2) = (x_1, x_2) \\
\phi\big(\psi_1(x_1)\big) & = & \phi(x_1, 0) = x_1 \otimes 1 \\
\phi\big(\psi_2(x_2)\big) & = & \phi(0, x_2) = 1 \otimes x_2.
\end{array}
$$

Hence ϕ and ψ are mutually inverse. $\square$

A deeper result which we shall not prove here is:

Proposition 6.11 *(Poincaré–Birkhoff–Witt)   If the R-Lie algebra L is free as an R-module then $i : L \longrightarrow \mathcal{U}(L)_L$ is injective.*

A Lie algebra L is called *commutative* when $\beta(x,y) = 0$ for all $x,y \in L$. So an algebra A is commutative iff A_L is commutative.

Notice that, for any module M, we can make M into a commutative Lie algebra. Then the universal enveloping algebra of M is precisely the same as the symmetric algebra of M; that is, $\mathcal{U}(M) = \mathcal{S}(M)$. In particular, we have (Proposition 6.10):

$$\mathcal{S}(M \oplus M') \;\cong\; \mathcal{S}(M) \otimes_R \mathcal{S}(M') \,.$$

Exercise 6.1 *Let R be a commutative ring and G be a group. Consider left modules M, N, L over the group algebra $R(G)$.*

(a) *Show that $M \otimes_R N$ becomes an $R(G)$-module on defining:*
$g(m \otimes n) = (gm) \otimes (gn)$ *for $g \in G, m \in M, n \in N$.*

(b) *Show that $\mathrm{Hom}_R(M, L)$ becomes an $R(G)$-module on defining:*
$(gu)(n) = gu(g^{-1}m)$ *for $g \in G, u \in \mathrm{Hom}_R(M, L), m \in M$.*

(c) *Show that evaluation $\mathrm{ev}_M : M \otimes_R \mathrm{Hom}_R(M, L) \longrightarrow L$ is an $R(G)$-module morphism.*

(d) *Prove that evaluation induces an isomorphism of R-modules*

$$\mathrm{Hom}_{R(G)}\big(N, \mathrm{Hom}_R(M, L)\big) \cong \mathrm{Hom}_{R(G)}\big(M \otimes_R N, L\big) \,.$$

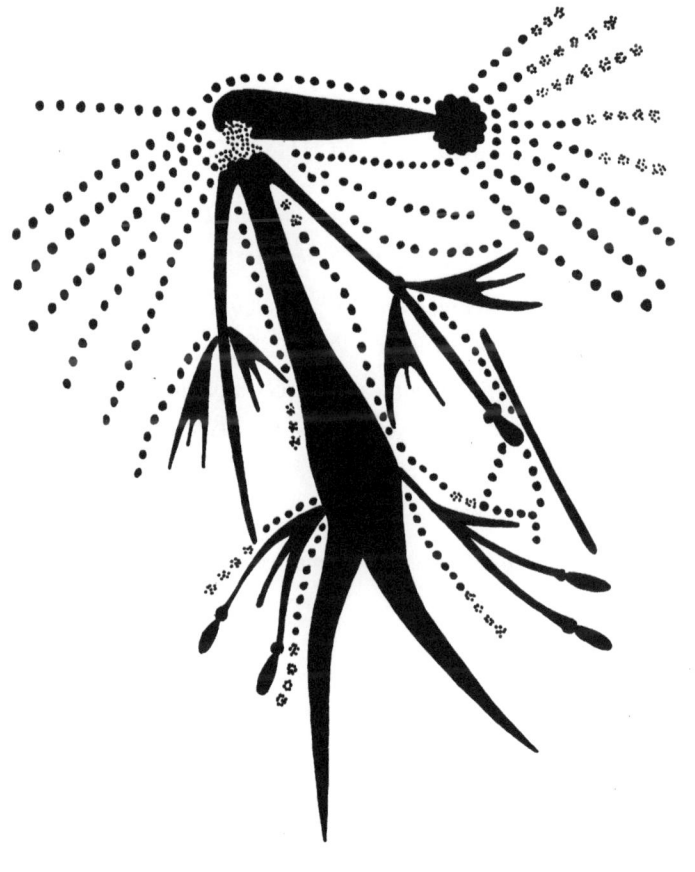

Bradshaw: Stylized Bradshaw Group, [Wal94, Plate 40].

7

Coalgebras and bialgebras

Let R be any ring. By a *coalgebra* over R (or *R-coalgebra*) we mean a module $C : R \rightarrowtail R$ together with module morphisms

$$\delta : C \longrightarrow C \otimes_R C \qquad \text{and} \qquad \varepsilon : C \longrightarrow R$$

such that

$$C \xrightarrow{\ \delta\ } C \otimes_R C \underset{1_X \otimes \delta}{\overset{\delta \otimes 1_X}{\rightrightarrows}} C \otimes_R C \otimes_R C$$

$$C \xrightarrow{\ \delta\ } C \otimes_R C \underset{1_X \otimes \varepsilon}{\overset{\varepsilon \otimes 1_X}{\rightrightarrows}} C \ .$$

$$1_C$$

We call δ the *comultiplication* and ε the *counit*. This structure provides a module with "formal diagonals". There is a uniquely determined

$$\delta : C \longrightarrow C \otimes_R C \otimes_R \cdots \otimes_R C$$

where for each $c \in C$ we have $\delta(c) = \sum_i c_{1i} \otimes \cdots \otimes c_{ni}$. The notation

$$\delta(c) = \sum_{(c)} c_{(1)} \otimes \cdots \otimes c_{(n)}$$

is sometimes used even though the representation of $\delta(c)$ in the tensor product is not uniquely determined — we act as though a choice of this representation has been made for each $c \in C$. Given a multilinear function $f : C \times \cdots \times C \longrightarrow A$ we also write

$$f\big(\delta(c)\big) = \sum_{(c)} f(c_{(1)}, \ldots, c_{(n)}) \ .$$

In terms of this notation the axioms can be rewritten as

$$\sum_{(c)} \delta(c_{(1)}) \otimes c_{(2)} = \sum_{(c)} c_{(1)} \otimes c_{(2)} \otimes c_{(3)} = \sum_{(c)} c_{(1)} \otimes \delta(c_{(2)})$$

$$c = \sum_{(c)} \varepsilon(c_{(1)}) \otimes c_{(2)} = \sum_{(c)} c_{(1)} \otimes \varepsilon(c_{(2)}) .$$

Suppose C and D are coalgebras. A *coalgebra morphism* $f : C \longrightarrow D$ is a module morphism such that

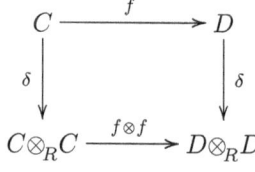   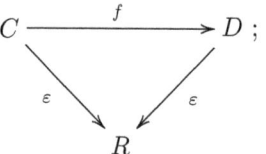

that is,

$$\sum_{(c)} f(c_{(1)}) \otimes f(c_{(2)}) = \sum_{(f(c))} f(c)_{(1)} \otimes f(c)_{(2)}$$

$$\varepsilon(f(c)) = \varepsilon(c) .$$

We write $\mathbf{Cog}_R(C,D)$ for the set of coalgebra morphisms from C to D.

Suppose R is commutative. A coalgebra C over R is *cocommutative* when

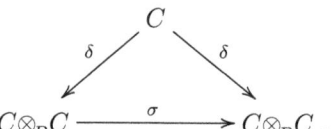

Return now to a general ring R. Suppose that A is an R-algebra and C is an R-coalgebra. Then $\mathrm{Hom}_R(C,A)$ becomes an R-algebra under the following *convolution structure*:

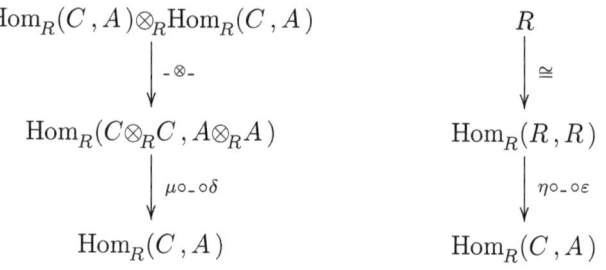

In terms of elements, for left R-module morphisms $f, g : C \longrightarrow A$ their *convolution product* is given by

$$f * g = \mu \circ (f \otimes g) \circ \delta \qquad \text{and} \qquad 1 = \eta \circ \varepsilon \,.$$

Using the notation for comultiplication this becomes the formula

$$(f * g)(c) = \sum_{(c)} f(c_{(1)}) \, g(c_{(2)}) \,.$$

In particular, with $A = R$ each R-coalgebra C gives rise to a convolution R-algebra structure on the dual $C^* = \operatorname{Hom}_R(C, R)$. However, we prefer to regard C^* as an R-algebra via the multiplication $\mu : C^* \otimes_R C^* \longrightarrow C^*$ defined by the following diagram.

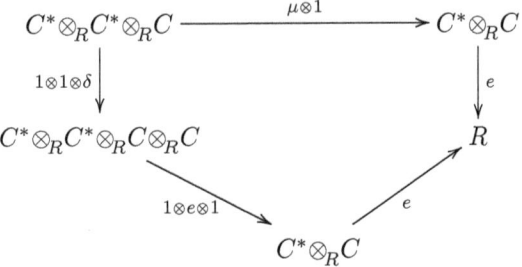

(This works even for non-commutative R.)

Example 7.1 *Suppose $\mathcal{X}$ is any category which admits finite products. Suppose $F : \mathcal{X} \longrightarrow \mathbf{Mod}_R^R$ is a functor into the category of modules from R to R. Suppose there are natural module morphisms*

$$\phi_{X_1, \dots, X_n} : F(X_1 \times \cdots \times X_n) \longrightarrow FX_1 \otimes_R FX_2 \otimes_R \cdots \otimes_R FX_n$$

compatible with the canonical associativity isomorphisms for product and tensor product. Then for each object X of $\mathcal{X}$ we obtain a coalgebra FX, with comultiplication and counit

$$FX \xrightarrow{\ F\delta\ } F(X \times X) \xrightarrow{\ \phi_{X,X}\ } FX \otimes_R FX$$

$$FX \xrightarrow{\ F\varepsilon\ } F1 \xrightarrow{\ \phi_0\ } R \,.$$

If furthermore R is commutative and F is compatible with the switches, then this coalgebra is cocommutative.

Sub-example (a). (R commutative) The free R-module construction gives a functor

$$\mathcal{F}_R : \mathbf{Set} \longrightarrow \mathbf{Mod}_R$$

from the category of sets to $\mathbf{Mod}_R$. We have isomorphisms

$$\phi : \ \mathcal{F}_R(X_1 \times \cdots \times X_n) \ \xrightarrow{\ \cong\ } \ \mathcal{F}_R X_1 \otimes_R \cdots \otimes_R \mathcal{F}_R X_n$$
$$(x_1, \ldots, x_n) \longmapsto \qquad x_1 \otimes \cdots \otimes x_n \ .$$

Proof. ($n = 2$)

$$
\begin{aligned}
\operatorname{Hom}_R\big(\mathcal{F}_R(X \times Y), M\big) \ &\cong \ M^{X \times Y} \ \cong \ (M^X)^Y \\
&\cong \ \operatorname{Hom}_R(\mathcal{F}_R Y, M^X) \\
&\cong \ \operatorname{Hom}_R\big(\mathcal{F}_R Y, \operatorname{Hom}(\mathcal{F}_R X, M)\big) \\
&\cong \ \operatorname{Hom}_R(\mathcal{F}_R X \otimes \mathcal{F}_R Y, M) \ .
\end{aligned}
$$

So *each $\mathcal{F}_R X$ becomes an R-coalgebra.* □

Sub-example (b). The universal enveloping algebra provides a functor

$$\mathcal{U} : \mathbf{Lie}_R \longrightarrow \mathbf{Mod}_R$$

and we have already observed

$$\mathcal{U}(L \oplus L') \ \cong \ \mathcal{U}(L) \otimes_R \mathcal{U}(L') \qquad \text{and} \qquad \mathcal{U}(\{0\}) \ \cong \ R$$
$$(x, x') \longmapsto \ x \otimes 1 + 1 \otimes x' \ .$$

Since direct sum is product in $\mathbf{Lie}_R$ we have another standard example. Thus *each universal enveloping algebra $\mathcal{U}(L)$ becomes an R-coalgebra.* The comultiplication here is determined by

$$L \xrightarrow{\ i\ } \mathcal{U}(L) \xrightarrow{\ \delta\ } \mathcal{U}(L) \otimes \mathcal{U}(L)$$
$$x \longmapsto \qquad\qquad\qquad x \otimes 1 + 1 \otimes x \ .$$

Sub-examples (a), (b) suggest two definitions that we can make for any coalgebra C.

- Say that $c \in C$ is *set-like* when $\delta(c) = c \otimes c$ and $\varepsilon(c) = 1$. (In the case of $C = \mathcal{F}_R(X)$ the set-like elements are precisely the elements of X.) Write $\mathcal{D}(C)$ for the set of set-like elements of C.

- Say that $c \in C$ is *primitive* when

$$\delta(c) = c \otimes 1 + 1 \otimes c.$$

(In the case of $C = \mathcal{U}(L)$ each element of L is primitive.)
Write $\mathcal{P}(C)$ for the submodule of primitive elements of C.

Proposition 7.2 *(with R a field.)* *The set-like elements of any coalgebra C form a linearly independent subset $\mathcal{D}(C)$.*

Proof. Suppose $\mathcal{D}(C)$ is linearly dependent. Let $n+1$ be the first natural number for which there is a linearly dependent subset of $\mathcal{D}(C)$ with that many elements. Then any set of n elements of $\mathcal{D}(C)$ must necessarily be linearly independent, while there exist distinct $g, g_1, \ldots, g_n \in \mathcal{D}(C)$ which are linearly dependent. Then we can write

$$g = \lambda_1 g_1 + \cdots + \lambda_n g_n$$

with the $\lambda_i \in R$ all non-zero. Then

$$\sum_{i=1}^{n} \lambda_i \, g_i \otimes g_i \; = \; \sum_{i=1}^{n} \lambda_i \, \delta(g_i) \; = \; \delta(g) \; = \; g \otimes g$$

$$= \; \sum_{i,j=1}^{n} \lambda_i \lambda_j \, g_i \otimes g_j \, .$$

Since $\{g_1, \ldots, g_n\}$ is linearly independent in C then $\{g_i \otimes g_j\}$ is linearly independent in $C \otimes C$, so we can equate coefficients: $\lambda_i \lambda_j = 0$ for $i \neq j$ and $\lambda_i = \lambda_i^2$. Since $\lambda_i \neq 0$ this means $n = 1$ and $\lambda_i = 1$. But $g = g_1$ was not allowed. □

We shall come back to set-like and primitive elements in the context of bialgebras.

Example 7.3 *(with R commutative.)* *Let $C = \mathcal{F}_R \mathbb{N}$ be the free R-module on the countable set $\mathbb{N}$. Define*

$$\delta(n) \; = \; \sum_{p+q=n} p \otimes q \quad \text{and} \quad \varepsilon(n) \; = \; \begin{cases} 1 & \text{for } n = 0 \\ 0 & \text{for } n > 0 \, . \end{cases}$$

This defines a cocommutative coalgebra structure on C.

Take an R-algebra A and look at an example of convolution with this coalgebra C. The convolution structure transports across the R-module isomorphism

$$\mathrm{Hom}_R(C, A) \; \cong \; A^{\mathbb{N}} \quad (\text{sequences in } A)$$

to give the multiplication

$$a\,b \;=\; \left(\sum_{p+q=n} a_p\, b_q \right)$$

for sequences $a = (a_n) = (a_0, a_1, \dots)$ *and* $b = (b_n) = (b_0, b_1, \dots)$ *in* A. *The unit sequence is* $(1, 0, 0, 0, \dots)$. *A precise definition of* indeterminate *can be taken to mean the sequence*

$$x \;=\; (0, 1, 0, 0, 0, 0, \dots) \in A^{\mathbb{N}}$$

in A. *Each* $u \in A$ *is identified with* $u\,1 = (u, 0, 0, 0, \dots) \in A^{\mathbb{N}}$. *Then each* $a \in A^{\mathbb{N}}$ *can be written as a* formal *(no convergence requirements!)* power series

$$a \;=\; \sum_{n=0}^{\infty} a_n\, x^n.$$

Write $A[\![x]\!]$ *for* $A^{\mathbb{N}}$ *with this algebra structure. It is the* R-algebra of formal *power series in* A. *If* A *is commutative so is* $A[\![x]\!]$. *In particular when* $A = R$ *we obtain the commutative* R-algebra $C^* = R[\![x]\!]$.

Example 7.4 *Let* $\mathbf{n} = \{1, 2, \dots, n\}$ *and put* $C = \mathcal{F}_R(\mathbf{n} \times \mathbf{n})$. *Then* C *becomes an* R-coalgebra on defining

$$\delta(i, j) \;=\; \sum_{k=1}^{n} (i, k) \otimes (k, j) \qquad \text{and} \qquad \varepsilon(i, j) \;=\; \begin{cases} 1 & \text{for } i = j \\ 0 & \text{otherwise.} \end{cases}$$

Given any R-algebra A, *the convolution structure simply transports across the* R-module isomorphism

$$\mathrm{Hom}_R(C, A) \;\cong\; A^{\mathbf{n} \times \mathbf{n}} \qquad (\, n \times n \text{ matrices in } A\,)$$

to give the usual matrix multiplication

$$(a_{ij})\,(b_{ij}) \;=\; \left(\sum_{k=1}^{n} a_{ik}\, b_{kj} \right).$$

In this way we obtain the R-algebra $\mathbf{Mat}(n, A) \cong \mathrm{End}_R(A^n)$ *of* $n \times n$ *matrices with entries in* A.

Suppose that R is a commutative ring. An R-*bialgebra* is an R-module B together with algebra and coalgebra structures

$$\mu : B \otimes_R B \longrightarrow B \quad \text{and} \quad \eta : R \longrightarrow B$$
$$\delta : B \longrightarrow B \otimes_R B \quad \text{and} \quad \varepsilon : B \longrightarrow R$$

satisfying the conditions

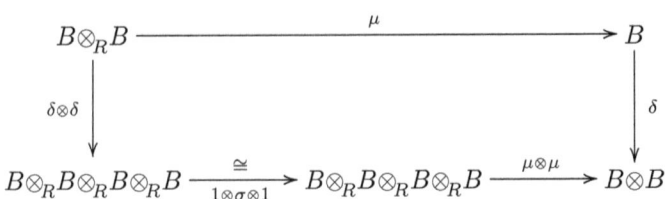

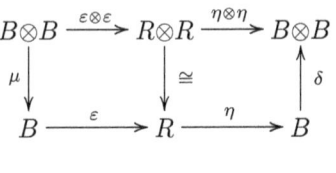

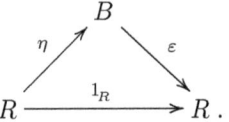

Notice the complete duality between (μ, η) and (δ, ε). When expressed in terms of elements the duality is not so apparent:

$$\delta(x, y) = \sum_{(x)} \sum_{(y)} x_{(1)} y_{(1)} \otimes x_{(2)} y_{(2)}$$

$$\varepsilon(x\,y) = \varepsilon(x)\,\varepsilon(y) \quad \text{and} \quad \delta(1) = 1 \otimes 1 \quad \text{and} \quad \varepsilon(1) = 1.$$

For R commutative, the tensor product $A \otimes_R A'$ of R-algebras A and A' becomes an R-algebra via the multiplication

$$(A \otimes_R A') \otimes_R (A \otimes_R A') \xrightarrow[\substack{1 \otimes \sigma \otimes 1 \\ \sigma_{1324}}]{\cong} (A \otimes_R A) \otimes_R (A' \otimes_R A') \xrightarrow{\mu \otimes \mu} A \otimes_R A'$$

and unit

$$R \cong R \otimes_R R \xrightarrow{\eta \otimes \eta} A \otimes_R A'.$$

Also the tensor product $C \otimes_R C'$, of R-coalgebras C and C', becomes an R-coalgebra via the comultiplication

$$C \otimes_R C' \xrightarrow{\delta \otimes \delta} (C \otimes_R C) \otimes_R (C' \otimes_R C') \xrightarrow[1 \otimes \sigma \otimes 1]{\cong} (C \otimes_R C') \otimes_R (C \otimes_R C')$$

and counit

$$C \otimes_R C' \xrightarrow{\varepsilon \otimes \varepsilon} R \otimes_R R \cong R.$$

With this, we can make the observation:

Proposition 7.5 *Suppose (μ, η) and (δ, ε) are respectively, algebra and coalgebra structures on the R-module B. Then the following conditions are equivalent:*

   (i) *B is a bialgebra;*

  (ii) *$\mu : B \otimes_R B \longrightarrow B$ and $\eta : R \longrightarrow B$ are coalgebra morphisms;*

 (iii) *$\delta : B \longrightarrow B \otimes_R B$ and $\varepsilon : B \longrightarrow R$ are algebra morphisms.*

For bialgebras B and B' a *bialgebra morphism* $f : B \longrightarrow B'$ is a function which is both an algebra and coalgebra morphism. Write $\mathbf{Big}_R(B, B')$ for the set of such functions f.

Before giving examples of bialgebras we prove some extra results on the set-like and primitive elements for the bialgebra case.

Proposition 7.6 *If B is a bialgebra then the set-like elements are closed under multiplication: so $\mathcal{D}(B)$ becomes a monoid.*

Proof. $\delta(bb') = \delta(b)\,\delta(b') = (b \otimes b)(b' \otimes b') = bb' \otimes bb'$ for $b, b' \in \mathcal{D}(B)$; also $\varepsilon(bb') = \varepsilon(b)\,\varepsilon(b') = 1 \cdot 1 = 1$.   □

Proposition 7.7 *If B is a bialgebra then the set of primitive elements is closed under commutator, so $\mathcal{P}(B)$ becomes a Lie algebra. Also $\varepsilon(x) = 0$ for all $x \in \mathcal{P}(B)$.*

Proof. For $x, y \in \mathcal{P}(B)$ we have

$$
\begin{aligned}
\delta([x,y]) &= \delta(x)\,\delta(y) - \delta(y)\,\delta(x) \\
&= (x \otimes 1 + 1 \otimes x)(y \otimes 1 + 1 \otimes y) - (y \otimes 1 + 1 \otimes y)(x \otimes 1 + 1 \otimes x) \\
&= xy \otimes 1 + x \otimes y + y \otimes x + 1 \otimes xy - (yx \otimes 1 + y \otimes x + x \otimes y + 1 \otimes yx) \\
&= [x,y] \otimes 1 + 1 \otimes [x,y]
\end{aligned}
$$

so that $[x,y] \in \mathcal{P}(B)$. Also $x = (1 \otimes \varepsilon)\,\delta(x) = (1 \otimes \varepsilon)(x \otimes 1 + 1 \otimes x) = x + \varepsilon(x)$; hence $\varepsilon(x) = 0$.   □

Example 7.8 *Return to the situation of coalgebras in Example 7.1. There are two conditions on the functor $F : \mathcal{X} \longrightarrow \mathbf{Mod}_R$ which give rise to bialgebras FX.*

(a) When the morphisms ϕ are all invertible.
Then F takes each monoid G in $\mathcal{X}$ to a bialgebra FG. The multiplication and unit for G give an algebra structure

$$FG \otimes_R FG \cong F(G \times G) \xrightarrow{\ F\mu\ } FG \quad \text{and} \quad R \cong F1 \xrightarrow{\ F\eta\ } FG$$

on FG. These are coalgebra morphisms since all arrows in $\mathcal{X}$ "commute with diagonals". By Proposition 7.5, each FG becomes a bialgebra. This is the situation for the functor $\mathcal{F}_R : \mathbf{Set} \longrightarrow \mathbf{Mod}_R$, so for each monoid G the monoid algebra $R(G)$ is a cocommutative bialgebra. Notice here that $G \cong \mathcal{D}\big(R(G)\big)$ as monoids (see Proposition 7.6).

(b) When F lifts to $F : \mathcal{X} \longrightarrow \mathbf{Alg}_R$.
In this case each FX is clearly a bialgebra since the comultiplication and counit are algebra morphisms (Proposition 7.5). In particular, for the functor $\mathcal{U} : \mathbf{Lie}_R \longrightarrow \mathbf{Alg}_R$ this is indeed the situation. Thus we have that:
each universal enveloping algebra $\mathcal{U}(L)$ is a cocommutative bialgebra.

Example 7.9 *Return to Example 7.3 of a coalgebra. This time, to use the symbol $\mathbb{N}$ to denote our countable set would be confusing. Instead we denote it by $E = \{e_0, e_1, e_2, e_3, \dots\}$. Then the coalgebra structure on $\mathcal{F}_R(E)$ is*

$$\delta(e_n) = \sum_{p+q=n} e_p \otimes e_q \quad \text{and} \quad \varepsilon(e_n) = \begin{cases} 0 & \text{for } n > 0, \\ 1 & \text{for } n = 0. \end{cases}$$

We now make $\mathcal{F}_R(E)$ into an algebra via

$$e_p\, e_q = \frac{(p+q)!}{p!\, q!}\, e_{p+q} \quad \text{with} \quad e_0 = 1.$$

(The binomial coefficient is an integer and so "lives" in any ring R.) Then $\mathcal{F}_R(E)$ is a bialgebra. If R is a field of characteristic 0 (i.e., $1 + \cdots + 1 \neq 0$ in R for any non-zero number of terms) put $x = e_1$ so one easily sees that $e_n = \dfrac{1}{n!}\, x^n$. Hence, as an algebra, $\mathcal{F}_R(E)$ is isomorphic to the polynomial algebra $R[x]$ in one variable. For general R we can think of $\mathcal{F}_R(E)$ as the algebra of Hurwitz polynomials in one indeterminate:

$$\sum_{n=0}^{k} \frac{a_n\, x^n}{n!} \quad \text{with each } a_n \in R.$$

Example 7.10 *Return to Example 7.4 and form the symmetric algebra $\mathcal{S}(C)$ of the coalgebra $C = \mathcal{F}_R(\mathbf{n} \times \mathbf{n})$. Since using $\mathbf{n} \times \mathbf{n}$ can be confusing we replace it by any set $X = \{x_{ij} \mid i, j \in \mathbf{n}\}$ of cardinality n^2. Then we*

identify $S(\mathcal{F}_R(X))$ with the polynomial R-algebra $R[(x_{ij})]$ *in n^2 commuting indeterminates x_{ij} for $i, j \in \mathbf{n}$. In Example 7.1 we saw that this becomes a bialgebra by virtue of the fact that it is the universal enveloping algebra of a commutative Lie algebra $\mathcal{F}_R(X)$, but this is not the structure of interest here. The coalgebra C induces the bialgebra structure*

$$\delta(x_{ij}) = \sum_k x_{ik} \otimes x_{kj} \quad \text{and} \quad \varepsilon(x_{ij}) = \begin{cases} 1 & \text{for } i = j \\ 0 & \text{for } i \neq j \end{cases}$$

which we call the matrix bialgebra $\mathrm{M}(n)$ *over R. This must not be confused with the matrix algebra*

$$\mathbf{Mat}(n, R) \cong \mathbf{Alg}_R(\mathrm{M}(n), R)$$

(which is the algebra of "points" of $\mathrm{M}(n)$).

Exercise 7.1 *For any R-coalgebra C, prove the following identities:*

(a) $\delta(c) = \displaystyle\sum_{(c)} \varepsilon(c_{(2)}) \otimes \delta(c_{(1)})$

$\qquad = \displaystyle\sum_{(c)} \delta(c_{(2)}) \otimes \varepsilon(c_{(1)}) \sum_{(c)} c_{(1)} \otimes \varepsilon(c_{(3)}) \otimes c_{(2)}$

(b) $\displaystyle\sum_{(c)} \varepsilon(c_{(1)}) \otimes c_{(3)} \otimes c_{(2)} = \sum_{(c)} c_{(2)} \otimes c_{(1)}$

(c) $\displaystyle\sum_{(c)} \varepsilon(c_{(1)}) \otimes \varepsilon(c_{(3)}) \otimes c_{(2)} = c$.

Bradshaw: "Transitionary Figures", Clothes Peg Figure Period, [Wal94, Plate 66].

8

Dual coalgebras of algebras

We have seen that the dual C^* of a coalgebra has a natural structure of an algebra. One might expect the dual A^* of an algebra to be a coalgebra in an obvious way, but this is not true because of the failure of the canonical morphism

$$M^* \otimes_R N^* \longrightarrow (M \otimes_R N)^*$$

to be always invertible. If M is Cauchy the morphism is invertible since

$$
\begin{aligned}
(M \otimes_R N)^* &= \operatorname{Hom}_R(M \otimes_R N, R) \\
&\cong \operatorname{Hom}_R(M, N^*) \cong M^* \otimes_R N^* .
\end{aligned}
$$

So for an algebra A which is Cauchy (as a module) we obtain a coalgebra, denoted by A^*, via

$$\delta : \quad A^* \xrightarrow{\mu^*} (A \otimes_R A)^* \cong A^* \otimes_R A^*$$

$$\varepsilon : \quad A^* \xrightarrow{\eta^*} R^* \cong R .$$

However, instead of restricting A, which is unsatisfactory since many of the examples are not Cauchy, we modify the definition of the dual A^*.

Let's call an ideal I of an algebra A *coCauchy* when the quotient algebra A/I is Cauchy (as a module). Define

$$A^0 = \{\, u \in A^* \mid u \text{ is zero on some coCauchy ideal of } A \,\} .$$

Proposition 8.1 *(with R a field.)*

(a) A^0 *is a submodule of* A^*.

(b) *If* $f \in \mathbf{Alg}_R(A, B)$ *then* $f^* : B^* \longrightarrow A^*$, *given by composition with f, takes B^0 into A^0.*

(c) *For any R-algebra B the canonical morphism* $A^* \otimes B^* \longrightarrow (A \otimes B)^*$ *induces an isomorphism* $A^0 \otimes B^0 \xrightarrow{\cong} (A \otimes B)^0$.

(d) *There exists a unique* $\delta : A^0 \longrightarrow A^0 \otimes A^0$ *satisfying*

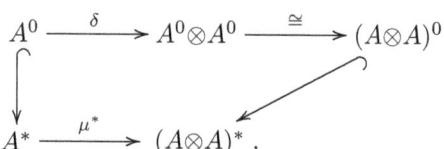

Proof. (a) If $u \in A^0$ and $r \in R$ then $\ker(r\,u) \supseteq \ker u$, so that $r\,u \in A^0$.

Take $u, v \in A^0$ zero on coCauchy ideals I and J respectively. We can find subspaces U, V and W of A with $A = (I \cap J) \oplus U \oplus V \oplus W$ and $I = (I \cap J) \oplus U$ and $J = (I \cap J) \oplus V$. So $A/I \cong V \oplus W$ and $A/J \cong U \oplus W$ are finite dimensional. Thus $A/I \cap J \cong U \oplus V \oplus W$ is finite dimensional. Hence $I \cap J$ is a coCauchy ideal on which $u + v$ is zero.

(b) Take $v \in B^0$ zero on coCauchy J in B. Then $f^{-1}(J) \subseteq \ker(vf) = \ker f^*(v)$ is an ideal of A; but $f^{-1}(J)$ is the kernel of $A \xrightarrow{\,f\,} B \rightarrow B/J$ so that $A/f^{-1}(J)$ is isomorphic to a subspace of B/J. So $f^{-1}(J)$ is coCauchy.

(c) We shall use Exercise 8.1.

Before beginning the proof of Proposition 8.1(c) notice that for any co-Cauchy ideal K in $A \otimes B$ we have coCauchy ideals

$$
\begin{aligned}
I &= \{a \in A \mid a \otimes 1 \in K\} &= (A \xrightarrow{\,-\otimes 1\,} A \otimes B)^*(K) \\
J &= \{b \in B \mid 1 \otimes b \in K\} &= (B \xrightarrow{\,1 \otimes -\,} A \otimes B)^*(K) \\
A \otimes_R J + I \otimes_R B &= \ker\big(A \otimes_R B \longrightarrow (A/I) \otimes_R (B/J) \big)
\end{aligned}
$$

of A and B and $A \otimes B$ respectively.

Now take $w \in (A \otimes B)^0$ which is zero on some coCauchy K as above. Then w is zero on $A \otimes J + I \otimes B \subseteq K$. However, $A \otimes B / (A \otimes J + I \otimes B) \cong A/I \otimes B/J$ so there exists a unique $\bar{w}$:

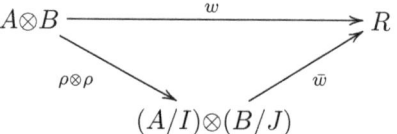

Furthermore since A/I and B/J are finite dimensional, so that we have $(A/I)^* \otimes (B/J)^* \xrightarrow{\cong} (A/I \otimes B/J)^*$ is invertible, there is some element $\sum \bar{h}_i \otimes \bar{k}_i \in (A/I)^* \otimes (B/J)^*$ corresponding to $\bar{w}$. In particular, for $x \in A/I$ and $y \in B/J$ we have that

$$
\bar{w}(x \otimes y) = \sum_i \bar{h}_i(x)\, \bar{k}_i(y) . \tag{$*$}
$$

Define the composite $h_i = \bar{h}_i \circ \rho : A \twoheadrightarrow A/I \twoheadrightarrow R$ and similarly define $k_i = \bar{k}_i \circ \rho : B \twoheadrightarrow B/J \twoheadrightarrow R$. These are in A^0 and B^0 since they are zero on I and J respectively. Hence we have

$$\sum_i h_i \otimes k_i \ \in \ A^0 \otimes B^0$$

which is the image of $w \in (A \otimes B)^0$ because of $(*)$.

Conversely, if $h \in A^0$ and $k \in B^0$ vanish on coCauchy I and J (ideals of A and B) then $h \otimes k$ vanishes on $A \otimes J + I \otimes B$ which is a coCauchy ideal of $A \otimes B$.

(d) Suppose $u \in A^0$ vanishes on a coCauchy ideal I. Then $\mu^*(u)(a \otimes b) = (u\mu)(a \otimes b) = u(ab)$, so $\mu^*(u)$ vanishes on $A \otimes I + I \otimes A$ which is a coCauchy ideal of $A \otimes A$. Hence μ^* takes A^0 into $(A \otimes A)^0$ and δ exists as desired. $\square$

Corollary 8.2 *(with R a field.) For each algebra A a coalgebra structure on A^0 is given by the δ in 8.1(d) and $\varepsilon = (A^0 \hookrightarrow A^* \xrightarrow{\eta^*} R^* \cong R)$. Also each algebra morphism $f : A \longrightarrow B$ induces a coalgebra morphism $f^0 : B^0 \longrightarrow A^0$ given by restriction of f^* (see Proposition 8.1(b)).*

Proof. Draw the diagrams expressing the axioms on μ, η and f. Simply apply $(\_)^*$ then restrict to $(\_)^0$. $\square$

For each algebra A we obtain a left and right A-module structure on A^* given as follows, for $a \in A$ and $u \in A^*$:

$$(au)(x) = u(xa) \quad , \quad (ua)(x) = u(ax) .$$

In fact $A^* : A \twoheadrightarrow A$. For any $f \in A^*$ write

$$Af \quad \text{and} \quad fA \quad \text{and} \quad AfA$$

for the R-submodules of A^* consisting of those elements of the form af and fa and afb respectively with $a, b \in A$.

Proposition 8.3 *(with R a field.) For $f \in A^*$ these are equivalent:*

(1) $f \in A^0$;

(2) $\mu^*(f)$ *is in the image of* $A^* \otimes A^* \longrightarrow (A \otimes A)^*$;

(3) $\mu^*(f)$ *is in the image of* $(A \otimes A)^0 \hookrightarrow (A \otimes A)^*$;

(4) Af *is Cauchy;* (5) fA *is Cauchy;* (6) AfA *is Cauchy.*

Proof. (1) $\Rightarrow$ (3) by Proposition 8.1(d). Also (3) $\Rightarrow$ (2) is trivial.

(2) $\Rightarrow$ (4). Let $\mu^*(f)$ be the image of $\sum_i u_i \otimes v_i \in A^* \otimes A^*$. Then we have that $f(ab) = \sum_i u_i(a) v_i(b)$ so $bf = \sum_i v_i(b) u_i \in A^*$. Thus bf is in the subspace of A^* spanned by the u_i. Hence Af is finite dimensional.

$(4) \Rightarrow (1)$. Suppose Af is finite dimensional. Then also $\mathrm{End}_R(Af)$ is finite-dimensional, so the kernel I of the morphism $A \longrightarrow \mathrm{End}_R(Af)$ given by $a \longmapsto (bf \longmapsto bfa)$, is a coCauchy ideal of A. But $a \in I$ implies $1f\,a = 0$, so $f(a) = 0$. Hence f is zero on I so that $f \in A^0$.

$(5) \Rightarrow (1)$ is similar to $(4) \Rightarrow (1)$ and $(6) \Rightarrow (5)$ is trivial.

$(1) \Rightarrow (6)$. Take $f \in A^0$ zero on the cofinite ideal I. Then for $c \in I$ we have $(a\,f\,b)(c) = f(b\,c\,a) = 0$. Thus $A\,fA \subseteq I^{\perp} = \{u \in A^* \mid u(I) = 0\} \cong (A/I)^*$ which is finite-dimensional since A/I is so. $\square$

Corollary 8.4 *For any coalgebra C the canonical injection $d : C \longrightarrow C^{**}$ given by $d(c)(u) = u(c)$, has image in $(C^*)^0$.*

Proof. Take $c \in C$. Then $C^*d(c) = \{u\,d(c) \mid u \in C^*\} \subseteq C^{**}$. Now $(u\,d(c))(v) = d(c)(v * u) = (v * u)(c) = (v \otimes u)\delta(c) = (v \otimes u) \sum_{(c)} c_{(1)} \otimes c_{(2)} = \sum_{(c)} v(c_{(1)})\,u(c_{(2)})$ using the definition of multiplication $v * u$ in C. Thus $u\,d(c) = \sum_{(c)} u(c_{(2)})\,d(c_{(1)})$, which is in the subspace of C^{**} spanned by the $d(c_{(1)})$. Hence $C^*\,d(c)$ has finite dimension and by using $(4) \Leftrightarrow (1)$ of Proposition 8.3 it follows that $d(c) \in (C^*)^0$. $\square$

Theorem 8.5 *(with R a field.)* *For all algebras A and coalgebras C there is a bijection*

$$\mathbf{Alg}_R(A, C^*) \quad \cong \quad \mathbf{Cog}_R(C, A^0)$$

$$\left(A \xrightarrow{\ f\ } C^* \right) \quad \longmapsto \quad \left(C \xrightarrow{\ d\ } (C^*)^0 \xrightarrow{\ f^0\ } A^0 \right).$$

Proof. The inclusion $i : A^0 \longrightarrow A^*$ induces $i^* : A^{**} \longrightarrow A^{0*}$, while the inverse to $f \longmapsto f^0 \circ d$ takes $g \in \mathbf{Cog}_R(C, A^0)$ to the composite

$$A \xrightarrow{\ d\ } A^{**} \xrightarrow{\ i^*\ } A^{0*} \xrightarrow{\ g^*\ } C^*.$$

The remaining details are left to the reader. $\square$

Exercise 8.1 *(with R a field.)* *When $f : M \longrightarrow N$ is an injective module morphism then $f \otimes_R 1 : M \otimes_R L \longrightarrow N \otimes_R L$ is injective. Furthermore we have that $M^* \otimes_R N^* \longrightarrow (M \otimes_R N)^*$ is injective. Show this.*

Exercise 8.2 *Show that an algebra morphism $f : A \longrightarrow R$ is a set-like element of A^0.*

Exercise 8.3 *Show that A^0 consists of those $u : A \longrightarrow R$ which factor (in the category of R-modules) as $u = w \circ f$, where f is surjective with Cauchy codomain. Use this to simplify some proofs in this chapter.*

9
Hopf algebras

Our base ring R will always be assumed commutative, and whenever $(\ )^0$ appears we happily suppose it to be a field.

An *R-Hopf algebra* is an R-bialgebra H together with R-module morphism

$$\nu : H \longrightarrow H$$

called the *antipode*, which satisfies the following diagram.

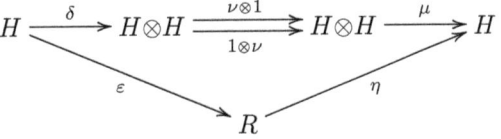

For any Hopf algebra H let H^{op} denote the Hopf algebra obtained by replacing μ with $\mu \circ \sigma :\ H \otimes H \xrightarrow{\ \sigma\ } H \otimes H \xrightarrow{\ \mu\ } H$ and replacing δ with $\sigma \circ \delta :\ H \xrightarrow{\ \delta\ } H \otimes H \xrightarrow{\ \sigma\ } H \otimes H$ while keeping the same η, ε and ν.

There is also a bialgebra H' obtained more simply by just replacing δ with $\sigma \circ \delta :\ H \xrightarrow{\ \delta\ } H \otimes H \xrightarrow{\ \sigma\ } H \otimes H$ while keeping the same μ, η, ε and ν. In general however, this H' is not a Hopf algebra.

Proposition 9.1 *Let H be a Hopf algebra. Then*

(a) *the antipode ν is uniquely determined;*

(b) *$\nu : H^{\mathrm{op}} \longrightarrow H$ is a bialgebra morphism;*

(c) *H' is a Hopf algebra if and only if ν is bijective (moreover the antipode for H' is the inverse for ν);*

(d) *if H is commutative or cocommutative then $\nu \circ \nu = 1_H$ (that is, ν is an involution).*

Proof. (a) Since H is a coalgebra and an algebra, we have the convolution algebra structure on $\mathrm{Hom}_R(H, H)$. An antipode is precisely an inverse for $1_H \in \mathrm{Hom}_R(H, H)$ under convolution. For any monoid, inverses are unique.

(b) To show $\nu : H^{\mathrm{op}} \longrightarrow H$ preserves multiplication we must show that

$$\left(H \otimes H \xrightarrow{\mu} H \xrightarrow{\nu} H \right) = \left(H \otimes H \xrightarrow{\sigma} H \otimes H \xrightarrow{\nu \otimes \nu} H \otimes H \xrightarrow{\mu} H \right).$$

We do this by showing that, under convolution, the left-hand side is a left inverse for $\mu \in \mathrm{Hom}_R(H \otimes H, H)$ while the right-hand side is a right inverse.

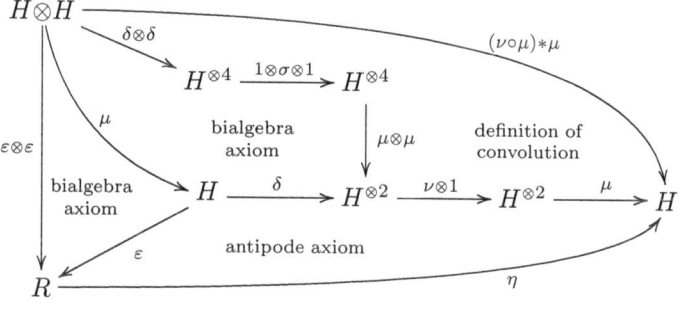

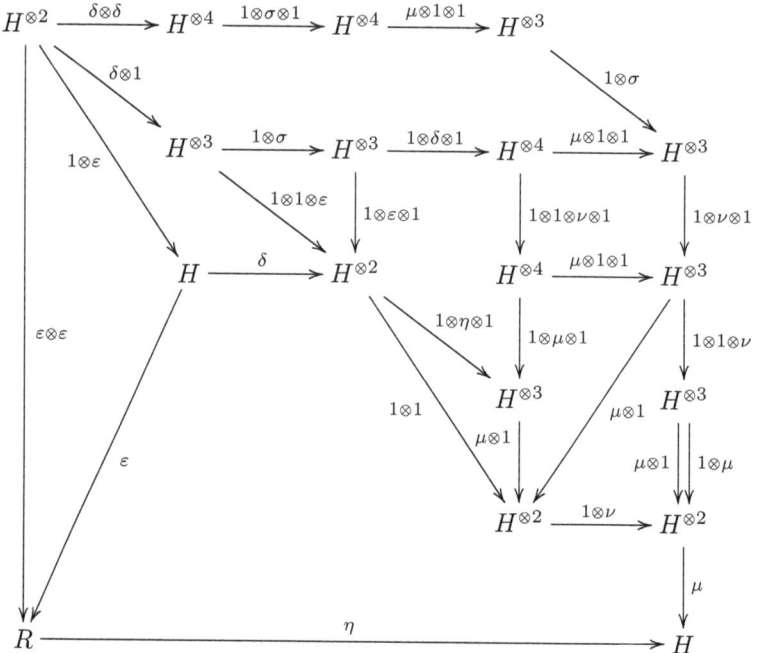

While the second commutativity is perhaps more easily seen by looking at elements, the bonus we get on using diagrams is that, formally reversing all the arrows and replacing μ and η by δ and ε, we have the proof that

$\nu : H^{\mathrm{op}} \longrightarrow H$ preserves comultiplication. The following diagram proves ν preserves unit, while the dual diagram proves ν preserves counit.

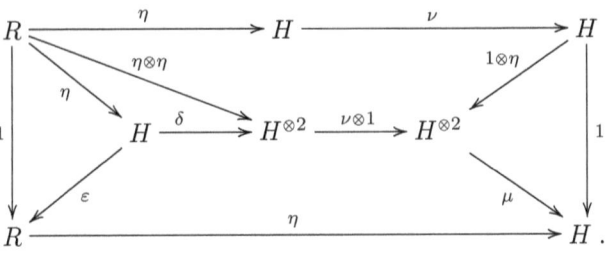

(c) ν' is a (composition) right inverse for ν $\Leftrightarrow$ $\nu \circ \nu' = 1_H$ $\Leftrightarrow$ $\nu \circ \nu'$ is a convolution left inverse for ν (since 1_H is the convolution inverse for ν)

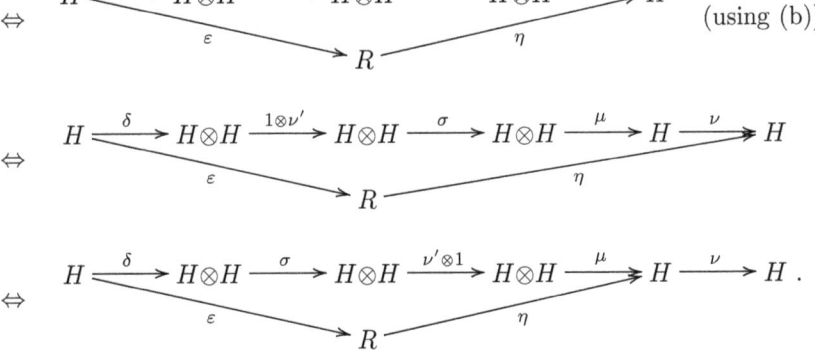

The last condition is the condition that ν' should be a left convolution inverse for $1_{H'}$ in $\mathrm{Hom}(H', H')$, except that ν is applied to the condition. Similarly, we get that ν' is a left (composition) inverse for ν if and only if ν' satisfies the condition to be a right convolution inverse for $1_{H'}$ in $\mathrm{Hom}(H', H')$ with ν applied to the condition. It follows then that ν and ν' are mutually (composition) inverse precisely when ν' and $1_{H'}$ are mutually convolution inverse; that is, if and only if ν' is an antipode for H'.

(d) If H is cocommutative then $H' = H$ so that H' is a Hopf algebra with antipode $\nu' = \nu$. So ν is its own (composition) inverse; that is, $\nu \circ \nu = 1_H$. For the commutative case replace H by H^{op}. $\qquad\square$

Remark. Proposition 9.1(d) can also be seen from the observation that *commutative Hopf algebras* are groups in the opposite of the category of commutative algebras, while cocommutative Hopf algebras are groups in the category of cocommutative coalgebras; the antipode is inversion so is clearly involutory.

Proposition 9.2 *Let H and K be any Hopf algebras. Then each bialgebra morphism $f : H \longrightarrow K$ preserves antipode.*

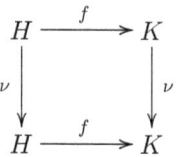

Proof. Clearly if $f : D \longrightarrow C$ and $g : A \longrightarrow B$ are coalgebra and algebra morphisms respectively, then

$$\mathrm{Hom}(\,C\,,A\,) \longrightarrow \mathrm{Hom}(\,D\,,B\,) \quad \text{whereby} \quad u \longmapsto g \circ u \circ f$$

is a monoid morphism for the convolution structures. In particular, here we have two monoid morphisms

$$\_ \circ f \quad \text{and} \quad f \circ \_ \; : \mathrm{Hom}(H, H) \longrightarrow \mathrm{Hom}(H, K)$$

that both take 1_H to f. Monoid morphisms take inverses to inverses. So

$$
\begin{aligned}
\nu \circ f \; &= \; (\text{ convolution inverse of } f \text{ in } \mathrm{Hom}(H, K)) \\
&= \; f \circ \nu \, .
\end{aligned}
$$

$\square$

Using other fancier words, the category $\mathbf{Hopf}_R$ of Hopf algebras is a full subcategory of the category $\mathbf{Big}_R$ of bialgebras.

For any algebra H we have seen that H^0 becomes a coalgebra. If H is a bialgebra then H^0 becomes a bialgebra using the multiplication

$$H^0 {\otimes} H^0 \; \cong \; (H {\otimes} H)^0 \xrightarrow{\;\delta^0\;} H^0$$

and unit

$$R \; \cong \; R^0 \xrightarrow{\;\varepsilon^0\;} H^0$$

(recall Proposition 8.1). Furthermore, if H is a Hopf algebra then so is H^0 with antipode

$$\nu^0 : H^0 \longrightarrow H^0 \, .$$

What we have here is a contravariant "self-adjoint" functor

$$(\_)^0 : \mathbf{Hopf}_R^{\mathrm{op}} \longrightarrow \mathbf{Hopf}_R \, .$$

What "self-adjoint" means in this context is that

$$\mathbf{Big}_R(H, K^0) \; \cong \; \mathbf{Big}_R(K, H^0) \, .$$

Proposition 9.3 *If H is any Hopf algebra then the monoid $\mathcal{D}(H)$ of set-like elements is a group.*

Proof. For $g \in \mathcal{D}(H)$ we have

$$
\begin{aligned}
\nu(g)\,g &= \big(\mu \circ (\nu \otimes 1)\big)(g \otimes g) \\
&= \big(\mu \circ (\nu \otimes 1)\big)\delta(g) \\
&= \big(\mu \circ (\nu \otimes 1) \circ \delta\big)(g) = \eta\big(\varepsilon(g)\big) = \eta(1) = 1 \ .
\end{aligned}
$$

$\square$

An *A-point* of a Hopf algebra H is an algebra morphism $f : H \longrightarrow A$.

Proposition 9.4 (a) *If $f, g : H \longrightarrow A$ are commuting A-points of H (meaning that $[\,f(h), g(k)\,] = 0$ for all $h, k \in H$) then $f * g : H \longrightarrow A$ is an A-point of H.*
(b) *If $f : H \longrightarrow A$ is an A-point of H then f has a convolution inverse $f \circ \nu : H^{\mathrm{op}} \longrightarrow A$ which is an A-point of H^{op}.*

Proof. (a) The commuting property yields that

$$
H \otimes H \xrightarrow{\ f \otimes g\ } A \otimes A \xrightarrow{\ \mu\ } A
$$

is an algebra morphism. But $\delta : H \longrightarrow H \otimes H$ is an algebra morphism since H is a bialgebra. So $f * g \in \mathbf{Alg}(H, A)$.
(b) Clear from Proposition 9.1(b). $\square$

Example 9.5 *For a monoid G, we have seen that the monoid algebra $R(G)$ is a bialgebra. If G is a group then the group algebra $R(G)$ becomes a Hopf algebra with antipode $\nu : R(G) \longrightarrow R(G)$ given by $\nu(g) = g^{-1}$. (The axioms for $(\_)^{-1} : G \longrightarrow G$ expressed diagrammatically in* **Set** *are taken by the functor $\mathcal{F}_R : \mathbf{Set} \longrightarrow \mathbf{Mod}_R$ into the axioms which define the antipode.)*

Example 9.6 *For a Lie algebra L, write L^{op} for the Lie algebra with the same module L but with Lie bracket β^{op} given by $\beta^{\mathrm{op}}(x, y) = \beta(y, x)$. For any algebra A we have $(A^{\mathrm{op}})_L = (A_L)^{\mathrm{op}}$. It follows (why?) that we have a canonical algebra isomorphism*

$$
\mathcal{U}(L^{\mathrm{op}}) \;\cong\; \mathcal{U}(L)^{\mathrm{op}} \ .
$$

We have a Lie algebra isomorphism $L \longrightarrow L^{\mathrm{op}}$ taking x to $-x$ (note that $[-x, -y] = [x, y] = -[\,y, x\,]$). So we define $\nu : \mathcal{U}(L) \longrightarrow \mathcal{U}(L)^{\mathrm{op}}$ by

$$L \xrightarrow{\quad - \quad} L^{\mathrm{op}}$$

$$i \downarrow \qquad\qquad \downarrow i$$

$$\mathcal{U}(L) \xrightarrow{\quad \nu \quad} \mathcal{U}(L)^{\mathrm{op}} \cong \mathcal{U}(L^{\mathrm{op}}) \,.$$

One easily checks that for $x_1 , \dots , x_n \in L$

$$\nu(i(x_1) \cdots i(x_n)) = (-1)^n \, i(x_1) \cdots i(x_n) \,.$$

With this antipode $\mathcal{U}(L)$ *becomes a Hopf algebra.*

Example 9.7 *The matrix bialgebra* $\mathrm{M}(n)$ *(Example 7.10 of a bialgebra) is not a Hopf algebra. We need to "adjoin an inverse for the determinant". Recall that* $\mathrm{M}(n) = R[\,X\,] = \mathcal{S}(\mathcal{F}_R(X))$ *where* $X = \{x_{ij} \mid i,j = 1, \dots, n\}$ *has cardinality* n^2. *Define*

$$\det(X) = \sum_{\xi \in \mathcal{S}_n} (-1)^{|\xi|} \, x_{1\,\xi(1)} \, x_{2\,\xi(2)} \cdots x_{n\,\xi(n)}$$

where $|\xi|$ *is the least number of simple transpositions required to obtain the permutation* ξ. *Form the following commutative polynomial* R-*algebra:* $R[\,X \cup \{t\}\,] = R[\,(x_{ij}), t\,] = \mathcal{S}(\mathcal{F}_R(X \cup \{t\}))$, *in* $n^2 + 1$ *(commuting) indeterminates* t *and* x_{ij} *with* $(1 \le i,j \le n)$. *Put*

$$\mathrm{GL}(n) = R[\,X \cup \{t\}\,]/(\,t \det(X) - 1\,)$$

as a commutative R-*algebra. We make* $\mathrm{GL}(n)$ *into a bialgebra by defining*

$$\delta(x_{ij}) = \sum_{k=1}^{n} x_{ik} \otimes x_{kj} \qquad \delta(t) = t \otimes t$$

$$\varepsilon(x_{ij}) = \delta_{ij} \qquad\qquad \varepsilon(t) = 1$$

modulo $(\,t \det(X) - 1\,)$. *Put* $X_{ij} = \{x_{rs} \mid r \ne i, \, s \ne j\}$. *Now define the morphism* $\nu : \mathrm{GL}(n) \longrightarrow \mathrm{GL}(n)$ *by*

$$\nu(x_{ij}) = t \det(X_{ji})$$
$$\nu(t) = \det(X)$$

modulo $(\,t \det(X) - 1\,)$. *Then* $\mathrm{GL}(n)$ *becomes a Hopf algebra.*

For any commutative R-*algebra* A *we have a canonical isomorphism of groups*

$$\mathbf{Alg}_R(\mathrm{GL}(n), A) \cong \mathrm{GL}(n, A) \,.$$

Examples 9.5 and 9.6 above exhibit cocommutative Hopf algebras $R(G)$ and $\mathcal{U}(L)$, while Example 9.7 is a commutative Hopf algebra $\mathrm{GL}(n)$. It is only recently that the importance of Hopf algebras which are neither commutative nor cocommutative has been properly understood.

Example 9.8 *We now describe a "quantum deformation" of Example 9.7. This is a generalization to $n \times n$, from the 2×2 case discussed in Chapter 3.*

Take $X = \{x_{ij} \mid i, j = 1, \ldots, n\}$ as in Example 9.7. First we form the free algebra $R\langle X \rangle = T(\mathcal{F}_R(X))$ on the (non-commuting) indeterminates x_{ij}. Let $\mathrm{M}_q(n)$ denote the quotient of $R\langle X \rangle$ by the ideal generated by the following elements:

$$
\begin{array}{ll}
x_{ir}\, x_{jk} - x_{jk}\, x_{ir} & \text{for } i < j \text{ and } k < r \\
x_{ir}\, x_{jk} - x_{jk}\, x_{ir} - (q - q^{-1})\, x_{ik}\, x_{jr} & \text{for } i < j \text{ and } r < k \\
x_{ik}\, x_{jk} - q\, x_{jk}\, x_{ik} & \text{for } i < j \\
x_{ik}\, x_{ir} - q\, x_{ir}\, x_{ik} & \text{for } k < r \, .
\end{array}
$$

This becomes a coalgebra with comultiplication

$$
\delta(x_{ij}) = \sum_{r=1}^{n} x_{ir} \otimes x_{rj} \quad (\text{ modulo the ideal })
$$

and counit

$$
\varepsilon(x_{ij}) = \delta_{ij} \quad (\text{ Kronecker delta }).
$$

Define the "quantum determinant" by

$$
\det_q(X) = \sum_{\xi \in \mathcal{S}_n} (-q)^{|\xi|}\, x_{1\,\xi(1)}\, x_{2\,\xi(2)} \cdots x_{n\,\xi(n)}
$$

which is a central element of $\mathrm{M}_q(n)$ (that is, it commutes with all other elements). The quantum general linear group *is defined by*

$$
\mathrm{GL}_q(n) = \mathrm{M}_q(n)[t]/(t \det_q(X) - 1).
$$

We adjust the comultiplication and counit of $\mathrm{M}_q(n)$ by defining $\delta(t) = t \otimes t$ and $\varepsilon(t) = 1$. Then we have a bialgebra epimorphism

$$
\rho : \mathrm{M}_q(n) \longrightarrow \mathrm{GL}_q(n) .
$$

Define $\nu : \mathrm{GL}_q(n) \longrightarrow \mathrm{GL}_q(n)$ by

$$
\nu(x_{ij}) = t \det_q(X_{ji}) \quad \text{and} \quad \nu(t) = \det_q(X) .
$$

Then $\mathrm{GL}_q(n)$ becomes a Hopf algebra. Notice that $\mathrm{GL}_q(n)^{\mathrm{op}} = \mathrm{GL}_{q^{-1}}(n)$.

Many claims have been made in this section. For $n = 2$ the calculations in Theorem 3.1 prove them all. (This should be compared with Proposition 9.4 in the present section.) The general case can be verified similarly, but will follow from later work.

Exercise 9.1 [Swe69] *Assume our base ring R is a field and write $\otimes$ for $\otimes_R$. An ideal in an algebra A is a submodule I such that $\mu(I\otimes A + A\otimes I) \subseteq I$. We know that A/I becomes an algebra. A* coideal *of a coalgebra C is a submodule I such that $\delta(I) \subseteq I\otimes C + C\otimes I$ and $\varepsilon(I) \subseteq 0$.*

(a) *If I is a coideal of a coalgebra C, describe a coalgebra structure on C/I for which $\rho : C \longrightarrow C/I$ becomes a coalgebra morphism. If C is a bialgebra and I is also an ideal, show that C/I is a bialgebra. What condition on I ensures C/I has an antipode if C has? I is called a* Hopf ideal *when this holds.*

(b) *Verify that the polynomial R-algebra $B = R\langle x, y, z\rangle$ on three non-commuting indeterminates becomes a bialgebra with*

$$\delta(x) = x\otimes x \ , \quad \delta(y) = y\otimes y \ , \quad \delta(z) = 1\otimes z + z\otimes x$$
$$\varepsilon(x) = \varepsilon(y) = 1 \ , \quad \varepsilon(z) = 0 \ .$$

(c) *Verify that the ideal $(xy - 1, yx - 1)$ is a coideal in B. Let H denote the quotient bialgebra.*

(d) *Show that H is a Hopf algebra with antipode ν given by*

$$\nu(x) = y, \ \nu(y) = x, \ \nu(z) = -zy \ \text{(modulo the ideal of (c))}.$$

Show further that $\nu^{2n}(z) = x^n z y^n$, $\nu^{2n+1}(z) = -x^n z y^{n+1}$. Hence, this antipode has infinite order.

(e) i. *Show that the ideals $I_n = (x^n z y^n - z)$, $J_n = (x^n - 1)$ are Hopf ideals in H.*

   ii. *Show that the antipodes of both H/I_n and H/J_n have order $2n$.*

Bradshaw: Elegant Action Figure Group, [Wal94, Plate 84].

10

Representations of quantum groups

We mentioned in Example 6.3 that a representation of a group G was an $R(G)$-module. One kind of representation for a Hopf algebra H therefore suggests itself: a module over H. We begin by discussing modules over bialgebras.

First note that if $f : E \longrightarrow A$ is a ring morphism then each (left) A-module M becomes a (left) E-module via the action

$$e\,m = f(e)m \qquad \text{for } e \in E \, , \; m \in M \, .$$

This is called *restriction of scalars along f*.

Let A be an R-algebra. Then each module is automatically an R-module via restriction of scalars along the unit $\eta : R \longrightarrow A$. Alternatively, we can view an A-module as an R-module M with a ring morphism $\hat{\mu} : A \longrightarrow \mathrm{End}_R(M)$. Later, we want to look at "comodules", and so we want a definition of A-module which dualizes. The good version is: an R-module M with a module morphism $\mu : A \otimes_R M \longrightarrow M$, called the *action* of A on M, satisfying

$$A \otimes_R A \otimes_R M \underset{1 \otimes \mu}{\overset{\mu \otimes 1}{\rightrightarrows}} A \otimes_R M \overset{\mu}{\longrightarrow} M$$

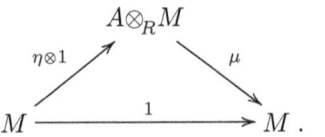

We write $\mathbf{Mod}_R(A)$ for $\mathbf{Mod}_A$ just to emphasize that we build it up from $\mathbf{Mod}_R$.

Suppose M and N are (left) modules over the R-algebra A. Regard $M : A \nrightarrow R$ and $N : R \nrightarrow A^{\mathrm{op}}$. We see that $M \otimes_R N : A \nrightarrow A^{\mathrm{op}}$, which means $M \otimes_R N$ becomes an $A \otimes A$-module. If A is a bialgebra then we

can restrict scalars along $\delta : A \longrightarrow A \otimes_R A$ to obtain an A-module structure on $M \otimes_R N$. Explicitly, the action is the composite

$$A \otimes_R M \otimes_R N \xrightarrow{\delta \otimes 1 \otimes 1} A \otimes_R A \otimes_R M \otimes_R N \xrightarrow{1 \otimes \sigma \otimes 1} A \otimes_R M \otimes_R A \otimes_R N \xrightarrow{\mu \otimes \mu} M \otimes_R N .$$

This generalizes to multiple tensor products (over R) of A-modules. In particular, the empty tensor product R becomes an A-module by restricting scalars along the counit $\varepsilon : A \longrightarrow R$.

With M and N left A-modules as before, we can regard $M : R \rightarrowtail A^{\mathrm{op}}$ and $N : R \rightarrowtail A^{\mathrm{op}}$, so that $\mathrm{Hom}_R(M, N) : A^{\mathrm{op}} \rightarrowtail A^{\mathrm{op}}$; or in other words $\mathrm{Hom}_R(M, N)$ becomes an $A^{\mathrm{op}} \otimes A$-module. Thus if $A = H$ is a Hopf algebra, we can restrict scalars along the R-algebra morphism

$$H \xrightarrow{\delta} H \otimes_R H \xrightarrow{\nu \otimes 1} H^{\mathrm{op}} \otimes_R H$$

to make $\mathrm{Hom}_R(M, N)$ into an H-module. Explicitly, the action of H on $\mathrm{Hom}_R(M, N)$ is the composite

$$H \otimes_R \mathrm{Hom}_R(M,N) \xrightarrow{\delta \otimes 1} H \otimes_R H \otimes_R \mathrm{Hom}_R(M,N) \xrightarrow{1 \otimes \sigma}$$
$$H \otimes_R \mathrm{Hom}_R(M,N) \otimes_R H \xrightarrow{\mu_1 \otimes \nu} \mathrm{Hom}_R(M,N) \otimes_R H \xrightarrow{\mu_2} \mathrm{Hom}_R(M,N)$$

where μ_1 and μ_2 are the left and right actions:

$$H \otimes_R \mathrm{Hom}_R(M,N) \xrightarrow{\hat{\mu} \otimes 1} \mathrm{Hom}_R(N,N) \otimes_R \mathrm{Hom}_R(M,N) \xrightarrow{\circ} \mathrm{Hom}_R(M,N)$$
$$\mu_1 : h \otimes f \longmapsto \qquad\qquad\qquad\qquad\qquad\qquad (m \longmapsto h(f\,m))$$
$$\mathrm{Hom}_R(M,N) \otimes_R H \xrightarrow{1 \otimes \hat{\mu}} \mathrm{Hom}_R(M,N) \otimes_R \mathrm{Hom}_R(M,M) \xrightarrow{\circ} \mathrm{Hom}_R(M,N)$$
$$\mu_2 : f \otimes h \longmapsto \qquad\qquad\qquad\qquad\qquad\qquad (m \longmapsto f(h\,m)) .$$

Proposition 10.1 *For left modules M and N over the Hopf algebra H, the canonical R-module morphisms*

$$e : \mathrm{Hom}_R(M, N) \otimes_R M \longrightarrow N \quad \text{where} \quad f \otimes m \longmapsto f(m)$$
$$d : M \longrightarrow \mathrm{Hom}_R(N, M \otimes_R N) \quad \text{where} \quad m \longmapsto (n \longmapsto m \otimes n)$$

are left H-module morphisms.

Proof. Omitting $\otimes_R$ and Hom_R from the notation, we obtain the first of

these from the following diagram. The second we leave to the reader.

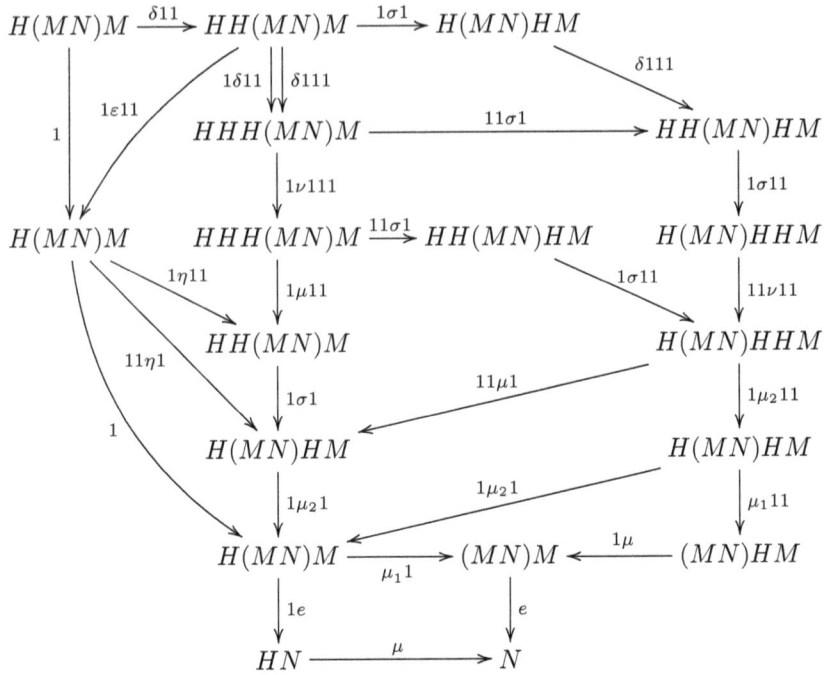

Corollary 10.2 *For modules M, N and L over a Hopf algebra H, the canonical isomorphism*

$$\mathrm{Hom}_R(M \otimes_R N, L) \;\cong\; \mathrm{Hom}_R(M, \mathrm{Hom}_R(N, L))$$

restricts to an isomorphism

$$\mathrm{Hom}_H(M \otimes_R N, L) \;\cong\; \mathrm{Hom}_H(M, \mathrm{Hom}_R(N, L)) \,.$$

Proof. The canonical isomorphism is obtained from the evaluation e and the canonical d of Proposition 10.1. □

In other words, we have a nice *tensor–hom* situation for the category $\mathbf{Mod}_R(H)$ of (left) H-modules. Both the tensor and the *hom* are preserved by the functor

$$\mathbf{Mod}_R(H) \longrightarrow \mathbf{Mod}_R$$

given by ignoring the H-action.

Although modules over the group algebra are representations of the group, so that the study of modules over a Hopf algebra does suggest itself, the point of view of Chapter 2 (i.e. space–algebra duality) leads more naturally to "comodules". For here, it is the comultiplication $\delta : H \longrightarrow H \otimes H$ of the Hopf algebra which corresponds to the spatial multiplication.

Suppose C is an R-coalgebra. A (left) C-*comodule* is an R-module M with a module morphism $\delta : M \longrightarrow C \otimes_R M$, called the *coaction* of C on M, satisfying

$$M \xrightarrow{\ \delta\ } C \otimes_R M \underset{1 \otimes \delta}{\overset{\delta \otimes 1}{\rightrightarrows}} C \otimes_R C \otimes_R M$$

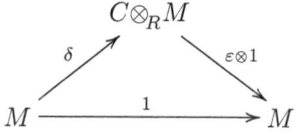

We write $\mathbf{Com}_R(C)$ for the category whose objects are C-comodules and whose arrows are C-*comodule morphisms*; that is, R-module morphisms $f : M \longrightarrow N$ such that

$$
\begin{array}{ccc}
M & \xrightarrow{\ \delta\ } & C \otimes_R M \\
\downarrow{\scriptstyle f} & & \downarrow{\scriptstyle 1 \otimes f} \\
N & \xrightarrow{\ \delta\ } & C \otimes_R N \ .
\end{array}
$$

Each C-comodule M becomes a C^*-module with the action

$$C^* \otimes_R M \xrightarrow{\ 1 \otimes \delta\ } C^* \otimes_R C \otimes_R M \xrightarrow{\ e \otimes 1\ } M \ .$$

See Chapter 7 for the algebra structure on C^*.

By the fundamental theorem of Morita theory (Theorem 5.2), if C is Cauchy (as an R-module) then this gives a bijection between C-coactions δ and C^*-actions μ on each R-module M: recover δ as the composite

$$M \xrightarrow{\ d \otimes 1\ } C \otimes_R C^* \otimes_R M \xrightarrow{\ 1 \otimes \mu\ } C \otimes_R M \ .$$

So for C Cauchy, we have an isomorphism of categories

$$\mathbf{Com}_R(C) \ \cong \ \mathbf{Mod}_R(C^*) \ .$$

If C is an R-bialgebra not necessarily Cauchy we obtain, in a manner dual to that for modules, a coaction on the tensor product (over R) of

C-comodules. Explicitly for C-modules M and N, the coaction for $M \otimes_R N$ is given by the composite

$$M \otimes_R N \xrightarrow{\delta \otimes \delta} C \otimes_R M \otimes_R C \otimes_R N \xrightarrow{\sigma_{1324}} C \otimes_R C \otimes_R M \otimes_R N \xrightarrow{\mu \otimes 1 \otimes 1} C \otimes_R M \otimes_R N \,.$$

The empty tensor product R has the coaction $\eta : R \longrightarrow C \otimes_R R$.

When it comes to Hom our formal duality fails: in reversing arrows we have maintained $\otimes$, yet Hom does not maintain its universal property. However, if M is Cauchy, $\mathrm{Hom}_R(M, N)$ does have the *reverse-arrow universal property*: there is a bijection between R-module morphisms

$$L \longleftarrow \mathrm{Hom}_R(M, N)$$

and R-module morphisms

$$M \otimes L \longleftarrow N$$

since $\mathrm{Hom}_R(M, N) \cong M^* \otimes_R N$ and $M \otimes_R L \cong \mathrm{Hom}_R(M^*, L)$.

Proposition 10.3 *Each Cauchy R-module M gives rise to an R-coalgebra $M \otimes_R M^*$ with counit $e : M \otimes_R M^* \longrightarrow R$ and comultiplication*

$$1 \otimes d \otimes 1 : M \otimes_R M^* \longrightarrow M \otimes_R M^* \otimes_R M \otimes_R M^*$$

(see Theorem 5.2). For any R-coalgebra C, the assignment

$$\hat{\delta} = (M \otimes_R M^* \xrightarrow{\delta \otimes 1} C \otimes_R M \otimes_R M^* \xrightarrow{1 \otimes e} C)$$

determines a bijection between coactions

$$\delta : M \longrightarrow C \otimes_R M$$

of C on M and coalgebra morphisms

$$\hat{\delta} : M \otimes_R M^* \longrightarrow C \,.$$

Proof. $M \otimes_R M^* \xrightarrow[\sigma]{\approx} M^* \otimes_R M \xrightarrow[\rho]{\approx} \mathrm{Hom}_R(M, M)$ has the universal property of Hom under reversal of arrows; so the diagrammatic proof that $\mathrm{End}_R(M)$ is an algebra and that an action is an algebra morphism of the kind $A \longrightarrow \mathrm{End}_R(M)$, dualizes. $\square$

Take $M = R^n$ in the above proposition and let $e_1, \ldots, e_n$ be the standard basis. Now let $e_1^*, \ldots, e_n^*$ be the dual basis for R^{n*} so $e_i^*(e_j) = \delta_{ij}$

(Kronecker-δ). A coaction of C on R^n thus amounts to a coalgebra morphism $\hat{\delta} : R^n \otimes_R R^{n*} \longrightarrow C$, and this is determined by its values on the basis elements $e_i \otimes e_j^*$ of $R^n \otimes_R R^{n*}$:

$$\hat{\delta}(e_i \otimes e_j^*) \;=\; x_{ij} \;\in C \,.$$

So C-comodule structures on R^n are in bijection with *multiplicative matrices* in C; that is, matrices $\vec{\mathbf{x}} = (x_{ij})$ in C satisfying

$$\delta(x_{ij}) = \sum_k x_{ik} \otimes x_{kj} \quad, \quad \varepsilon(x_{ij}) = \delta_{ij} \,.$$

Following Manin [Man88], we write the last two equations as

$$\delta(\vec{\mathbf{x}}) = \vec{\mathbf{x}} \otimes \vec{\mathbf{x}} \quad, \quad \varepsilon(\vec{\mathbf{x}}) = \vec{\mathbf{i}}$$

where $\vec{\mathbf{i}}$ is the identity matrix and $\vec{\mathbf{x}} \otimes \vec{\mathbf{y}} = (\sum_k x_{ik} \otimes y_{kj})$ is *not* the usual tensor product of matrices.

Example 10.4 *In the situation of Example 9.8, $\vec{\mathbf{x}} = (x_{ij})$ and $\begin{pmatrix} \vec{\mathbf{x}} & 0 \\ 0 & t \end{pmatrix}$ are multiplicative matrices for* $\mathrm{M}_q(n)$ *and* $\mathrm{GL}_q(n)$, *respectively.*

Now suppose $C = H$ is a Hopf algebra and M is a Cauchy R-module. By applying Proposition 10.1 to M^* (and using the canonical $M^{**} \cong M$), we see that $M^* \otimes_R M$ becomes a coalgebra with counit

$$M^* \otimes_R M \xrightarrow{\;\sigma\;} M \otimes_R M^* \xrightarrow{\;e\;} R$$

and comultiplication

$$M^* \otimes_R M \xrightarrow{\;1 \otimes d \otimes 1\;} M^* \otimes_R M^* \otimes_R M \otimes_R M \xrightarrow{\;1 \otimes \sigma \otimes 1\;} M^* \otimes_R M \otimes_R M^* \otimes_R M \,.$$

Proposition 10.5 (a) $\hat{\delta} : M \otimes_R M^* \longrightarrow H$ *is a coalgebra morphism if and only if the composite*

$$M^* \otimes_R M \xrightarrow{\;\sigma\;} M \otimes_R M^* \xrightarrow{\;\hat{\delta}\;} H^{\mathrm{op}}$$

is a coalgebra morphism.

   (b) *Suppose M is an H-comodule and put*

$$\hat{\delta}_k = \left(M \otimes_R M^* \xrightarrow{\;\hat{\delta}\;} H \xrightarrow{\;\nu^k\;} H \right) \,.$$

For k even, $\hat{\delta}_k$ is a coalgebra morphism and has convolution inverse $\hat{\delta}_{k+1}$ in $\mathrm{Hom}_R(M \otimes_R M^*, H)$.

Proof. (a)

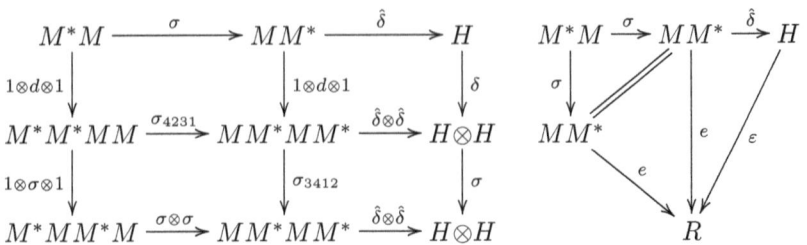

(b) Since $\nu : H \longrightarrow H^{\mathrm{op}}$ is a coalgebra morphism, by Proposition 9.1(b); then it is also true that $\nu^k : H \longrightarrow H$ is a coalgebra morphism for k even; so $\hat{\delta}_k = \nu^k \circ \hat{\delta}$ is a coalgebra morphism, as required. This also means that right composition with $\hat{\delta}_k$ preserves convolution. Since 1_H and ν are convolution inverses, so are $1_H \circ \hat{\delta}_k$ and $\nu \circ \hat{\delta}_k$; that is, so are $\hat{\delta}_k$ and $\hat{\delta}_{k+1}$. □

Proposition 10.6 *Suppose that M is a comodule over the Hopf algebra H and that M is Cauchy as an R-module. Then M^* becomes an H-comodule via*

$$\hat{\delta} = \left(M^* \otimes_R M \xrightarrow{\sigma} M \otimes_R M^* \xrightarrow{\hat{\delta}} H \xrightarrow{\nu} H \right).$$

Moreover, the R-module morphisms

$$e : M \otimes_R M^* \longrightarrow R$$
$$d : R \longrightarrow M^* \otimes_R M$$

become H-comodule morphisms. One might therefore say that M becomes a Cauchy H-comodule.

Proof. By Propositions 10.5(a) and 9.1(b), the stated $\hat{\delta}$ is a coalgebra morphism. So by Proposition 10.5, it determines a coaction of H on M^*. Tracing through, one sees that this is dual to the situation for $\mathrm{Hom}_R(M, R)$ as in Proposition 10.1; so the proof dualizes, but it can also be shown directly that e and d are comodule morphisms. □

Remark. To obtain results as in Proposition 10.5(b) for k odd, apply Proposition 10.5(b) to the M^* of Proposition 10.6; compare with [Man88, p.14].

Exercise 10.1 *Give a direct proof of Proposition 10.3, concerning the coalgebra structure on $M \otimes_R M^*$ where M is a Cauchy R-module.*

11

Tensor categories

It is clear that specific categories have entered explicitly into the above discussion, but we have made little use of them as categories apart from diagrams and duality. For what follows it is hard to imagine how to express the results without categories.

A *tensor category* (also called "monoidal category" [EK66]) is a category $\mathcal{V}$ together with functor $\otimes : \mathcal{V} \times \mathcal{V} \longrightarrow \mathcal{V}$ called *tensor product*, an object I of $\mathcal{V}$ called the *unit object*, and natural families of isomorphisms

$$a_{A,B,C} : (A \otimes B) \otimes C \longrightarrow A \otimes (B \otimes C)$$

$$r_A : A \otimes I \longrightarrow A \quad , \quad l_A : I \otimes A \longrightarrow A$$

called respectively the *associativity* constraint, the *right unit* constraint and the *left unit* constraint, subject to the two conditions:

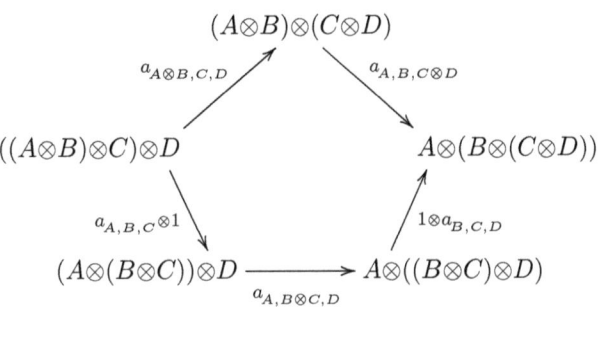

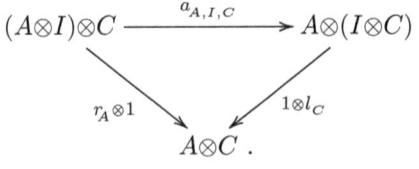

Define $A_1 \otimes \cdots \otimes A_n$ to be the object obtained by inserting brackets in some chosen preassigned way, such as from the left $((\cdots(A_1 \otimes A_2) \otimes \cdots) \otimes A_n)$.

It is an important fact (Mac Lane's coherence theorem) that, in general, the only automorphism which is a composite of isomorphisms of the form $1 \otimes (x \otimes 1)$ or $(1 \otimes x) \otimes 1$, where x is a component of a, r, l or their inverses, is the identity arrow of $A_1 \otimes \cdots \otimes A_n$. This essentially allows one to work as if the a, r, l are all identities. If all the a, r, l are indeed identities, then the tensor category is called *strict*. The *opposite* $\mathcal{V}^{\mathrm{op}}$ of a tensor category $\mathcal{V}$ consists of the opposite category of $\mathcal{V}$ (obtained by reversing the direction of arrows of $\mathcal{V}$) and the reverse tensor product, so that $A \otimes B$ in $\mathcal{V}^{\mathrm{op}}$ is just $B \otimes A$ in $\mathcal{V}$.

A *braiding* for a tensor category $\mathcal{V}$ is a natural family of isomorphisms

$$c_{A,B} : A \otimes B \longrightarrow B \otimes A$$

subject to the conditions

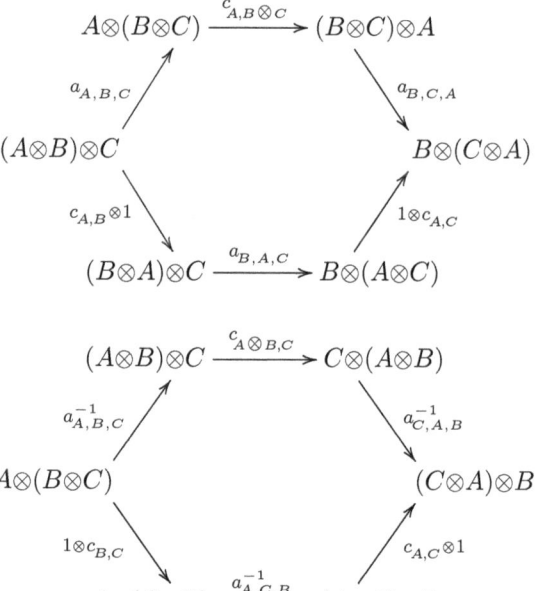

A *braided tensor category* is a tensor category with a chosen braiding (see [JS93]).

A *symmetry* for a tensor category is a braiding which satisfies the following extra condition:

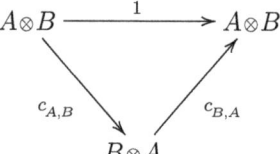

A *symmetric tensor category* is a tensor category with a chosen symmetry.

Example 11.1 *The* braid category $\mathcal{B}$ *has as objects the natural numbers* $0, 1, 2, \ldots$ *and as arrows* $\alpha : n \longrightarrow n$ *the braids on* n *strings; there are no arrows* $m \longrightarrow n$ *for* $m \neq n$. *A braid* α

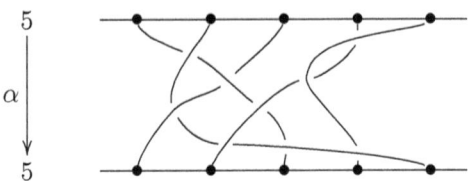

on n *strings can be regarded as an element of the Artin braid group* $\mathcal{B}_n$ *with generators* $s_1, \ldots, s_{n-1}$ *subject to the relations*

$$s_i s_j = s_j s_i \qquad for \; j < i - 1$$
$$s_{i+1} s_i s_{i+1} = s_i s_{i+1} s_i$$

where s_i *is the braid depicted as:*

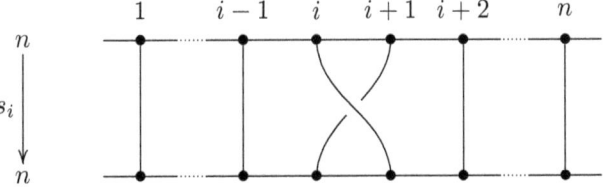

Composition of braids is just multiplication in this group, represented diagrammatically by vertical stacking of braids with the same number of strings.

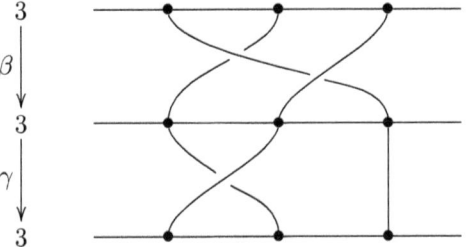

Tensor product of braids adds the number of strings by placing one braid next to the other longitudinally.

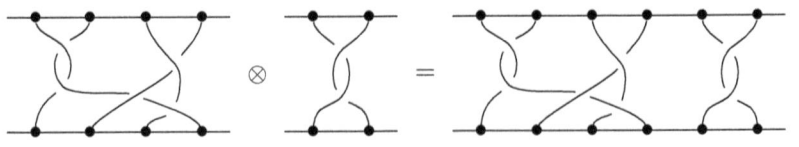

This makes $\mathcal{B}$ *a strict tensor category. A braiding* $c_{m,n} : m+n \longrightarrow n+m$ *is given by crossing the first* m *strings over the remaining* n.

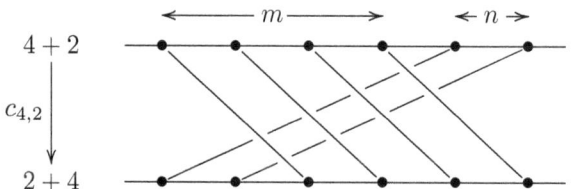

The axioms that show $\mathcal{B}$ *is braided are easily checked diagrammatically.*

Example 11.2 *The category* $\mathbf{Mod}_R$ *of modules over a commutative ring* R *is a symmetric tensor category with tensor product* $\otimes_R$, *with the canonical constraints, and with symmetry* $\sigma : A\otimes_R B \longrightarrow B\otimes_R A$.

Example 11.3 *Let* A *be an* R-*bialgebra. If* M *and* N *are* A-*modules, we have an* A-*module structure on* $M\otimes_R N$ *given by*

$$a \cdot (m\otimes n) = \sum_{(a)} a_{(1)} m \otimes a_{(2)} n$$

as seen in the last chapter. So $\mathbf{Mod}_R(A)$ *becomes a tensor category with tensor product* $\otimes_R$.

If A *is cocommutative, the switch morphism* $\sigma : M\otimes_R N \longrightarrow N\otimes_R M$ *is a symmetry for* $\mathbf{Mod}_R(A)$. *However, as in the rest of this book, we are more interested in non-cocommutative* A.

We ask: *what are the possible braidings on the tensor category* $\mathbf{Mod}_R(A)$?

A braiding $c_{M,N} : M\otimes_R N \longrightarrow N\otimes_R M$ *gives, for each* A, *a morphism* $c_{A,A} : A\otimes_R A \longrightarrow A\otimes_R A$ *which gives an element* $\gamma = c_{A,A}(1\otimes 1) \in A\otimes A$.

Conversely, each element $\gamma = \sum_i u_i \otimes v_i \in A\otimes A$ *determines a natural morphism* $c_{M,N} : M\otimes_R N \longrightarrow N\otimes_R M$ *via the formula*

$$c_{M,N}(m\otimes n) = \sum_i (u_i n)\otimes(v_i m).$$

This is a bijection, as can be seen from the following diagram in which $\hat{m} : A \longrightarrow M$ *is the unique module morphism with* $\hat{m}(1) = m$.

$$
\begin{array}{ccc}
R\otimes_R R & \xrightarrow{\ \eta\otimes\eta\ } & A\otimes_R A & \xrightarrow{\ c_{A,A}\ } & A\otimes_R A \\
{\scriptstyle \hat{m}\otimes\hat{n}}\big\downarrow & & & & \big\downarrow{\scriptstyle \hat{n}\otimes\hat{m}} \\
& M\otimes_R N & \xrightarrow{\ c_{M,N}\ } & N\otimes_R M. &
\end{array}
$$

In order for each $c_{M,N}$ to be an isomorphism it is necessary for $\gamma \in A \otimes_R A$ to be invertible. In order for each $c_{M,N}$ to be a module morphism we need

$$c\big(a \cdot (m \otimes n)\big) = c\Big(\sum_{(a)} (a_{(1)}m) \otimes (a_{(2)}n)\Big)$$

$$= \sum_{(a)} \sum_i (u_i a_{(2)} n) \otimes (v_i a_{(1)} m)$$

to be equal to

$$a \cdot c(m \otimes n) = a \cdot \sum_i (u_i n) \otimes (v_i m)$$

$$= \sum_{(a)} \sum_i (a_{(1)} u_i n) \otimes (a_{(2)} v_i m).$$

This is equivalent to the requirement

$$\sum_{i,(a)} (u_i a_{(2)}) \otimes (v_i a_{(1)}) = \sum_{i,(a)} (a_{(1)} u_i) \otimes (a_{(2)} v_i).$$

Regarding $\gamma \in A \otimes_R A$ as a morphism $\gamma : R \longrightarrow A \otimes_R A$ whose value at $1 \in R$ is the given γ, we can express this condition diagrammatically as

(B0) $$A \xrightarrow{\gamma \otimes \delta} A^{\otimes 4} \underset{\sigma_{3142}}{\overset{\sigma_{1423}}{\rightrightarrows}} A^{\otimes 4} \xrightarrow{\mu \otimes \mu} A^{\otimes 2}.$$

For a braiding, we require two more conditions:

$$c_{M, N \otimes L}(m \otimes n \otimes l) = (1_N \otimes c_{M,L})(c_{M,N} \otimes 1_L)(m \otimes n \otimes l),$$

$$c_{M \otimes N, L}(m \otimes n \otimes l) = (c_{M,L} \otimes 1_N)(1_M \otimes c_{N,L})(m \otimes n \otimes l);$$

that is,

$$\sum_{i,(u_i)} u_{i(1)} n \otimes u_{i(2)} l \otimes v_i m = \sum_{i,j} u_i n \otimes u_j l \otimes v_j v_i m,$$

$$\sum_{i,(v_i)} u_i l \otimes v_{i(1)} m \otimes v_{i(2)} n = \sum_{i,j} u_j u_i l \otimes v_j m \otimes v_i n.$$

These are equivalent to the two conditions

$$\sum_{i,(u_i)} u_{i(1)} \otimes u_{i(2)} \otimes v_i = \sum_{i,j} u_i \otimes u_j \otimes v_j v_i,$$

$$\sum_{i,(v_i)} u_i \otimes v_{i(1)} \otimes v_{i(2)} = \sum_{i,j} u_j u_i \otimes v_j \otimes v_i.$$

Diagrammatically, these conditions become:

(B1)

$$
\begin{array}{ccccc}
R & \xrightarrow{\gamma\otimes\gamma} & A^{\otimes 4} & \xrightarrow{\sigma_{1342}} & A^{\otimes 4} \\
\gamma \downarrow & & & & \downarrow 1\otimes 1\otimes\mu \\
A^{\otimes 2} & & \xrightarrow{\delta\otimes 1} & & A^{\otimes 3}
\end{array}
$$

(B2)

$$
\begin{array}{ccccc}
R & \xrightarrow{\gamma\otimes\gamma} & A^{\otimes 4} & \xrightarrow{\sigma_{3142}} & A^{\otimes 4} \\
\gamma \downarrow & & & & \downarrow \mu\otimes 1\otimes 1 \\
A^{\otimes 2} & & \xrightarrow{1\otimes\delta} & & A^{\otimes 3} \;.
\end{array}
$$

Hence, we define a braiding element *for a bialgebra A to be an invertible element $\gamma \in A\otimes_R A$ which satisfies (B0), (B1), (B2). We have proved above that* braiding elements for A are in bijection with braidings *on the tensor category* $\mathbf{Mod}_R(A)$.

A braided bialgebra *(also called "quasitriangular bialgebra")* is a bialgebra equipped with a braiding element $\gamma \in A\otimes_R A$. A braiding element γ is called a symmetry element *when $\gamma^2 = 1 \in A\otimes_R A$; these are in bijection with symmetries on* $\mathbf{Mod}_R(A)$. A symmetric bialgebra *(also known as a "triangular algebra")* is a bialgebra equipped with a symmetry element.

Before leaving this example, we point out that conditions (B1), (B2) can be put in a more familiar form in the case where A is Cauchy as an R-module. For in this case, elements $\gamma = \sum_i u_i\otimes v_i \in A\otimes_R A$ are in bijection with R-module morphisms $g : A^* \longrightarrow A$ via the formula

$$
\gamma = \left(R \xrightarrow{\;d\;} A^*\otimes_R A \xrightarrow{\;g\otimes 1_A\;} A\otimes_R A \right) .
$$

Condition (B1) precisely says that g preserves comultiplication, while condition (B2) says that g reverses multiplication. In fact, if γ is a braiding element, $g : A^* \longrightarrow A'^{\mathrm{op}}$ is a bialgebra morphism; preservation of unit and counit follows from $c_{M,I} = c_{I,M} = \mathbf{1}_M$.

We shall just look at the translation of (B2) to g. Begin with the defining diagram

$$
\begin{array}{ccccc}
R & \xrightarrow{\;d\;} & A^*\otimes_R A & \xrightarrow{1\otimes d\otimes 1} & A^*\otimes_R A^*\otimes_R A\otimes_R A \\
d \downarrow & & & & \downarrow \delta^*\otimes 1\otimes 1 \\
A^*\otimes_R A & & \xrightarrow{\;\;1\otimes\delta\;\;} & & A^*\otimes_R A\otimes_R A
\end{array}
$$

for δ^*, which is the multiplication for A^*. To prove g reverses multiplication is to prove

$$A^* \otimes_R A^* \xrightarrow{\ g \otimes g\ } A \otimes_R A \xrightarrow{\ \sigma\ } A \otimes_R A$$

with d^* down on the left to A^*, and μ down on the right to A, and $A^* \xrightarrow{\ g\ } A$.

This is equivalent to proving the legs are equal after applying $\_ \otimes_R A \otimes_R A$ and composing with

$$R \xrightarrow{\ d\ } A^* \otimes_R A^* \xrightarrow{\ 1 \otimes d \otimes 1\ } A^* \otimes_R A^* \otimes_R A \otimes_R A.$$

From the defining diagram for δ^*, this amounts to

$$R \xrightarrow{d} A^* \otimes_R A^* \xrightarrow{1 \otimes d \otimes 1} A^* \otimes_R A^* \otimes_R A \otimes_R A \xrightarrow{g \otimes g \otimes 1 \otimes 1} A^{\otimes 4} \xrightarrow{\sigma} A^{\otimes 4}$$

with $1 \otimes \delta$ down on the left to $A^* \otimes_R A \otimes_R A$, and $\mu \otimes 1 \otimes 1$ down on the right to $A^{\otimes 3}$, and $A^* \otimes_R A \otimes_R A \xrightarrow{g \otimes 1 \otimes 1} A^{\otimes 3}$.

Using $\gamma = (g \otimes 1_A) \circ d$, we easily see that this is equivalent to (B2).

Although a braiding is as useful as a symmetry for most purposes, there is sometimes further structure on a braiding which makes it even more like a symmetry without actually forcing it to be one.

Suppose $\mathcal{V}$ is a braided tensor category. A *twist* for $\mathcal{V}$ is a natural family of isomorphisms

$$\theta_A : A \longrightarrow A$$

such that $\theta_I = 1_I$ and

$$A \otimes B \xrightarrow{\ c_{A,B}\ } B \otimes A$$

with $\theta_{A \otimes B}$ down on the left to $A \otimes B$, and $\theta_B \otimes \theta_A$ down on the right to $B \otimes A$, and $A \otimes B \xleftarrow{\ c_{B,A}\ } B \otimes A$.

A *balanced tensor category* is a braided tensor category with a chosen twist. (A braiding is a symmetry if and only if the identity arrows provide a twist.)

Example 11.4 *The braid category* $\mathcal{B}$ *is canonically balanced. The twist* $\theta_n : n \longrightarrow n$ *is obtained by taking n vertical parallel strings with ends tied to two horizontal parallel rods, and rotating the bottom rod through a full* 2π *twist in the right-hand screw direction with thumb vertical. Then* θ_0, θ_1 *are identities, while* θ_2 *(which can be written as* $(s_1)^2$ *using the notation from Example 11.1) is:*

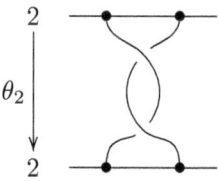

Example 11.5 *There is a tensor category* $\tilde{\mathcal{B}}$ *which is defined similarly to* $\mathcal{B}$, *except that the arrows are braids on ribbons (instead of on strings) and it is permissible to twist the ribbons through full* 2π *turns (as in the following diagram).*

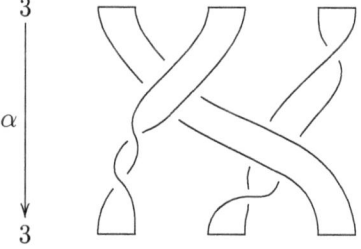

The homsets $\tilde{\mathcal{B}}(n, n) = \tilde{\mathcal{B}}_n$ *are groups under composition. A presentation of this group* $\tilde{\mathcal{B}}_n$ *is given by generators* $s_1, \ldots, s_n$ *where* $s_1, \ldots, s_{n-1}$ *satisfy the relations as for* $\mathcal{B}_n$. *These are depicted by thickened versions of the diagrams in Example 11.1, along with the extra relation*

$$s_{n-1} s_n s_{n-1} s_n = s_n s_{n-1} s_n s_{n-1}$$

where s_n *is depicted as follows*

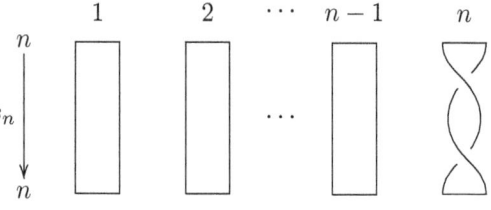

Composition in $\tilde{\mathcal{B}}$ *is vertical stacking of diagrams, and tensor product for* $\tilde{\mathcal{B}}$ *is horizontal placement of diagrams, much as for* $\mathcal{B}$. *The braiding*

$c_{m,n} : m+n \longrightarrow n+m$ *for* $\tilde{\mathcal{B}}$ *is obtained by placing the first m ribbons over the remaining n without introducing any twists. Now the twist* $\theta_n : n \longrightarrow n$ *for* $\tilde{\mathcal{B}}$ *is obtained by regarding the two boundary edges of the ribbons as extra strings and taking* $\theta_{2n} : 2n \longrightarrow 2n$ *in* $\mathcal{B}$. *Then in* $\tilde{\mathcal{B}}$ *we have*

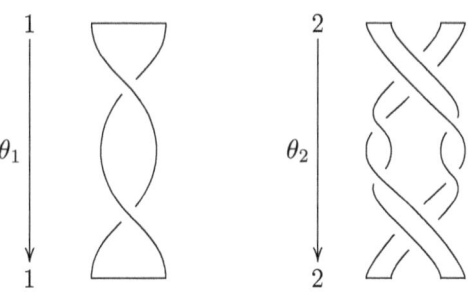

Example 11.6 *Let A and B be abelian groups and* $f : A \times A \longrightarrow B$ *be a bilinear function. There is a balanced strict tensor category* $\mathcal{C}_f$ *constructed as follows. The objects are the elements of A. The homset* $\mathcal{C}_f(x,y)$ *is empty unless* $x = y$, *in which case* $\mathcal{C}_f(x,x) = B$. *The tensor product is given by*

$$(x \xrightarrow{\alpha} x) \otimes (y \xrightarrow{\beta} y) \;=\; (x+y \xrightarrow{\alpha+\beta} x+y) \,.$$

The braiding is $c_{x,y} = f(x,y) : x+y \longrightarrow y+x$ *and the twist is given by* $\theta_X = f(x,x) : x \longrightarrow x$.

Example 11.7 *Let A be a braided R-bialgebra with braiding element* $\gamma = \sum_i u_i \otimes v_i \in A \otimes_R A$. *A twist element for A is an invertible central element* $\tau \in A$ *such that* $\varepsilon(\tau) = 1$ *and*

$$\delta(\tau) = \sum_{i,j} (u_i \tau v_j) \otimes (v_i \tau u_j) \,.$$

Diagrammatically the last equation becomes:

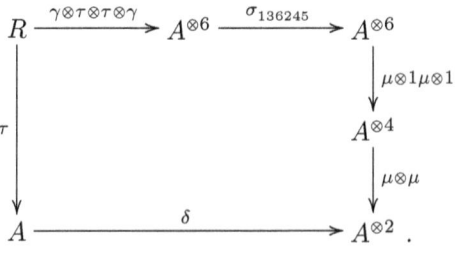

Twist elements τ *for A are in bijection with twists* θ *for the braided tensor category* $\mathbf{Mod}_R(A)$. *Naturality of* $\theta_M : M \longrightarrow M$ *means it has the form*

$\theta_M(m) = \tau m$ *for some* $\tau \in A$; *for* θ_M *to be an A-module morphism, τ needs to be central (meaning $\tau \cdot a = a \cdot \tau$ for all $a \in A$); for θ_M to be an isomorphism, τ needs to be invertible; for $\theta_M = 1_R$, the condition $\varepsilon(\tau) = 1$ is needed; and of course the remaining twist conditions correspond.*

A *balanced bialgebra* is a braided bialgebra with a twist.

Exercise 11.1

(a) *In a braided tensor category* $\mathcal{V}$ *show that (ignoring the constraints* a, l, r) $c_{A,I} = c_{I,A} = 1_A$ *and*

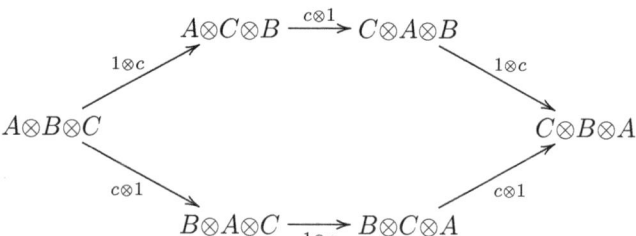

(b) *For a braided bialgebra A and $\mathcal{V} = \mathbf{Mod}_R(A)$, interpret the properties of c in (a) in terms of the braiding element $\gamma \in A \otimes_R A$.*

(c) *Draw diagrams of braids which express the hexagonal diagram of (a) in the braid category* $\mathcal{B}$.

Exercise 11.2
Define the centre $\mathcal{Z}_\mathcal{V}$ *of a tensor category $\mathcal{V}$ to be the category whose objects are pairs (A, a) where $A \in \mathcal{V}$ and $a : A \otimes_- \longrightarrow {}_- \otimes A$ is a natural isomorphism such that the following conditions hold:*

- $a_I = 1$ *(more precisely, a_I is the composite of the canonical isomorphisms $A \otimes I \cong A \cong I \otimes A$).*

- $a_{X \otimes Y} = (1 \otimes a_Y) \circ (a_X \otimes 1)$ *for all $X, Y \in \mathcal{V}$.*

An arrow $f : (A, a) \longrightarrow (B, b)$ in $\mathcal{Z}_\mathcal{V}$ is an arrow $f : A \longrightarrow B$ such that, for all $X \in \mathcal{V}$, we have $b_X \circ (f \otimes 1) = (1 \otimes f) \circ a_X$.

(a) *Show that $\mathcal{Z}_\mathcal{V}$ becomes a tensor category with:*

$$(A, a) \otimes (B, b) = \bigl(A \otimes B, (a \otimes 1) \circ (1 \otimes b)\bigr).$$

(b) *Show that the tensor category $\mathcal{Z}_\mathcal{V}$ is braided via*

$$c_{(A,a),(B,b)} = a_B : (A, a) \otimes (B, b) \longrightarrow (B, b) \otimes (A, a).$$

Exercise 11.3
Show that the first dot-point condition on objects of $\mathcal{Z}_\mathcal{V}$ is redundant.

12

Internal homs and duals

Suppose $\mathcal{V}$ is a tensor category with A and B being objects of $\mathcal{V}$. A (left) *internal hom* for A and B consists of an object $[A,B]$ of $\mathcal{V}$ together with an arrow

$$e_A : [A,B] \otimes A \longrightarrow B \qquad \text{(called } \textit{evaluation}\text{)}$$

such that, for all arrows $f : C \otimes A \longrightarrow B$, there exists a unique arrow $\hat{f} : C \longrightarrow [A,B]$ with

$$f \;=\; \left(C \otimes A \xrightarrow{\hat{f} \otimes 1_A} [A,B] \otimes A \xrightarrow{e_A} B \right).$$

Thus we have a natural bijection

$$\mathcal{V}(C, [A,B]) \qquad \cong \qquad \mathcal{V}(C \otimes A, B)$$

$$g \qquad \longleftrightarrow \qquad e_A \circ (g \otimes 1_A)\,.$$

A tensor category is called *left-closed* when each pair of objects has a left internal hom. If $f : C \longrightarrow A$ and $g : B \longrightarrow D$ are arrows of $\mathcal{V}$ then, provided the internal homs exist, there is a unique arrow

$$[f,g] : [A,B] \longrightarrow [C,D]$$

such that

$$
\begin{array}{ccc}
[A,B] \otimes C & \xrightarrow{\;[f,g] \otimes 1_C\;} & [C,D] \otimes C \\[2pt]
{\scriptstyle 1_{[A,B]} \otimes f} \big\downarrow & & \big\downarrow {\scriptstyle e_C} \\[2pt]
[A,B] \otimes A \xrightarrow{\;e_A\;} B & \xrightarrow{\;g\;} & D\,.
\end{array}
$$

In this way, when $\mathcal{V}$ is left-closed the internal hom becomes a functor

$$[-,-] : \mathcal{V}^{\mathrm{op}} \times \mathcal{V} \longrightarrow \mathcal{V}\,.$$

From the universal property, the internal hom for A, B is unique up to isomorphism. An internal hom for I, A always exists; namely

$$[I, A] = A \quad \text{with} \quad e_A = r_A : A \otimes I \longrightarrow A .$$

If B, C and $A \otimes B, C$ have internal homs then so do $A, [B, C]$; namely,

$$[A, [B, C]] = [A \otimes B, C] \quad \text{with} \quad e_A = \hat{e}_{A \otimes B} : [A \otimes B, C] \otimes A \longrightarrow [B, C] .$$

Usually the internal hom functor $[\,\text{-}\,, \text{-}\,]$ is given *a priori*; in which case all we have are canonical isomorphisms

$$[I, A] \cong A \quad \text{and} \quad [A, [B, C]] \cong [A \otimes B, C] .$$

There is a *composition arrow*

$$[B, C] \otimes [A, B] \longrightarrow [A, C]$$

in $\mathcal{V}$ (whenever the internal homs exist) which corresponds, by using the universal property of $[A, C]$, to the composite

$$[B, C] \otimes [A, B] \otimes A \xrightarrow{\ 1 \otimes e_A\ } [B, C] \otimes B \xrightarrow{\ e_B\ } C .$$

A *right internal hom* $[A, B]'$ for A, B comes equipped with an arrow

$$e_A' : A \otimes [A, B]' \longrightarrow B$$

which induces a bijection

$$\mathcal{V}(C, [A, B]') \cong \mathcal{V}(A \otimes C, B)$$

for all objects C. If $\mathcal{V}$ is braided, each left internal hom $[A, B]$ gives a right internal hom via $[A, B]' = [A, B]$ and

$$e_A' = \left(A \otimes [A, B] \xrightarrow{\ c_{A,[A,B]}\ } [A, B] \otimes A \xrightarrow{\ e_A\ } B \right) .$$

A tensor category is called *closed* when all left- and right-internal homs exist. (In the literature, "closed" is sometimes used for our left-closed, while "biclosed" is used for our closed.) When the internal homs exist, we have an arrow

$$w_A' : A \longrightarrow [[A, B], B]'$$

which corresponds to $e_A : [A, B] \otimes A \longrightarrow B$ via $e_{[A,B]}'$. Similarly, we have

$$w_A : A \longrightarrow [[A, B]', B]$$

corresponding to e_A'.

Example 12.1 *The symmetric tensor category* $\mathbf{Mod}_R$ *of modules over a commutative ring R is closed with internal homs given by*

$$[M, N] = [M, N]' = \mathrm{Hom}_R(M, N).$$

Example 12.2 *Suppose H to be an R-Hopf algebra. Then the tensor category $\mathbf{Mod}_R(H)$ of H-modules (with tensor product $\otimes_R$) is left-closed with internal hom given by*

$$[M, N] = \mathrm{Hom}_R(M, N);$$

see Proposition 10.1.

Suppose the antipode ν for H is invertible. Using Proposition c(c), H' becomes a Hopf algebra having antipode ν^{-1}. Write $\mathrm{Hom}'_R(M, N)$ for $\mathrm{Hom}_R(M, N)$ as a left H'-module. Clearly, $\mathbf{Mod}_R(H)$ is right-closed with $[M, N]' = \mathrm{Hom}'_R(M, N)$. Therefore $\mathbf{Mod}_R(H)$ becomes closed when ν is invertible. The forgetful functor

$$\mathbf{Mod}_R(H) \longrightarrow \mathbf{Mod}_R$$

preserves tensor product, and both left and right internal homs.

Example 12.3 *Let $\mathbf{Ban}$ denote the category of Banach spaces (over the complex numbers), where the arrows $f : A \longrightarrow B$ are linear functions for which*

$$\|f(a)\| \leq \|a\|.$$

(The analysts in the audience will think these fairly uninteresting functions.) We make $\mathbf{Ban}$ into a symmetric tensor category by taking tensor products as vector spaces, completing in the obvious way. The internal hom $[A, B]$ exists for all Banach spaces A, B; it is the Banach space of bounded linear functions from A to B with the usual norm. (These functions are of more interest to the analyst.) Thus $\mathbf{Ban}$ is a closed symmetric tensor category.

Suppose $\mathcal{V}$ is a tensor category. An object J is called *left dualizing* when, for all objects A, internal homs $[A, J]$, $[A, J]'$ exist, and the arrow

$$\omega_A : A \longrightarrow [[A, J]', J]$$

is invertible. It follows that $\mathcal{V}$ is left-closed with

$$[A, B] = [A \otimes [B, J]', J],$$

since

$$\begin{aligned}
\mathcal{V}(C, [A \otimes [B, J]', J]) &\cong \mathcal{V}(C \otimes A \otimes [B, J]', J) \\
&\cong \mathcal{V}(C \otimes A, [[B, J]', J]) \cong \mathcal{V}(C \otimes A, B).
\end{aligned}$$

The concept of *right dualizing object* is defined in the same way, with ω replaced by ω'. A *dualizing object* is one which is both left and right dualizing.

Example 12.4 *The category of finite-dimensional vector spaces over a field has a dualizing object, namely the field itself. In this case, the dualizing object is the unit for tensor product.*

Example 12.5 *Fix a field* **k**. *A* quadratic algebra *is a pair* (V, R) *where V is a finite-dimensional vector space and R is a subspace of $V \otimes V$. A* quadratic algebra morphism $f : (V, R) \longrightarrow (W, S)$ *is a linear function* $f : V \longrightarrow W$ *for which*

$$(f \otimes f)(R) \subseteq S .$$

Write $\mathcal{QA}$ for the category of quadratic algebras. Each quadratic algebra (V, R) *determines an actual algebra*

$$T(V)/R$$

where $T(V)$ is the tensor algebra on V (see Example 6.2). The category $\mathcal{QA}$ has a symmetric tensor product given by

$$(V, R) \otimes (W, S) \;=\; \big(V \otimes W, \sigma_{1324}(R \otimes S)\big) .$$

The unit object is $I = (\mathbf{k}, \mathbf{k} \otimes \mathbf{k})$.
   We claim that $J = (\mathbf{k}, 0)$ is a dualizing object. *It is easy to see that*

$$[(V, R), J] \;=\; (V^*, R^\perp)$$

where $V^* = \mathrm{Hom}_{\mathbf{k}}(V, \mathbf{k})$ and $R^\perp$ is the kernel of the composite surjection

$$V^* \otimes V^* \xrightarrow{\;\cong\;} (V \otimes V)^* \xrightarrow{\;i^*\;} R^*$$

with $i : R \longrightarrow V \otimes V$ being inclusion. It follows that R is the kernel of

$$V \otimes V \xrightarrow{\;\omega \otimes \omega\;} V^{**} \otimes V^{**} \xrightarrow{\;\cong\;} (V^* \otimes V^*)^* \xrightarrow{\;i^*\;} R^{\perp *}$$

so we have an isomorphism

$$\omega : (V, R) \longrightarrow (V^{**}, R^{\perp\perp}) .$$

   The quadratic algebra which we identify with the quantum plane $\mathbb{A}_q^{2|0}$ *(recall Chapter 3) is*

$$\mathbb{A}_q^{2|0} \;=\; \big(\mathbf{k}^2, (y \otimes x - q\, x \otimes y)\big)$$

where $x = (1, 0)$, $y = (0, 1) \in \mathbf{k}^2$. Let $\xi, \eta \in \mathbf{k}^{2*}$ be the dual basis given by $\xi(x) = \eta(y) = 1$, $\xi(y) = \eta(x) = 0$. Then

$$(y \otimes x - q\, x \otimes y)^{\perp} = (\xi \otimes \xi, \eta \otimes \eta, \xi \otimes \eta + q\, \eta \otimes \xi)$$

as a subspace of $\mathbf{k}^{2*} \otimes \mathbf{k}^{2*}$. Hence the quantum superplane arises as the quadratic algebra

$$
\begin{aligned}
\mathbb{A}_q^{0|2} &= \left(\mathbf{k}^{2*}, (\xi \otimes \xi, \eta \otimes \eta, \xi \otimes \eta + q\, \eta \otimes \xi) \right) \\
&= [\mathbb{A}_q^{2|0}, J].
\end{aligned}
$$

Notice also that

$$
\begin{aligned}
\mathbb{A}_q^{0|2} \otimes \mathbb{A}_q^{2|0} &= \left(\mathbf{k}^{2*} \otimes \mathbf{k}^2, (b \otimes a - q\, a \otimes b, d \otimes c - q\, c \otimes d, \right. \\
&\qquad\qquad \left. q^{-1} b \otimes c - q\, c \otimes b + d \otimes a - a \otimes d) \right)
\end{aligned}
$$

(which should be compared with equations (∗∗) in the proof of Theorem 3.2) where

$$a = \xi \otimes x, \quad b = \xi \otimes y, \quad c = \eta \otimes x, \quad d = \eta \otimes y$$

gives half the relations required for $\mathrm{M}_q(2)$.

Let $\mathcal{V}$ be a tensor category. Write $d, e : B \dashv A$, or briefly $B \dashv A$, for objects A, B of $\mathcal{V}$ and arrows

$$e : B \otimes A \longrightarrow I, \qquad d : I \longrightarrow A \otimes B$$

when the following diagrams commute:

We call B a *left dual* for A, and we call A a *right dual* for B. We call e the *counit* and d the *unit*. Duals are uniquely determined up to isomorphism.

The tensor category is called *left autonomous* when each object A has a left dual A^*. Each arrow $f : A \longrightarrow B$ determines a unique arrow $f^* : B^* \longrightarrow A^*$ given by the composite

$$B^* \xrightarrow{\;1 \otimes d\;} B^* \otimes A \otimes A^* \xrightarrow{\;1 \otimes f \otimes 1\;} B^* \otimes B \otimes A^* \xrightarrow{\;e \otimes 1\;} A^*.$$

This makes *left dual* into a functor

$$(\text{-})^* : \mathcal{V}^{\mathrm{op}} \longrightarrow \mathcal{V}.$$

We also have $(A \otimes B)^* \cong B^* \otimes A^*$ and $I^* \cong I$. The tensor category is called *autonomous* when each object A has both a left dual A^* and right dual $A^\vee$.

If $\mathcal{V}$ is a braided tensor category then each left dual A^* is also a right dual with counit and unit, respectively

$$A \otimes A^* \xrightarrow{\ c_{A,A^*}\ } A^* \otimes A \xrightarrow{\ e\ } I$$

$$I \xrightarrow{\ d\ } A \otimes A^* \xrightarrow{\ c_{A,A^*}^{-1}\ } A^* \otimes A \ .$$

This implies $A^{**} \cong A$.

A *tortile tensor category* is an autonomous balanced tensor category in which the twist is related to the dual via the condition

$$\theta_{A^*} \;=\; \theta_A{}^* : A^* \longrightarrow A^* \ .$$

A left dual A^* for A gives a left internal hom $[A, B] = B \otimes A^*$ with "evaluation":

$$e_A \;=\; 1_B \otimes e : B \otimes A^* \otimes A \longrightarrow B$$

for all objects B of $\mathcal{V}$. A right dual $A^\vee$ for A gives a right internal hom $[A, B]' = A^\vee \otimes B$. Hence a *left/right autonomous* tensor category is left-/right-closed.

Example 12.6 *For each commutative ring R, an object M of $\mathbf{Mod}_R$ has a (left) dual if and only if it is Cauchy (Theorem 5.2). Write $\mathbf{Prf}_R$ for the full subcategory of $\mathbf{Mod}_R$ consisting of Cauchy R-modules. Since $\mathbf{Prf}_R$ is closed under tensor, it is an autonomous (symmetric) tensor category.*

Example 12.7 *Let H be an R-Hopf algebra. An object M of the tensor category $\mathbf{Mod}_R(H)$ has a left dual precisely when it is Cauchy as an R-module; in this case, $M^* = \mathrm{Hom}_R(M, R)$ (Proposition 10.1). If H has an invertible antipode then each such M has a right dual $M^\vee = \mathrm{Hom}'_R(M, R)$. Write $\mathbf{Prf}_R(H)$ for the full subcategory of $\mathbf{Mod}_R(H)$ consisting of those H-modules M which are Cauchy when viewed as R-modules. For H with invertible antipode, $\mathbf{Prf}_R(H)$ is an autonomous tensor category.*

Example 12.8 Tangles on strings. *(This example was discovered by Freyd–Yetter [FY89].) Let P be a Euclidean plane. A geometric tangle T is a compact 1-dimensional oriented submanifold of $[0, 1] \times P$ which is tamely embedded and whose boundary ∂T is equal to $T \cap \partial([0, 1] \times P)$. Thus a geometric tangle T is a disjoint union of directed (topological) circles contained in $(0, 1) \times P$ and of directed paths connecting two points on the boundary $\partial([0, 1] \times P)$. The target of T is the subset $\partial T \cap (\{1\} \times P)$*

as an oriented 0-dimensional manifold. The source of T is the subset $\partial T \cap (\{0\} \times P)$ but with orientation reversed. A geometric tangle can be pictured as follows:

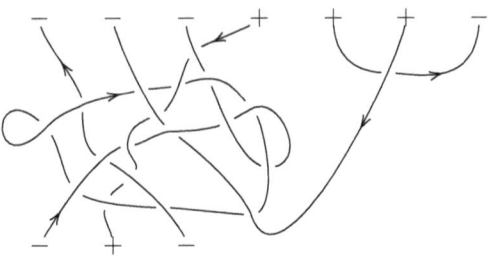

A tangle *is an isotopy class of geometric tangles where isotopies keep the boundaries fixed. The source and target of a tangle are regarded as signed subsets of* P. *Let* $1, 2, 3, \ldots$ *denote equally spaced collinear points on* P.

Now we can define the autonomous braided tensor category $\mathcal{T}$ *of tangles. The objects are functions* $A: \{1, 2, \ldots, n\} \longrightarrow \{+, -\}$ *for* $n \geq 0$, *called* signed sets. *The arrows of* $\mathcal{T}$ *are the tangles which have these signed sets as sources and targets. Composition and tensor-product are as for braids.*

The braiding is illustrated below.

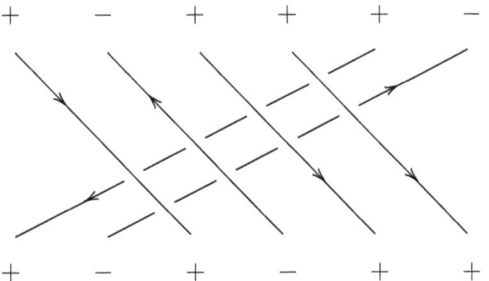

The left dual A^* *of a signed set* A *is given by reversing the order and the sign of the points; that is,* $A^*(i)$ *and* $A(n - i + 1)$ *are opposite signs for* $1 \leq i \leq n$. *The counit and unit are illustrated below for* $A = \{- + -\}$.

$$e: A^* \otimes A \longrightarrow I$$

$$d: I \longrightarrow A \otimes A^*$$

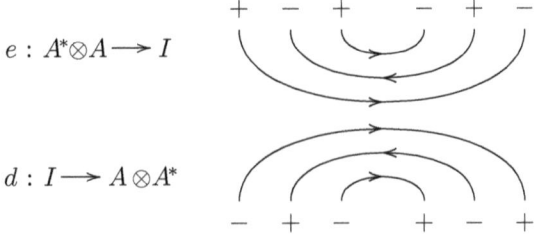

The next diagram proves one triangle for e and d; the other is similar.

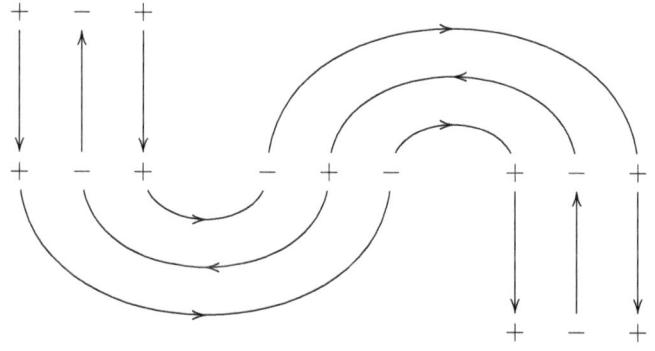

Example 12.9 Tangles on ribbons. *(The full details of this example appear in the thesis of Shum [Shu94].) The category $\widetilde{T}$ of tangles on ribbons is obtained from T just as we obtained $\widetilde{B}$ from B in Example 11.5. The directed strings of tangles are thickened into (directed) ribbons. Ribbons obtained from strings with boundary may be twisted through complete turns. Those which are thickenings of closed strings may have twists, as long as they remain 2-sided 2-manifolds; the Möbius ribbon is not allowed.*

Again we obtain an autonomous tensor category. The counits and units look like this:

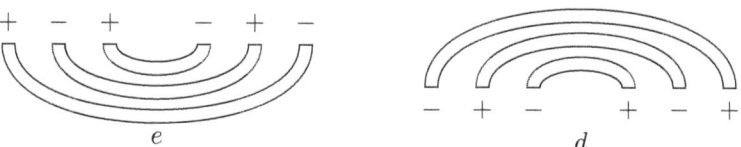

In fact, $\widetilde{T}$ is a tortile tensor category. The twist is as for $\widetilde{B}$ and the identity $\theta_{A^*} = \theta_A^*$ can be seen from the following diagram.

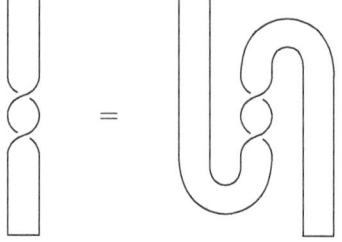

13

Tensor functors and Yang–Baxter operators

Suppose $\mathcal{C}$ and $\mathcal{V}$ are tensor categories. A *tensor functor* $F : \mathcal{C} \longrightarrow \mathcal{V}$ consists of a functor $F : \mathcal{C} \longrightarrow \mathcal{V}$ (denoted by the same symbol) together with a natural isomorphism $\phi_{A,B} : FA \otimes FB \stackrel{\cong}{\Longrightarrow} F(A \otimes B)$ and another isomorphism $\phi_0 : I \stackrel{\cong}{\Longrightarrow} FI$, such that

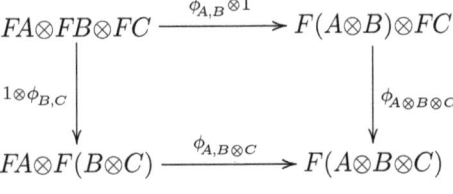

and

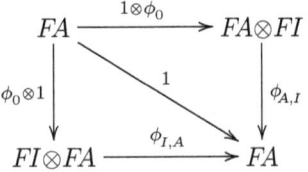

(where we have suppressed the constraints a, r, l as usual). If the condition that ϕ, ϕ_0 be invertible is dropped, we have a *weak tensor functor*. If in fact ϕ and ϕ_0 are identities, then F is called a *strict tensor functor*. (Weak tensor functors are also called "monoidal functors" and tensor functors called "strong monoidal functors".)

If $\mathcal{C}$ and $\mathcal{V}$ are braided, we describe a tensor functor as *braided* when

$$
\begin{array}{ccc}
FA \otimes FB & \xrightarrow{\ \phi_{A,B}\ } & F(A \otimes B) \\
\downarrow{\scriptstyle c_{FA,FB}} & & \downarrow{\scriptstyle F(c_{A,B})} \\
FB \otimes FA & \xrightarrow{\ \phi_{B,A}\ } & F(B \otimes A) \ .
\end{array}
$$

If $\mathcal{C}$ and $\mathcal{V}$ are symmetric, we say F is *symmetric* instead of "braided". If $\mathcal{C}$ and $\mathcal{V}$ are balanced, we say F is *balanced*, when it is braided and

$$F(\theta_A) \;=\; \theta_{FA} : FA \longrightarrow FA \;.$$

Suppose $F : \mathcal{C} \longrightarrow \mathcal{V}$ is a weak tensor functor and $\mathcal{C}$ and $\mathcal{V}$ are left-closed tensor categories. Then the composite

$$F[A,B]\otimes FA \xrightarrow{\;\phi_{[A,B],A}\;} F([A,B]\otimes A) \xrightarrow{\;F\,e_A\;} FB$$

corresponds, using the defining property of $[FA,FB]$, to an arrow

$$\tilde{\phi}_{A,B} : F[A,B] \longrightarrow [FA,FB] \;.$$

We call F a *left-closed tensor functor* when each $\tilde{\phi}_{A,B}$ is invertible; *right-closed* and *closed* are now defined in the obvious way. (This differs from the notion of "closed functor" in the literature.)

When it comes to duals, the situation is better: *tensor functors preserve duals*. More precisely, if $F : \mathcal{C} \longrightarrow \mathcal{V}$ is a tensor functor and $d,e : B \dashv A$ in $\mathcal{C}$, then $FB \dashv FA$ in $\mathcal{V}$ with unit

$$I \xrightarrow{\;\phi_0\;} FI \xrightarrow{\;Fd\;} F(B\otimes A) \xrightarrow{\;\phi_{[A,B],A}^{-1}\;} FA\otimes FB$$

and counit

$$FB\otimes FA \xrightarrow{\;\phi_{[B,A]}\;} F(B\otimes A) \xrightarrow{\;Fe\;} FI \xrightarrow{\;\phi_0^{-1}\;} I \;.$$

Hence, if $\mathcal{C}$ is (left-) autonomous and $\mathcal{V}$ is (left-) closed, then each tensor functor $\mathcal{C} \longrightarrow \mathcal{V}$ is (left-) closed (since $[A,B] = B\otimes A^*$ in $\mathcal{C}$).

Example 13.1 *The category* **Set** *of (small) sets is a tensor category using cartesian product as tensor product. For each commutative ring R, the "free module functor" (see Chapter 4)*

$$\mathcal{F}_R : \mathbf{Set} \longrightarrow \mathbf{Mod}_R$$

is a tensor functor. It is certainly not closed since we have

$$\mathrm{Hom}_R(\mathcal{F}_R X, \mathcal{F}_R Y) \cong (\mathcal{F}_R Y)^X \ncong Y^X \;.$$

The functor $|\,\_\,| : \mathbf{Mod}_R \longrightarrow \mathbf{Set}$ *which takes each module to its underlying set $|M|$ is a good example of a weak tensor functor: we have functions*

$$\phi_{M,N} : |M| \times |N| \longrightarrow |M\otimes N|$$

$$(m,n) \longmapsto m\otimes n$$

$$\phi_0 = \eta : \quad 1 \quad \longrightarrow |R|$$

which are not invertible.

Example 13.2 Universal algebra. *The "universal enveloping algebra" functor* $\mathcal{U} : \mathbf{Lie}_R \longrightarrow \mathbf{Alg}_R$ *is a tensor functor (see Proposition 6.10).*

Example 13.3 Yang–Baxter operators. *We want to examine what is involved in giving a tensor functor* $F : \mathcal{B} \longrightarrow \mathcal{V}$ *from the braid category into an arbitrary tensor category* $\mathcal{V}$.

A *Yang–Baxter (YB) operator on an object* A *of* $\mathcal{V}$ *is an invertible arrow* $y : A \otimes A \longrightarrow A \otimes A$ *such that the following hexagon commutes.*

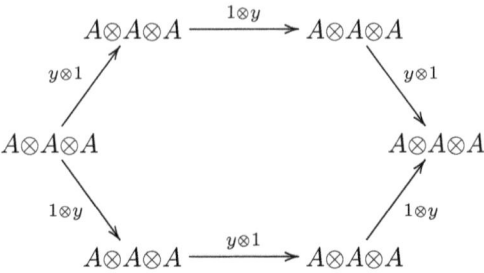

For example, the object 1 of $\mathcal{B}$ *admits the following YB-operator:*

$$c_{1,1} = s_1 : 1 + 1 \longrightarrow 1 + 1$$

which is the element of the braid group $\mathcal{B}_2$ *depicted by the next diagram.*

$s_1 :$

The YB-hexagon becomes the following simple identity.

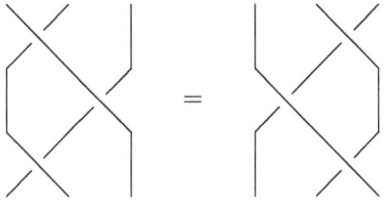

Since any tensor functor $F : \mathcal{B} \longrightarrow \mathcal{V}$ *preserves tensor products "up to coherent isomorphism", we obtain a YB-operator* y *on* $F(1) = A$; *namely,*

$$y : A \otimes A \xrightarrow{\phi_{1,1}} F(1+1) \xrightarrow{Fs_1} F(1+1) \xrightarrow{\phi_{1,1}^{-1}} A \otimes A .$$

Conversely, given a YB-operator y *on an object* A *of* $\mathcal{V}$, *we can determine a tensor functor* $F : \mathcal{B} \longrightarrow \mathcal{V}$ *such that* $F(1) = A$ *and* y *is the above composite. In fact,* F *is unique up to isomorphism (arising from the different*

possible choices for the n-fold tensor product $A^{\otimes n}$). Since F is to be a tensor functor, we are forced to have

$$F(n) = F(1 + 1 + \cdots + 1) \cong A^{\otimes n}.$$

Each generator s_i of $\mathcal{B}_\mathbf{n}$ can be written as $s_i = 1_{i-1} \otimes s_1 \otimes 1_{n-i-1}$ in $\mathcal{B}$, so the definition of $Fs_i : A^{\otimes n} \longrightarrow A^{\otimes n}$ is forced. We just need to check that this is compatible with the braid relations (Example 11.1); but this follows from the YB-hexagon and the functoriality of tensor product. Details are left as an exercise (which is worth doing).

Hence, up to the appropriate notion of isomorphism, tensor functors $F : \mathcal{B} \longrightarrow \mathcal{V}$ correspond to pairs (A, y) consisting of an object A of $\mathcal{V}$ and a YB-operator y on A. We can express this by saying:

> $(\mathcal{B}, 1, s_1)$ is the free tensor category having an object equipped with a YB-operator.

Example 13.4 *A tensor functor from ribbons* $F : \widetilde{\mathcal{B}} \longrightarrow \mathcal{V}$ *also determines a YB-operator y on $F1 = A$ as in Example 13.3 (compare with* [Tur88]). *This time the twist* $\theta_1 : 1 \longrightarrow 1$ *in* $\widetilde{\mathcal{B}}$ *gives an isomorphism* $z = F\theta_1 : A \longrightarrow A$.
Due to the equalities in $\widetilde{\mathcal{B}}$

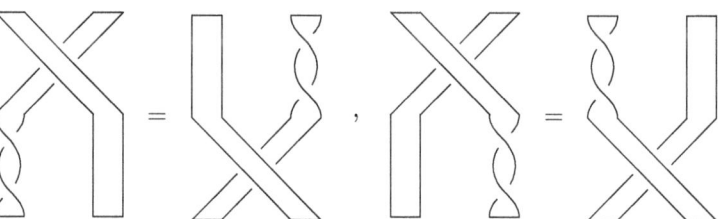

this gives an example of the following concept.

A YB-operator y on an object A of $\mathcal{V}$ is said to be balanced when it is equipped with an isomorphism $z : A \longrightarrow A$ such that these commute:

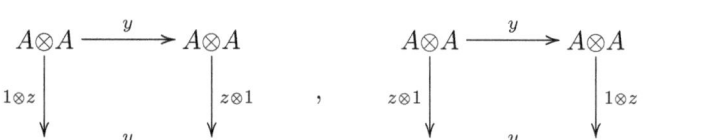

Each balanced YB-operator determines a unique (once the n-fold tensors $A^{\otimes n}$ are chosen) tensor functor $F : \widetilde{\mathcal{B}} \longrightarrow \mathcal{V}$ from which A, y, z are recovered as above. So we have that

> $(\widetilde{\mathcal{B}}, 1, c_{1,1}, \theta_1)$ is the free tensor category containing an object equipped with a balanced YB-operator.

The easier part of Example 13.3 can be obtained from two observations.

- Tensor functors take *YB*-operators into *YB*-operators.
  More precisely, if $F : \mathcal{C} \longrightarrow \mathcal{V}$ is a tensor functor and if y is a *YB*-operator on X in $\mathcal{C}$, then $Fy : F(X \otimes X) \longrightarrow F(X \otimes X)$ carries across the isomorphism $\phi_{X,X}$ to a *YB*-operator y on FX in $\mathcal{V}$.

- A braiding on a tensor category gives, on each object X, a *YB*-operator: $c_{X,X} : X \otimes X \longrightarrow X \otimes X$.

Moreover, tensor functors take balanced *YB*-operators into balanced *YB*-operators. Furthermore, in a balanced tensor category there is a balanced *YB*-operator $(c_{X,X}, \theta_X)$ on each object X. (See Example 13.4.)

We now look at compatibility of *YB*-operators with duals.

A *YB*-operator y on A is called *(left-) dualizable* when A has a left dual A^* and both the arrows $u, v : A^* \otimes A \longrightarrow A^*$ given by the composites:

$$A^* \otimes A \xrightarrow{1 \otimes 1 \otimes d} A^* \otimes A \otimes A \otimes A^* \underset{1 \otimes y^{-1} \otimes 1}{\overset{1 \otimes y \otimes 1}{\rightrightarrows}} A^* \otimes A \otimes A \otimes A^* \xrightarrow{e \otimes 1 \otimes 1} A \otimes A^*$$

are invertible. It follows that the composite w given by

$$A^* \otimes A^* \xrightarrow{1 \otimes 1 \otimes d} A^* \otimes A^* \otimes A \otimes A^* \xrightarrow{1 \otimes u \otimes 1} A^* \otimes A \otimes A^* \otimes A^* \xrightarrow{e \otimes 1 \otimes 1} A^* \otimes A^*$$

has inverse w^{-1} given by the composite:

$$A^* \otimes A^* \xrightarrow{1 \otimes 1 \otimes d} A^* \otimes A^* \otimes A \otimes A^* \xrightarrow{1 \otimes v \otimes 1} A^* \otimes A \otimes A^* \otimes A^* \xrightarrow{e \otimes 1 \otimes 1} A^* \otimes A^* \ .$$

Proposition 13.5 *In a braided tensor category, if an object A has a dual, then $y = c_{A,A}$ is a dualizable YB-operator on A with*

$$u = c_{A,A^*}^{-1} \ , \qquad v = c_{A^*,A}^{-1} \ , \qquad w = c_{A^*,A^*}^{-1} \ .$$

Proof. To prove $c_{A,A^*} \circ u = 1_{A^* \otimes A}$, it suffices (by the property of duals) to show that equality holds after applying $A \otimes \_$ to both sides and composing with $d \otimes 1_A$. Thus the following diagram gives the first equation.

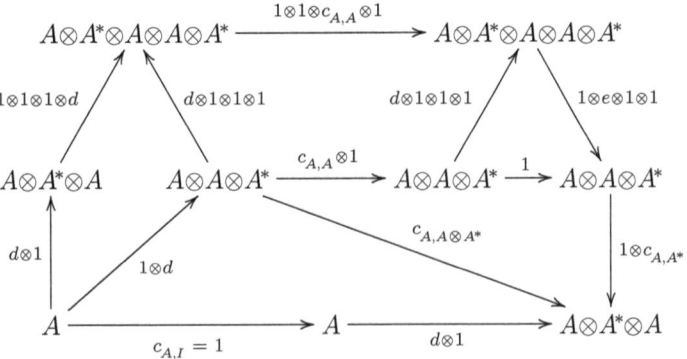

The result for v follows by using the braiding $c_{B,A}^{-1}$ in place of $c_{A,B}$. For w, consider A^* in place of A. $\qquad\qquad\square$

Tensor functors $F : \mathcal{C} \longrightarrow \mathcal{V}$ preserve dualizability. So if $\mathcal{C}$ is braided and X has a dual in $\mathcal{V}$, we obtain a dualizable YB-operator on FX in $\mathcal{V}$.

A YB-operator on A is called *tortile* when it is balanced, dualizable, and

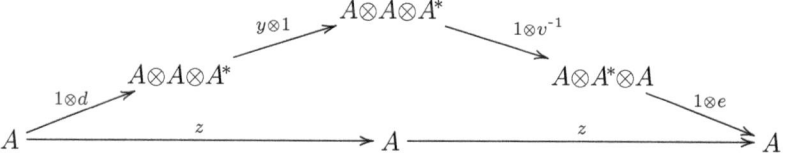

Proposition 13.6 *In a balanced tensor category, if an object A has a dual then the pair $(c_{A,A}, \theta_A)$ is a tortile YB-operator precisely when $\theta_{A^*} = (\theta_A)^*$.*

Proof. The following diagram proves the equation

$$(\theta_{A^*})^* \theta_A = (1 \otimes e) \circ (1 \otimes v^{-1}) \circ (y \otimes 1) \circ (1 \otimes d).$$

So the balanced YB-operator (y, z) is tortile if and only if $(\theta_{A^*})^* \theta_A = z^2$; that is, if and only if $(\theta_{A^*})^* \theta_A = (\theta_A)^2 \Leftrightarrow (\theta_{A^*})^* = \theta_A \Leftrightarrow \theta_{A^*} = (\theta_A)^*$. $\qquad\square$

It follows that, in a tortile tensor category, each object A is equipped with a tortile YB-operator $(c_{A,A}, \theta_A)$.

Example 13.7 *Since $\widetilde{T}$ is a tortile tensor category, we obtain a tortile YB-operator $(c_{+,+}, \theta_+)$ on the object $+$ of $\widetilde{T}$. Thus, each tensor functor $F : \widetilde{T} \longrightarrow V$ yields a tortile YB-operator on $F(+)$ in V.*

In fact, $\left(\widetilde{T}, +, c_{+,+}, \theta_+\right)$ is the free tensor category equipped with a tortile YB-operator ([Shu94] together with [JS91c]). This means that, given a tortile YB-operator (y, z) on an object A in a tensor category V, there exists a (unique up to isomorphism) tensor functor $F : \widetilde{T} \longrightarrow V$ which takes $\left(+, c_{+,+}, \theta_+\right)$ to (A, y, z). We do not intend to prove this here; after all, our geometric description of $\widetilde{T}$ was incomplete. We hope the result is believable. All we really need is that such a free $\widetilde{T}$ should exist; but the description of the realistic model is too pretty to omit.

It should be clear how to define F in terms of A, y, z. For example,

$$
\begin{aligned}
F(+ - - - + -) &= A \otimes A^* \otimes A^* \otimes A^* \otimes A \otimes A^* \\
F(c_{+,-}) &= u^{-1} \\
F(c_{-,-}) &= w \\
F(\theta_+) &= z \\
F(\theta_-) &= z^*
\end{aligned}
$$

and so on. Any tangle of ribbons can be decomposed, using composition and tensor product in $\widetilde{T}$, into single crossings $(c_{+,+}, c_{+,-}, c_{-,+}, c_{-,-},$ and their inverses), turnings (e and d) and twists (θ_+, θ_-, and their inverses). So the value of F on the tangle is forced. The hard part, which we shall not include in these notes, is to show that this value is independent of the decomposition.

It is instructive to see in this example what is meant by the equation $z^2 = (1 \otimes e) \circ (1 \otimes v^{-1}) \circ (y \otimes 1) \circ (1 \otimes d)$. It is expressed by the following diagram (which can be tested by taking off your belt).

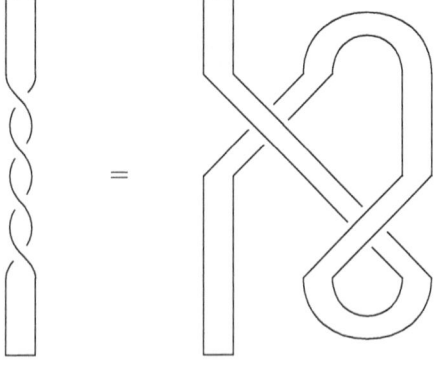

Bradshaw: "Barred Variation", Sash Bradshaw Group, [Wal94, Plate 58].

14

A tortile Yang–Baxter operator for each finite-dimensional vector space

Let $\mathbf{k}$ be a field and let $q \in \mathbf{k}$ be a fixed non-zero element. Let V be a vector space over $\mathbf{k}$ with basis $\varepsilon_1, \varepsilon_2, \ldots, \varepsilon_n$. Define a linear function

$$y : V \otimes V \longrightarrow V \otimes V$$

on the basis elements $\varepsilon_i \otimes \varepsilon_j$ of $V \otimes V$ by

$$y(\varepsilon_i \otimes \varepsilon_j) = \begin{cases} \varepsilon_j \otimes \varepsilon_i & \text{for } i > j \\ \varepsilon_j \otimes \varepsilon_i + (q - q^{-1}) \, \varepsilon_i \otimes \varepsilon_j & \text{for } i < j \\ q \, \varepsilon_i \otimes \varepsilon_i & \text{for } i = j \, . \end{cases}$$

In order to check the YB-hexagon for y, we look at $(y \otimes 1)(1 \otimes y)(y \otimes 1)$, $(1 \otimes y)(y \otimes 1)(1 \otimes y)$ at each $\varepsilon_i \otimes \varepsilon_j \otimes \varepsilon_k$. There are thirteen of these cases to check to account for all possible relative positions of i, j and k. We shall only give three of these cases as an illustration: put $\rho = q - q^{-1}$ and omit the ε and $\otimes$ symbols from the notation.

$$i < j < k \; : \; (ijk) \xmapsto{\;y \otimes 1\;} (jik) + \rho\,(ijk)$$
$$\xmapsto{\;1 \otimes y\;} (jki) + \rho\,(jik) + \rho\,(ikj) + \rho^2 (ijk)$$
$$\xmapsto{\;y \otimes 1\;} (kji) + \rho\,(jki) + \rho\,(ijk) + \rho\,(kij) + \rho^2 (ikj)$$
$$+ \rho^2 (jik) + \rho^3 (ijk)$$

$$(ijk) \xmapsto{\;1 \otimes y\;} (ikj) + \rho\,(ijk)$$
$$\xmapsto{\;y \otimes 1\;} (kij) + \rho\,(ikj) + \rho\,(jik) + \rho^2 (ijk)$$
$$\xmapsto{\;1 \otimes y\;} (kji) + \rho\,(kij) + \rho\,(ijk) + \rho\,(jki) + \rho^2 (jik)$$
$$+ \rho^2 (ijk) + \rho^3 (ijk) \, ,$$

93

$$j < i < k \; : \; (ijk) \xmapsto{\;y\otimes 1\;} (jik)$$
$$\xmapsto{\;1\otimes y\;} (jki) + \rho\,(jik)$$
$$\xmapsto{\;y\otimes 1\;} (kji) + \rho\,(jki) + \rho\,(ijk) + \rho^2(jik)$$

$$(ijk) \xmapsto{\;1\otimes y\;} (ikj) + \rho\,(ijk)$$
$$\xmapsto{\;y\otimes 1\;} (kij) + \rho\,(ikj) + \rho\,(jik)$$
$$\xmapsto{\;1\otimes y\;} (kji) + \rho\,(ijk) + \rho\,(jki) + \rho^2(jik)\,,$$

$$i = j < k \; : \; (iik) \xmapsto{\;y\otimes 1\;} q\,(iik)$$
$$\xmapsto{\;1\otimes y\;} q\,(iki) + \rho q\,(iik)$$
$$\xmapsto{\;y\otimes 1\;} q\,(kii) + \rho q\,(iki) + \rho q^2(iik)$$

$$(iik) \xmapsto{\;1\otimes y\;} (iki) + \rho\,(iik)$$
$$\xmapsto{\;y\otimes 1\;} (kii) + \rho\,(iki) + \rho q\,(iik)$$
$$\xmapsto{\;1\otimes y\;} q\,(kii) + \rho\,(iik) + \rho q\,(iki) + \rho^2 q\,(iik)\,.$$

(Note that $q^2\,\rho = \rho + q\,\rho^2$ since $\rho = q - q^{-1}$.)

Clearly y is invertible with inverse given by:

$$y^{-1}(\varepsilon_i \otimes \varepsilon_j) \;=\; \begin{cases} \varepsilon_j \otimes \varepsilon_i & \text{for } i < j \\ \varepsilon_j \otimes \varepsilon_i + (q^{-1} - q)\,\varepsilon_i \otimes \varepsilon_j & \text{for } i > j \\ q^{-1}\varepsilon_i \otimes \varepsilon_i & \text{for } i = j\,. \end{cases}$$

Hence, y *is a YB-operator on the object V of* $\mathbf{Mod_k}$.

It is now possible to calculate the operators u, v, w and their inverses (see the definition of dualizable *YB*-operator in Chapter 13). For this, let $\varepsilon_1^*, \ldots, \varepsilon_n^* \in V^*$ be the dual basis for $\varepsilon_1, \ldots, \varepsilon_n \in V$; this means

$$\varepsilon_i^*(\varepsilon_j) = \delta_{ij}\,.$$

Recall that $e : V^* \otimes V \longrightarrow \mathbf{k}$ is the evaluation functor and $d = \sum_k \varepsilon_k \otimes \varepsilon_k^*$. Now we obtain

$$
\begin{aligned}
u(\varepsilon_i^* \otimes \varepsilon_j) &= (e \otimes 1 \otimes 1)(1 \otimes y \otimes 1)(1 \otimes 1 \otimes d)(\varepsilon_i^* \otimes \varepsilon_j) \\
&= \sum_k (e \otimes 1 \otimes 1)(1 \otimes y \otimes 1)(\varepsilon_i^* \otimes \varepsilon_j \otimes \varepsilon_k \otimes \varepsilon_k^*) \\
&= \sum_{k>j} (e \otimes 1 \otimes 1) \left(\varepsilon_i^* \otimes \varepsilon_k \otimes \varepsilon_j \otimes \varepsilon_k^* + (q - q^{-1}) \, \varepsilon_i^* \otimes \varepsilon_j \otimes \varepsilon_k \otimes \varepsilon_k^* \right) \\
&\quad + q \, (e \otimes 1 \otimes 1) \left(\varepsilon_i^* \otimes \varepsilon_j \otimes \varepsilon_j \otimes \varepsilon_j^* \right) + \sum_{k<j} (e \otimes 1 \otimes 1) \left(\varepsilon_i^* \otimes \varepsilon_k \otimes \varepsilon_j \otimes \varepsilon_k^* \right) \\
&= \sum_{k>j} \left(\delta_{ik} \, \varepsilon_j \otimes \varepsilon_k^* + (q - q^{-1}) \, \delta_{ij} \, \varepsilon_k \otimes \varepsilon_k^* \right) + q \, \delta_{ij} \left(\varepsilon_j \otimes \varepsilon_j^* \right) \\
&\quad + \sum_{k<j} \delta_{ik} \, \varepsilon_j \otimes \varepsilon_k^* \\
&= \begin{cases} \varepsilon_j \otimes \varepsilon_i^* & \text{for } j < i \\ \varepsilon_j \otimes \varepsilon_i^* & \text{for } i < j \\ q^{-1} \, \varepsilon_i \otimes \varepsilon_i^* + \sum_{k>i} (q - q^{-1}) \, \varepsilon_k \otimes \varepsilon_k^* & \text{for } i = j \, . \end{cases}
\end{aligned}
$$

The other operators are calculated similarly. We record the results below:

$$
u(\varepsilon_i^* \otimes \varepsilon_j) = \begin{cases} \varepsilon_j \otimes \varepsilon_i^* & \text{for } i \neq j \\ q \, \varepsilon_i \otimes \varepsilon_i^* + \sum_{k>i} (q - q^{-1}) \, \varepsilon_k \otimes \varepsilon_k^* & \text{for } i = j \end{cases}
$$

$$
u^{-1}(\varepsilon_i \otimes \varepsilon_j^*) = \begin{cases} \varepsilon_j^* \otimes \varepsilon_i & \text{for } i \neq j \\ q^{-1} \varepsilon_i^* \otimes \varepsilon_i + \sum_{k>i} (q^{-1} - q) q^{-2(k-i)} \, \varepsilon_k^* \otimes \varepsilon_k & \text{for } i = j \end{cases}
$$

$$
v(\varepsilon_i^* \otimes \varepsilon_j) = \begin{cases} \varepsilon_j \otimes \varepsilon_i^* & \text{for } i \neq j \\ q^{-1} \, \varepsilon_i \otimes \varepsilon_i^* + \sum_{k<i} (q^{-1} - q) \, \varepsilon_k \otimes \varepsilon_k^* & \text{for } i = j \end{cases}
$$

$$
v^{-1}(\varepsilon_i \otimes \varepsilon_j^*) = \begin{cases} \varepsilon_j^* \otimes \varepsilon_i & \text{for } i \neq j \\ q \, \varepsilon_i^* \otimes \varepsilon_i + \sum_{k<i} (q - q^{-1}) q^{2(i-k)} \, \varepsilon_k^* \otimes \varepsilon_k & \text{for } i = j \end{cases}
$$

$$
w(\varepsilon_i^* \otimes \varepsilon_j^*) = \begin{cases} \varepsilon_j^* \otimes \varepsilon_i^* & \text{for } i < j \\ \varepsilon_j^* \otimes \varepsilon_i^* + (q - q^{-1}) \, \varepsilon_i^* \otimes \varepsilon_j^* & \text{for } i > j \\ q \, \varepsilon_i^* \otimes \varepsilon_i^* & \text{for } i = j \end{cases}
$$

$$
w^{-1}(\varepsilon_i^* \otimes \varepsilon_j^*) = \begin{cases} \varepsilon_j^* \otimes \varepsilon_i^* & \text{for } i > j \\ \varepsilon_j^* \otimes \varepsilon_i^* + (q^{-1} - q) \, \varepsilon_i^* \otimes \varepsilon_j^* & \text{for } i < j \\ q^{-1} \varepsilon_i^* \otimes \varepsilon_i^* & \text{for } i = j \, . \end{cases}
$$

Hence y is dualizable. It enriches to a balanced YB-operator on defining $z : V \longrightarrow V$ simply to be the homothety

$$z(x) = q^n x .$$

Proposition 14.1 *The YB-operator (y, z) defined above is tortile.*

Proof. First observe that, for $i \neq j$, the value of $e\,v^{-1}$ at $\varepsilon_i \otimes \varepsilon_j^*$ is 0; while for $i = j$ the value is

$$q + \sum_{k<i}(q - q^{-1})q^{2(i-k)}$$

$$= q + (q - q^{-1})\big(q^{2(i-1)} + q^{2(i-2)} + \cdots + q^{2(i-(i-1))}\big)$$

$$= q + q(q^2 - 1)\,\frac{(q^2)^{i-1} - 1}{q^2 - 1}$$

$$= q + q(q^2)^{i-1} - q = q^{2i-1} .$$

Hence, we have the following remarkable calculation:

$$(1 \otimes e)(1 \otimes v^{-1})(y \otimes 1)(1 \otimes d)(\varepsilon_i)$$

$$= \sum_{j}(1 \otimes (e\,v^{-1}))(y \otimes 1)(\varepsilon_i \otimes \varepsilon_j \otimes \varepsilon_j^*)$$

$$= \sum_{j<i}(1 \otimes (e\,v^{-1}))(\varepsilon_j \otimes \varepsilon_i \otimes \varepsilon_j^*) + (1 \otimes (e\,v^{-1}))\,q\,(\varepsilon_i \otimes \varepsilon_i \otimes \varepsilon_i^*)$$

$$\quad + \sum_{i<j}(1 \otimes (e\,v^{-1}))\big((\varepsilon_j \otimes \varepsilon_i \otimes \varepsilon_j^*) + (q - q^{-1})\,(\varepsilon_i \otimes \varepsilon_j \otimes \varepsilon_j^*)\big)$$

$$= 0 + q\,q^{2i-1}\,\varepsilon_i + 0 + \sum_{i<j}(q - q^{-1})q^{2j-1}\,\varepsilon_i$$

$$= \big(q^{2i} + (q - q^{-1})(q^{2i+1} + q^{2i+3} + \cdots + q^{2n-1})\big)\,\varepsilon_i$$

$$= \big(q^{2i} + (q^2 - 1)\,q^{2i}(1 + q^2 + \cdots + q^{2(n-i-1)})\big)\,\varepsilon_i$$

$$= \Big(q^{2i} + (q^2 - 1)\,q^{2i}\,\frac{(q^2)^{n-i} - 1}{q^2 - 1}\Big)\,\varepsilon_i$$

$$= \big(q^{2i} + q^{2i}(q^{2n-2i} - 1)\big)\,\varepsilon_i$$

$$= q^{2n}\,\varepsilon_i$$

$$= z\big(z(\varepsilon_i)\big) .$$

□

15

Monoids in tensor categories

A *monoid* in a tensor category $\mathcal{V}$ consists of an object A and arrows

$$\mu : A \otimes A \longrightarrow A \qquad \text{and} \qquad \eta : I \longrightarrow A$$

which satisfy the usual identity and associativity conditions (see Chapters 1 and 6). A *monoid arrow* $f : A \longrightarrow B$ is an arrow in $\mathcal{V}$ which preserves μ and η, in the diagrammatically expressed sense.

It is also useful to consider "arrows between monoid arrows". Suppose that $f, g : A \longrightarrow B$ are monoid arrows. A *2-cell*

$$\xi : f \Rightarrow g : A \longrightarrow B$$

is defined to be an arrow $\xi : I \longrightarrow B$ in $\mathcal{V}$ such that

$$A \overset{\xi \otimes f}{\underset{g \otimes \xi}{\rightrightarrows}} B \otimes B \overset{\mu}{\longrightarrow} B \ .$$

The more 2-dimensional notation

$$A \underset{g}{\overset{f}{\rightrightarrows}}\Downarrow\xi \ B$$

is also used. (In the case where $\mathcal{V} = \mathbf{Set}$, such a 2-cell amounts to an element $\xi \in B$ for which $\xi f(a) = g(a)\xi$ for all $a \in A$.)

There are two basic *pasting operations* for 2-cells. Given the situation

$$A' \overset{h}{\longrightarrow} A \underset{g}{\overset{f}{\rightrightarrows}}\Downarrow\xi \ B \overset{k}{\longrightarrow} B'$$

where the arrows are monoid arrows and ξ is a 2-cell, there is a 2-cell

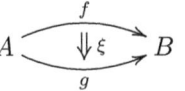

obtained since we have the following factorization:

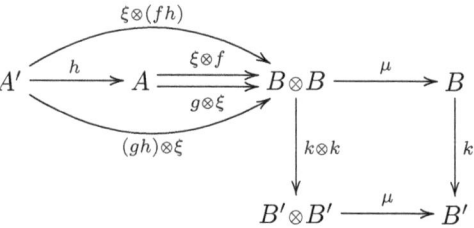

This is called *whiskering* ξ by h and k.

The other basic pasting operation is *vertical composition*, which takes a pair of 2-cells ξ and ζ, as in the following situation

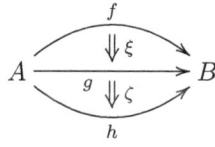

to a 2-cell

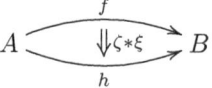

where $\zeta * \xi = \left(I \xrightarrow{\zeta \otimes \xi} B \otimes B \xrightarrow{\mu} B \right)$.

The *identity 2-cell* $\eta : f \Rightarrow f : A \longrightarrow B$ is an identity for the operation of vertical composition.

When we write a diagram such as

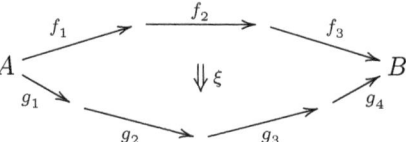

it is intended that $\xi : f_3 f_2 f_1 \Rightarrow g_4 g_3 g_2 g_1 : A \longrightarrow B$. This allows us to define a more general *pasting* operation, which assigns to a diagram like

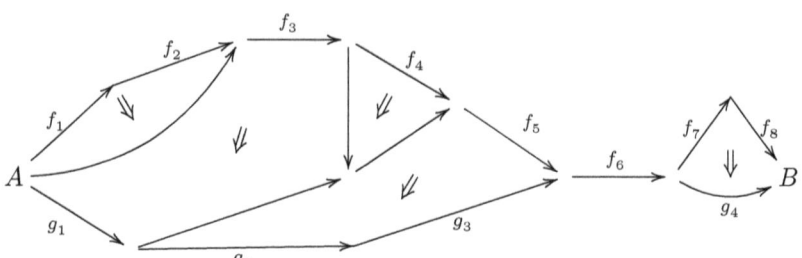

a 2-cell $f_8 f_7 f_6 f_5 f_4 f_3 f_2 f_1 \Rightarrow g_4 f_6 g_3 g_2 g_1 : A \longrightarrow B$, obtained (quite possibly in several different ways) by first whiskering the 2-cells in the diagram to be of the form

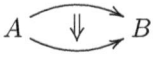

as well as being vertically composable, and then composing vertically.

As an example of this pasting, consider the diagram

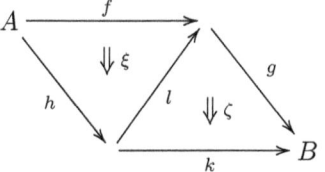

First whisker ξ and ζ appropriately, as in

to obtain two vertically composable 2-cells

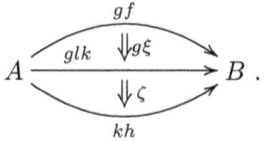

Then vertically compose to obtain a 2-cell

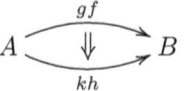

which is called the *pasted composite* of the original diagram. In this case there was only one way of performing the pasting.

As another example, consider the diagram

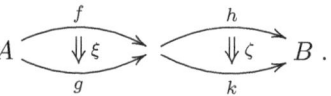

On the one hand we can whisker ξ and ζ as in

and then vertically compose; while on the other hand we can whisker ξ and ζ as in

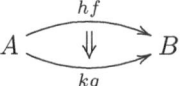

and then vertically compose. The reader should verify that the resultant 2-cells of the form

$$A \xrightarrow{\quad hf \quad \Downarrow \quad kg \quad} B$$

are actually equal.

It is a general fact that the result of pasting is independent of the way it is broken down into basic pasting operations. In fact, all ambiguities in the method can be traced back to instances of the last example. For the particular diagrams we shall use here, it is easily shown that they have a uniquely determined pasted composite.

Write $\mathbf{Mon}(\mathcal{V})$ to denote the category of monoids in $\mathcal{V}$; the arrows are monoid arrows. With the extra structure of 2-cells, $\mathbf{Mon}(\mathcal{V})$ is an example of a "2-category" (see [KS74]).

(For a commutative ring R, we have that $\mathbf{Mon}(\mathbf{Mod}_R) = \mathbf{Alg}_R$ and also $\mathbf{Mon}(\mathbf{Mod}_R^{\mathrm{op}}) = \mathbf{Cog}_R^{\mathrm{op}}$ where $\mathbf{Mod}_R$ and $\mathbf{Mod}_R^{\mathrm{op}}$ have the same tensor product $\otimes_R$.)

- *Weak tensor functors take monoids to monoids.*

More precisely, if $F : \mathcal{V} \longrightarrow \mathcal{W}$ is a weak tensor functor, each monoid A in $\mathcal{V}$ gives a monoid $F(A)$ in $\mathcal{W}$ with multiplication

$$F(A) \otimes F(A) \xrightarrow{\ \phi_{A,A}\ } F(A \otimes A) \xrightarrow{\ F(\mu)\ } F(A)$$

and unit

$$I \xrightarrow{\ \phi_0\ } F(I) \xrightarrow{\ F(\eta)\ } F(A) \ .$$

In fact, we obtain a functor

$$\mathbf{Mon}(F) : \mathbf{Mon}(\mathcal{V}) \longrightarrow \mathbf{Mon}(\mathcal{W})$$

which preserves the basic pasting operations of 2-cells (and so is an example of a "2-functor").

For each monoid A in $\mathcal{V}$ there is a category, called $\mathbf{Mod}_{\mathcal{V}}(A)$, of (left) A-modules. An A-*module* consists of an object M of $\mathcal{V}$ and an arrow $\mu : A \otimes M \longrightarrow M$, called the *action*, which satisfies all the usual defining diagrams for a module (see Chapter 9). An A-*module arrow* $\mu : M \longrightarrow N$ is an arrow in $\mathcal{V}$ such that

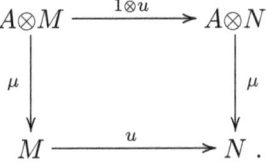

There is a "forgetful" functor $U_A : \mathbf{Mod}_{\mathcal{V}}(A) \longrightarrow \mathcal{V}$ which simply forgets the module action.

Each monoid arrow $f : A \longrightarrow B$ determines a functor

$$\mathbf{Mod}(f) : \mathbf{Mod}_{\mathcal{V}}(B) \longrightarrow \mathbf{Mod}_{\mathcal{V}}(A)$$

given by "restriction of scalars" along f. That is, for a B-module M, we take $\mathbf{Mod}(f)(M)$ to be M with A-action

$$A \otimes M \xrightarrow{\ f \otimes 1\ } B \otimes M \xrightarrow{\ \mu\ } M \ .$$

Each B-module arrow becomes an A-module arrow, thereby giving the following commutative triangle of categories and functors:

$$\mathbf{Mod}_{\mathcal{V}}(B) \xrightarrow{\ \mathbf{Mod}(f)\ } \mathbf{Mod}_{\mathcal{V}}(A)$$

$$U_B \searrow \qquad \swarrow U_A$$

$$\mathcal{V}$$

Furthermore, each 2-cell $\xi : f \Rightarrow g : A \longrightarrow B$ between monoid arrows f and g in $\mathcal{V}$ determines a natural transformation

$$\mathbf{Mod}(\xi) : \mathbf{Mod}(f) \longrightarrow \mathbf{Mod}(g)$$

whose component at the B-module M is the composite

$$M \xrightarrow{\xi \otimes 1} B \otimes M \xrightarrow{\mu} M$$

which is an A-module arrow, as can be seen from the diagram

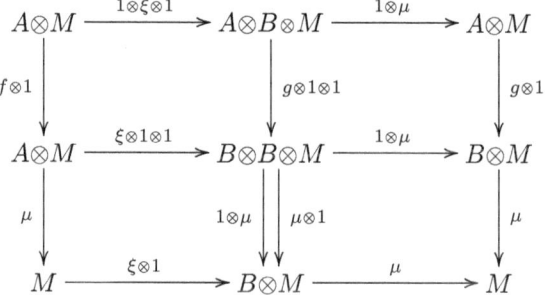

Naturality follows from the following diagram involving a B-module arrow $u : M \longrightarrow N$.

The assignment $\xi \longmapsto \mathbf{Mod}(\xi)$ turns the two basic pasting operations into corresponding familiar operations on natural transformations. This gives an example of a "2-functor"

$$\mathbf{Mod}_{\mathcal{V}} : \mathbf{Mon}(\mathcal{V})^{\mathrm{op}} \longrightarrow \mathbf{Cat}$$

where $\mathbf{Cat}$ is some appropriate "2-category" of categories.

For our purposes, it is important to remember the forgetful functors $U_A : \mathbf{Mon}_{\mathcal{V}}(A) \longrightarrow \mathcal{V}$. So, rather than $\mathbf{Cat}$, we consider $\mathbf{Cat}/\mathcal{V}$, whose objects are functors $F : \mathcal{C} \longrightarrow \mathcal{V}$, whose arrows

$$T : (\mathcal{C}, F) \longrightarrow (\mathcal{D}, G)$$

are functors $T : \mathcal{C} \longrightarrow \mathcal{D}$ such that $GT = F$, and for which the 2-cells $\alpha : T \longrightarrow T'$ are arbitrary natural transformations from T to T' (that is, no

condition relating it to F and G). Observe that $\mathbf{Mod}_\mathcal{V}$ really lands in $\mathbf{Cat}/\mathcal{V}$ by taking $A \in \mathbf{Mon}(\mathcal{V})^{\mathrm{op}}$ to $\left(\mathbf{Mod}_\mathcal{V}(A), U_A\right)$. So we have that

$$\mathbf{Mod}_\mathcal{V} : \mathbf{Mon}(\mathcal{V})^{\mathrm{op}} \longrightarrow \mathbf{Cat}/\mathcal{V}$$

is a 2-functor.

There is an obvious candidate for a tensor product on $\mathbf{Cat}/\mathcal{V}$, namely

$$(\mathcal{C}, F) \otimes (\mathcal{D}, G) \;=\; \left(\mathcal{C} \times \mathcal{D}, \; \mathcal{C} \times \mathcal{D} \xrightarrow{F \times G} \mathcal{V} \times \mathcal{V} \xrightarrow{\otimes} \mathcal{V}\right).$$

This tensor product respects all of the pasting operations for natural transformations. Ignoring the 2-cells, $\mathbf{Cat}/\mathcal{V}$ becomes a tensor category with unit given by $\left(1, \; I : 1 \longrightarrow \mathcal{V}\right)$.

Suppose now that $\mathcal{V}$ is braided. For monoids A and B in $\mathcal{V}$, we enrich $A \otimes B$ with the multiplication

$$A \otimes B \otimes A \otimes B \xrightarrow{1 \otimes c_{B,A} \otimes 1} A \otimes A \otimes B \otimes B \xrightarrow{\mu \otimes \mu} A \otimes B$$

and unit $\eta \otimes \eta : I \longrightarrow A \otimes B$; the braiding properties imply (Exercise 15.1) that this makes $A \otimes B$ into a monoid. Thus $\mathbf{Mon}(\mathcal{V})$ becomes a tensor category such that the forgetful functor

$$\mathbf{Mon}(\mathcal{V}) \longrightarrow \mathcal{V}$$

is a strict tensor functor. The tensor product on $\mathbf{Mon}(\mathcal{V})$ respects the basic pasting operations of 2-cells; see Exercise 15.2.

We shall now see that

$$\mathbf{Mod}_\mathcal{V} : \mathbf{Mon}(\mathcal{V})^{\mathrm{op}} \longrightarrow \mathbf{Cat}/\mathcal{V}$$

is *essentially* a weak tensor functor. For this, observe that $\mathbf{Mod}_\mathcal{V}(I) = \mathcal{V}$ and we have arrows

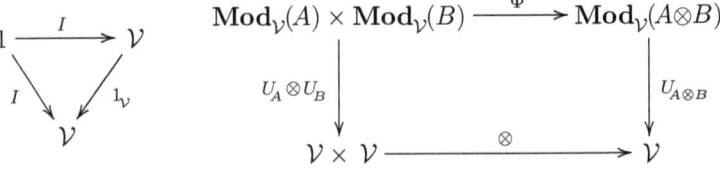

in $\mathbf{Cat}/\mathcal{V}$, where $\Phi(M, N) = M \otimes N$, with action

$$A \otimes B \otimes M \otimes N \xrightarrow{1 \otimes c_{B,M} \otimes 1} A \otimes M \otimes B \otimes N \xrightarrow{\mu \otimes \mu} M \otimes N \,.$$

The reason for the word "essentially" is that the axioms for a weak tensor functor (see the beginning of Chapter 12) hold only up to isomorphism (instead of equality); in fact, the isomorphisms are precisely provided by the associativity and unit constraints a, r, l for the tensor product.

Just as weak tensor functors take monoids to monoids, the 2-functor $\mathbf{Mod}_{\mathcal{V}}$ takes *tensor objects* in $\mathbf{Mon}(\mathcal{V})^{\mathrm{op}}$ to tensor objects in $\mathbf{Cat}/\mathcal{V}$.

• A tensor object is "essentially" a monoid.

• A tensor object in $\mathbf{Cat}$ is precisely a tensor category; whereas a monoid in $\mathbf{Cat}$ is a *strict* tensor category.

• A tensor object in $\mathbf{Cat}/\mathcal{V}$ is a pair $(\mathcal{C}, F)$ consisting of a tensor category $\mathcal{C}$ and a strict tensor functor $F : \mathcal{C} \longrightarrow \mathcal{V}$.

• A tensor object in $\mathbf{Mon}(\mathcal{V})^{\mathrm{op}}$ will be called a *quasi-bimonoid* in $\mathcal{V}$. This consists of a monoid A in $\mathcal{V}$, monoid arrows $\delta : A \longrightarrow A \otimes A$, $\epsilon : A \longrightarrow I$, and 2-cells:

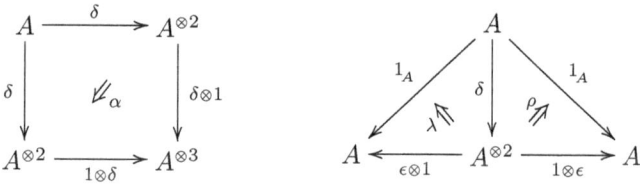

which are invertible under vertical composition and satisfy the following equalities between pasted diagrams.

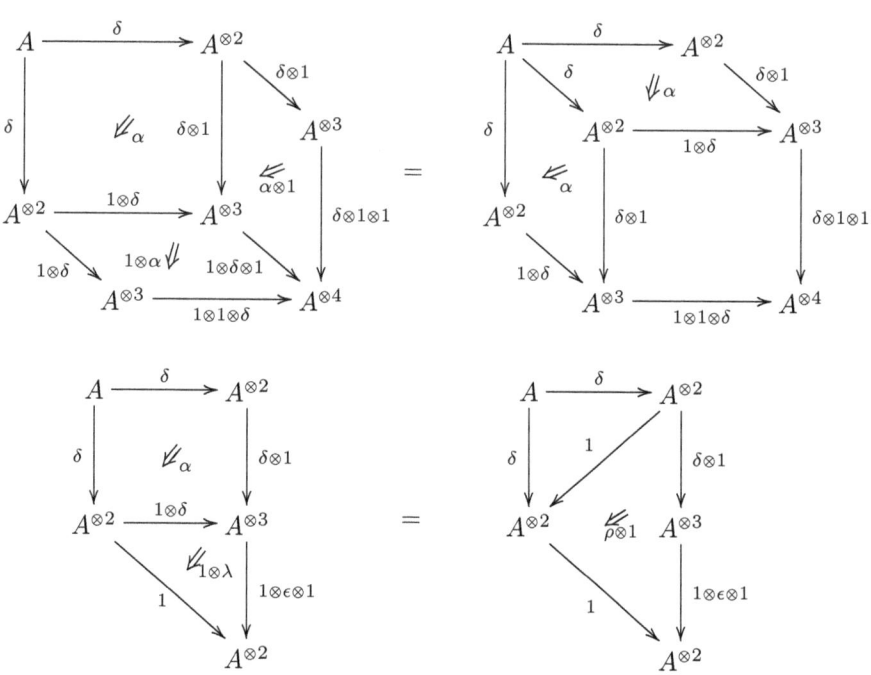

A *bimonoid* in $\mathcal{V}$ is a strict quasi-bimonoid; that is, one for which the 2-cells α, λ, ρ are actually identity 2-cells. (A bimonoid in $\mathbf{Mod}_R$ is an R-bialgebra; see Proposition 7.5.)

Hence we have that a (quasi-) bimonoid A in $\mathcal{V}$ determines the structure of a tensor category on $\mathbf{Mod}_\mathcal{V}(A)$ as well as a tensor functor structure on $U_A : \mathbf{Mod}_\mathcal{V}(A) \longrightarrow \mathcal{V}$ (see Chapter 10 for those cases when $\mathcal{V} = \mathbf{Mod}_R$ and when $\mathcal{V} = \mathbf{Mod}_R^{\mathrm{op}}$).

A *(quasi-) Hopf monoid* in $\mathcal{V}$ is a (quasi-) bimonoid H together with an arrow $\nu : H \longrightarrow H$, called the *antipode*, such that

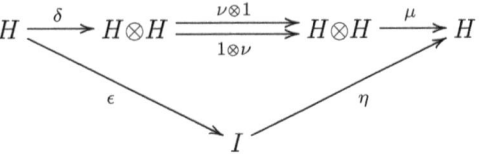

Drinfel'd [Dri89] obtained interesting examples of quasi-Hopf monoids in $\mathbf{Mod}_k$.

Proposition 15.1 *Suppose $\mathcal{V}$ is a braided tensor category and H is a Hopf monoid in $\mathcal{V}$. If M is an H-module which has a left dual M^* as an object of $\mathcal{V}$ then M^* becomes the left dual of M in $\mathbf{Mod}_\mathcal{V}(H)$ via the action*

$$H \otimes M^* \xrightarrow{\nu \otimes 1} H \otimes M^* \xrightarrow{c_{H,M^*}} M^* \otimes H \xrightarrow{1 \otimes d} M^* \otimes H \otimes M \otimes M^*$$

$$\xrightarrow{1 \otimes \mu \otimes 1} M^* \otimes M \otimes M^* \xrightarrow{e \otimes 1} M^* .$$

Proof. This is a matter of proving that e and d are module arrows. For this, compare with Propositions 10.1 and 10.5. □

A *braiding* for a (quasi-) bimonoid A in $\mathcal{V}$ is a 2-cell

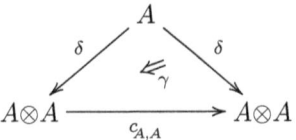

which is invertible (with respect to vertical composition) and satisfies the two equalities below.

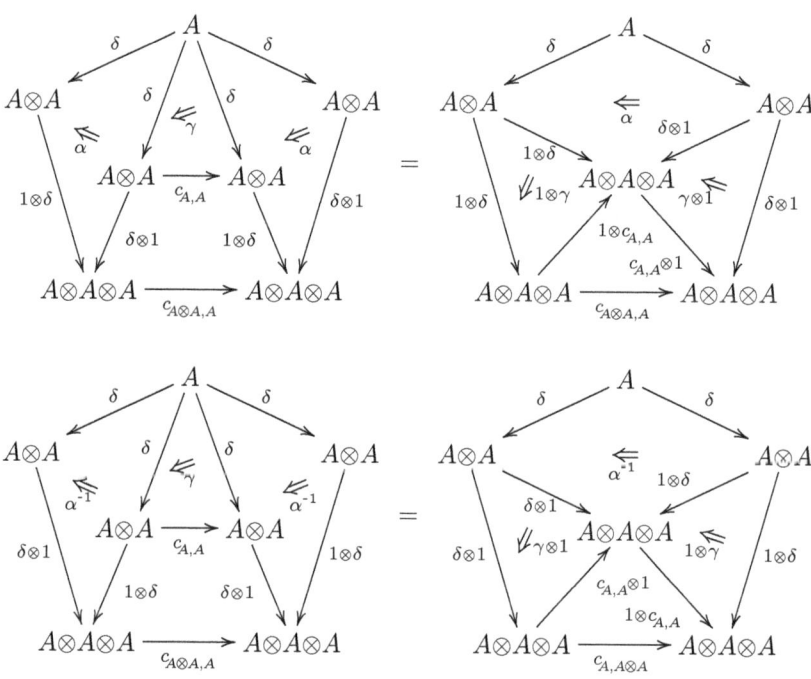

Now suppose $\mathcal{V}$ is symmetric. A *twist* for A is a 2-cell

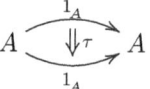

which is invertible, with respect to $*$, and satisfies

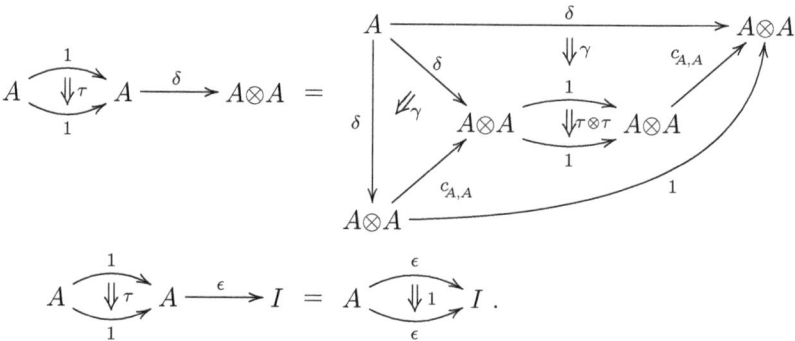

$$A \underset{1}{\overset{1}{\Downarrow \tau}} A \xrightarrow{\epsilon} I = A \underset{\epsilon}{\overset{\epsilon}{\Downarrow 1}} I \;.$$

A (quasi-) bimonoid with a braiding and twist is called *balanced*.

The reader should interpret the above pasting diagrams in the special case where $\mathcal{V} = \mathbf{Mod}_k$ to see that these definitions agree with the definitions of braiding element and twist for bialgebras as in Examples 11.3 and 11.7. We now have a conceptual version of the calculations in those examples.

Proposition 15.2 *Suppose $\mathcal{V}$ is a symmetric tensor category and A is a bimonoid in $\mathcal{V}$. There is a bijection between braidings γ for A and braidings c for $\mathbf{Mod}_{\mathcal{V}}(A)$ determined by $c_{M,N}$ as the composite:*

$$M \otimes N \xrightarrow{\gamma \otimes 1 \otimes 1} A \otimes A \otimes M \otimes N \xrightarrow{1 \otimes c_{A \otimes M, N}} A \otimes N \otimes A \otimes M \xrightarrow{\mu \otimes \mu} N \otimes M \ .$$

There is a bijection between twists τ for A and twists θ for $\mathbf{Mod}_{\mathcal{V}}(A)$ determined by

$$\theta_M = \left(M \xrightarrow{\tau \otimes 1} A \otimes M \xrightarrow{\mu} M \right) .$$

Proof. Apply the 2-functor $\mathbf{Mod}_{\mathcal{V}}$ to the triangle containing γ, and paste below it a square containing a natural isomorphism whose components are the symmetry of $\mathcal{V}$ (we omit the subscripts $\mathcal{V}$ on maps in the diagram):

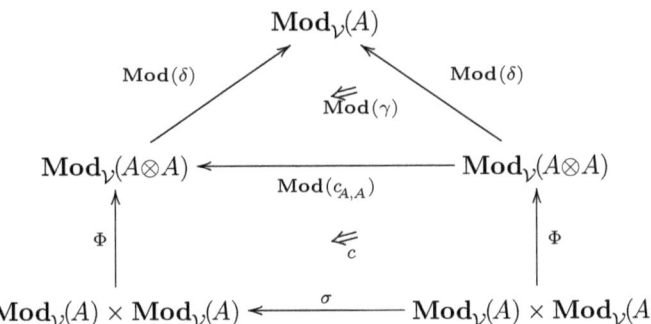

The result is a natural isomorphism, whose component at the pair $(M, N) \in \mathbf{Mod}_{\mathcal{V}}(A) \times \mathbf{Mod}_{\mathcal{V}}(A)$ is the $c_{M,N}$ as stated in the proposition. The axioms on γ convert to the braiding axioms for c.

Conversely, to recapture γ from c, take the composite

$$\gamma = \left(I \xrightarrow{\eta \otimes \eta} A \otimes A \xrightarrow{c_{A,A}} A \otimes A \right) ,$$

where we regard A as an object of $\mathbf{Mod}_{\mathcal{V}}(A)$ with action μ. The proof of the twist bijection is similar. $\qquad\square$

A *tortile bimonoid* in $\mathcal{V}$ is a balanced Hopf monoid H such that

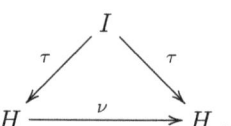

Proposition 15.3 *Suppose that $\mathcal{V}$ is a symmetric autonomous tensor category. If H is a tortile bimonoid in $\mathcal{V}$ then $\mathbf{Mod}_{\mathcal{V}}(H)$ is a tortile tensor category.*

Proof. By Propositions 15.1 and 15.2, $\mathbf{Mod}_{\mathcal{V}}(H)$ becomes an autonomous balanced tensor category. All that remains to see is that $\theta_{M^*} = (\theta_M)^*$, which follows from $\nu \circ \tau = \tau$. ◻

Consider the replacement of the tensor category $\mathcal{V}$ by its opposite tensor category $\mathcal{V}^{\mathrm{op}}$. Monoids become comonoids, but bimonoids and Hopf monoids are unchanged. For a bimonoid H in $\mathcal{V}$, a *cotwist* $\tau : H \longrightarrow I$ on H in $\mathcal{V}$ is defined to be a twist on H in $\mathcal{V}^{\mathrm{op}}$; a *cobraiding* $\gamma : H \otimes H \longrightarrow I$ on H in $\mathcal{V}$ is defined to be a braiding on H in $\mathcal{V}^{\mathrm{op}}$.

Thus we have the corresponding notions of *cobraided, cobalanced* and *cotortile* bimonoid in $\mathcal{V}$.

Our proposal for a definition of a *quantum group over R* is that it should be a cotortile bimonoid in $\mathbf{Mod}_R$. In Chapter 18 we shall see that our main example, the quantum general linear group, does indeed give an instance of this concept.

Exercise 15.1 *Check that the tensor product of monoids in a braided tensor category is a monoid.*

Exercise 15.2 *Show that the natural tensor product on $\mathbf{Mon}(\mathcal{V})$ preserves whiskering and vertical composition of 2-cells in each variable.*

Exercise 15.3 *Verify that braidings and twists in $\mathbf{Mod}_k$, in the sense of this chapter, include those of Examples 11.3 and 11.7.*

16

Tannaka duality

Given a compact group G, the set **Rep** G of isomorphism classes of *appropriate representations* admits various operations; for example direct sum and tensor product. Tannaka's duality theorem (1939) provided a recipe for recovering a compact group **Gp** R from a structure R such as **Rep** G whereby **Gp Rep** $G \cong G$.

For algebraic groups, Saavedra Rivano [SR72] considered the category of appropriate representations together with the tensor structure and the underlying functor into vector spaces. He gave criteria on a tensor functor into vector spaces under which it should be equivalent to such an underlying functor. A non-commutative generalization of this was given by Ulbrich [Ulb89]. We shall lead into this Hopf algebra version by examining the 2-functor **Mod**$_\mathcal{V}$ of the previous chapter.

For simplicity of exposition we suppose our tensor category $\mathcal{V}$ is strict. This loses no generality in fact since every tensor category is equivalent to a strict one (Mac Lane's coherence theorem). We also suppose that $\mathcal{V}$ is symmetric (but we cannot suppose the tensor product is strictly commutative). A consequence of this simplification is that we really do have a weak tensor functor

$$\mathbf{Mod}_\mathcal{V} : \mathbf{Mon}_\mathcal{V}^{\mathrm{op}} \longrightarrow \mathbf{Cat}/\mathcal{V}.$$

We are now interested in going back from $\mathbf{Cat}/\mathcal{V}$ to $\mathbf{Mon}(\mathcal{V})^{\mathrm{op}}$. A possible way to do this is via a left adjoint to $\mathbf{Mod}_\mathcal{V}$, if it exists. Under reasonable conditions, a left adjoint $E_F \in \mathbf{Mon}(\mathcal{V})$ does exist at an object $(\mathcal{C}, F)$ of $\mathbf{Cat}/\mathcal{V}$. It is constructed as follows.

If each internal hom $[FX, FX]$ exists in $\mathcal{V}$ and if $\mathcal{V}$ is suitably complete, we put

$$E_F = \int_{X \in \mathcal{C}} [FX, FX]$$

where the integral sign denotes the *end* (see Mac Lane [Mac71]) of the functor $\mathcal{C}^{\mathrm{op}} \times \mathcal{C} \longrightarrow \mathcal{V}$ taking (X, Y) to $[FX, FY]$; it is the equalizer of the obvious pair of arrows (see Exercise 16.1)

$$\prod_X [FX, FX] \Longrightarrow \prod_{f : X \to Y} [FX, FY].$$

There are projection arrows

$$\pi_X : E_F \longrightarrow [FX, FX]$$

for each object $X \in \mathcal{C}$. These correspond, using the definition of *internal hom*, to arrows

$$\mu_X : E_F \otimes FX \longrightarrow FX .$$

The universal properties of *end* and *internal hom* show that there exists a bijection between the arrows $f : A \longrightarrow E_F$ in $\mathcal{V}$ and natural families of arrows $\theta_X : A \otimes FX \longrightarrow FX$, given by

$$\theta_X = \mu_X \circ (f \otimes 1_X) .$$

The natural families

$$E_F \otimes E_F \otimes FX \xrightarrow{\;1 \otimes \mu_X\;} E_F \otimes FX \xrightarrow{\;\mu_X\;} FX , \qquad FX \xrightarrow{\;1_{FX}\;} FX$$

induce, under such bijections, the monoid structure on E_F :

$$\mu : E_F \otimes E_F \longrightarrow E_F \quad , \quad \eta : I \longrightarrow E_F .$$

Example 16.1 *Take* $\mathcal{V} = \mathbf{Mod}_R$ *for some commutative ring* R. *Then we have that* $\mathbf{Mon}(\mathcal{V}) = \mathbf{Alg}_R$. *Now for any functor* $F : \mathcal{C} \longrightarrow \mathcal{V}$, *the algebra* E_F *has as elements the natural families* $\theta = (\theta_X)_{X \in G}$ *of* R-*linear morphisms* $\theta_X : FX \longrightarrow FX$ *; addition and multiplication by scalars are done componentwise, while multiplication is componentwise composition. In particular, for any* R-*algebra* A, *if we have*

$$F = U_A : \mathbf{Mod}_R(A) \longrightarrow \mathbf{Mod}_R$$

then there is a natural isomorphism of algebras

$$E_F \cong A .$$

To see this, notice that each element m *of an* A-*module* M *determines a unique* $\hat{m} : A \longrightarrow M$ *in* $\mathbf{Mod}_R(A)$ *with* $\hat{m}(1) = m$ *; so for a natural* $\theta : U_A \longrightarrow U_A$, *we have*

$$
\begin{array}{ccc}
A & \xrightarrow{\;\theta_A\;} & A \\
{\scriptstyle \hat{m}} \downarrow & & \downarrow {\scriptstyle \hat{m}} \\
M & \xrightarrow{\;\theta_M\;} & M
\end{array}
$$

which implies $\theta_M(m) = \theta_A(1)\, m$ *; so* θ *is determined by* $\theta_A(1) \in A$.

Suppose A is a monoid in $\mathcal{V}$. The two axioms which are required for an arrow $f : A \longrightarrow E_F$ to be in $\mathbf{Mon}(\mathcal{V})$ translate to the two conditions on the corresponding family of arrows $\theta_X : A \otimes FX \longrightarrow FX$ which say that each θ_X is an *action* of A on FX. This is precisely what is needed to lift F to a functor $T : \mathcal{C} \longrightarrow \mathbf{Mod}_{\mathcal{V}}(A)$ such that $U_A T = F$; just put $TX = (FX, \theta_X)$. This gives a natural bijection between hom sets

$$\mathbf{Mon}(\mathcal{V})(A, E_F) \cong (\mathbf{Cat}/\mathcal{V})\big((\mathcal{C}, F), (\mathbf{Mod}_{\mathcal{V}}(A), U_A)\big),$$

whereby $(\mathcal{C}, F) \longmapsto E_F$ is *left adjoint to* $\mathbf{Mod}_{\mathcal{V}} : \mathbf{Mon}(\mathcal{V})^{\mathrm{op}} \longrightarrow \mathbf{Cat}/\mathcal{V}$. In fact, the above bijection becomes an isomorphism of categories, since it extends to 2-cells:

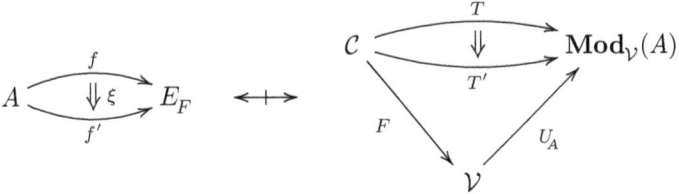

This is expressed by saying that $(\mathcal{C}, F) \longmapsto E_F$ is *left 2-adjoint to* $\mathbf{Mod}_{\mathcal{V}}$.

Taking $A = E_F$ in the above bijection and looking at the image of the identity arrow, we obtain

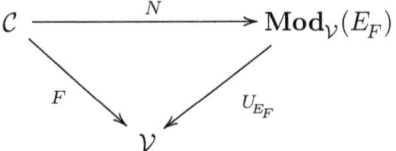

where $NX = (FX, \mu_X)$. We obtain a (partial) 2-functor

$$E_- : \mathbf{Cat}/\mathcal{V} \longrightarrow (\mathbf{Mon}(\mathcal{V}))^{\mathrm{op}}$$

by taking the 2-cell $\alpha : T \Rightarrow T' : (\mathcal{C}, F) \longrightarrow (\mathcal{D}, G)$ in $\mathbf{Cat}/\mathcal{V}$ into the 2-cell $E_\alpha : E_T \Rightarrow E_{T'} : E_G \longrightarrow E_F$ in $\mathbf{Mon}(\mathcal{V})$ corresponding (under the 2-adjunction) to the 2-cell in $\mathbf{Cat}/\mathcal{V}$:

$$N\alpha : NT \Rightarrow NT' : (\mathcal{C}, F) \longrightarrow (\mathbf{Mod}_{\mathcal{V}}(E_G), U_{E_G}).$$

Remark 16.2 *The* Formal Tannaka Duality *criteria on* $F : \mathcal{C} \longrightarrow \mathcal{V}$ *are that* $N : \mathcal{C} \longrightarrow \mathbf{Mod}_{\mathcal{V}}(E_F)$ *should be faithful and that every "appropriate" E_F-module should be isomorphic to some NX.*

We can equally well regard E_- as a 2-functor

$$E_- : (\mathbf{Cat}/\mathcal{V})^{\mathrm{op}} \longrightarrow \mathbf{Mon}(\mathcal{V})$$

whereupon (for general reasons as an adjoint to $\mathbf{Mod}_{\mathcal{V}}$) it is a weak tensor functor. It preserves the unit in the sense that $E_I \cong I$, while we have a canonical arrow ϕ such that

$$
\begin{array}{ccc}
E_F \otimes F_G & \xrightarrow{\ \ \phi_{F,G}\ \ } & E_{F \otimes G} \\
{\scriptstyle \pi_X \otimes \pi_Y} \downarrow & & \downarrow {\scriptstyle \pi_{X \otimes Y}} \\
[FX, FX] \otimes [GY, GY] & \xrightarrow{\ \ -\otimes-\ \ } & [FX \otimes GY, FX \otimes GY]
\end{array}
$$

where the bottom arrow corresponds to the composite

$$[FX, FX] \otimes [GY, GY] \otimes FX \otimes GY \xrightarrow{\ 1 \otimes c \otimes 1\ } [FX, FX] \otimes FX \otimes [GY, GY] \otimes GY$$

$$\xrightarrow{\ e \otimes e\ } FX \otimes GY \ .$$

So E_- takes monoids in $(\mathbf{Cat}/\mathcal{V})^{\mathrm{op}}$ to monoids in $\mathbf{Mon}(\mathcal{V})$, the latter being the commutative monoids in $\mathcal{V}$, but this is of no interest to us here.

Our real interest is in to what extent

$$E_- : (\mathbf{Cat}/\mathcal{V}) \longrightarrow \mathbf{Mon}(\mathcal{V})^{\mathrm{op}}$$

takes monoids to monoids. This will be true of those monoids $(\mathcal{C}, F)$ in $\mathbf{Cat}/\mathcal{V}$ for which $\phi_{F,F} : E_F \otimes E_F \longrightarrow E_{F \otimes F}$ is invertible.

Is this a reasonable condition? At first glance, invertibility of

$$\phi_{F,G} : \int_X [FX, FX] \otimes \int_Y [GY, GY] \longrightarrow \int_{X,Y} [FX \otimes GY, FX \otimes GY]$$

looks unlikely. It would be implied by the two conditions:

(a) *each* $A \otimes _- : \mathcal{V} \longrightarrow \mathcal{V}$ *preserves ends*; and

(b) *each* $[A, B] \otimes [C, D] \xrightarrow{\ -\otimes-\ } [A \otimes C, B \otimes D]$ *is invertible, for every* $A = FX$ *and every* $C = GY$.

However, these look unlikely too, if we think in terms of Example 16.1.

We shall look at the conditions (a) and (b) more closely. If $\mathcal{V}$ is a closed tensor category then $A \otimes_-$ preserves colimits (since it has a right adjoint

$[A, \_]$; see Mac Lane [Mac71, Chapter V §5]. But *end* is a limit, not a colimit. So (a) can be ensured by taking $\mathcal{V}$ to be the opposite of a complete closed tensor category. We need to be careful here since we still need the internal homs of the form $[FX, FX]$ in $\mathcal{V}$, not in $\mathcal{V}^{\mathrm{op}}$.

Condition (b) is true, for example for finite-dimensional vector spaces. What is needed is that A and C should have duals; then we have canonical isomorphisms

$$
\begin{aligned}
[A, B] \otimes [C, D] & \cong A^* \otimes B \otimes C^* \otimes D \\
& \cong (A \otimes C)^* \otimes (B \otimes D) \cong [A \otimes C, B \otimes D] .
\end{aligned}
$$

Hence, conditions (a) and (b) are not unreasonable after all. They are satisfied when $\mathcal{V}$ is the opposite of a closed symmetric tensor category which is cocomplete enough for coends over $\mathcal{C}$ to exist, and when each FX and GY has a dual.

Suppose then that $\mathcal{V}^{\mathrm{op}}$ is a closed symmetric (strict) tensor category which is (small) cocomplete. Suppose $\mathcal{C}$ is a left autonomous small (strict) tensor category and $F : \mathcal{C} \longrightarrow \mathcal{V}$ is a (strict) tensor functor. Then each FX has a dual FX^*. Since $(\mathcal{C}, F)$ is a monoid in $\mathbf{Cat}/\mathcal{V}$, we obtain a monoid E_F in $\mathbf{Mon}(\mathcal{V})^{\mathrm{op}}$; that is, a bimonoid E_F in $\mathcal{V}$. This gives a factorization

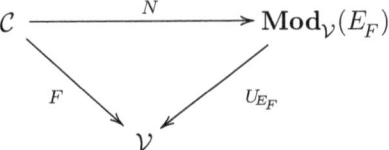

of our tensor functor F into tensor functors N and U_{E_F}.

In fact, E_F is a *Hopf monoid*. To see this, define $F^* : \mathcal{C}^{\mathrm{op}} \longrightarrow \mathcal{V}$ by $F^* X = FX^*$. We obtain a monoid arrow

$$
(\ )^* : (\mathcal{C}, F) \longrightarrow (\mathcal{C}^{\mathrm{op}}, F^*)
$$

in $\mathbf{Cat}/\mathcal{V}$. This induces a monoid arrow ν with

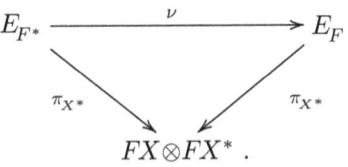

It is easy to see that $E_{F^*} = E_F{}^{\mathrm{op}}$ as bimonoids in $\mathcal{V}$ (that is, E_{F^*} is just E_F with switched multiplication and switched comultiplication), and ν is an antipode for the bimonoid E_F.

Now suppose $\mathcal{C}$ is braided. The braiding can be regarded as an invertible 2-cell:

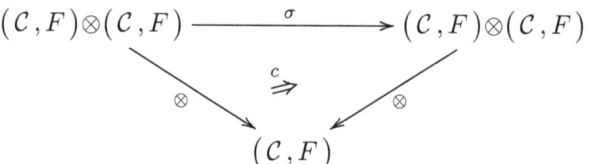

in $\mathbf{Cat}/\mathcal{V}$. Applying E_-, we obtain an invertible 2-cell in $\mathbf{Mon}(\mathcal{V})$:

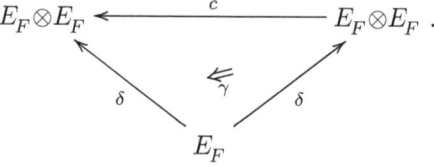

The braiding arrows for c on $\mathcal{C}$ carry over precisely to those for γ on E_F. Moreover, $N : \mathcal{C} \longrightarrow \mathbf{Mod}_{\mathcal{V}}(E_F)$ becomes a braided tensor functor.

Next suppose $\mathcal{C}$ is balanced. The twist on $\mathcal{C}$ can be regarded as being an invertible 2-cell:

$$(\mathcal{C},F) \underset{1_C}{\overset{1_C}{\rightrightarrows}} \Downarrow\theta \ (\mathcal{C},F)$$

in $\mathbf{Cat}/\mathcal{V}$, and, applying E_-, we obtain a twist

$$E_F \underset{1}{\overset{1}{\rightrightarrows}} \Downarrow\tau \ E_F$$

for the braided bimonoid E_F. So E_F becomes a balanced Hopf monoid and $N : \mathcal{C} \longrightarrow \mathbf{Mod}_{\mathcal{V}}(E_F)$ becomes a balanced tensor functor.

Finally, if $\mathcal{C}$ is a tortile tensor category, E_F is a tortile bimonoid in $\mathcal{V}$.

To obtain Ulbrich's [Ulb89] setting, we take $\mathcal{V} = \mathbf{Mod}_R^{\mathrm{op}}$ for a commutative ring R. For each R-coalgebra C, we have

$$\mathbf{Mod}_{\mathcal{V}}(C)^{\mathrm{op}} \ = \ \mathbf{Comod}_R(C) .$$

We use the notation $\mathbf{Comod}_R(C)_c$ to denote the full subcategory consisting of C-comodules M for which the underlying R-module $U_C M$ is Cauchy.

Consider a small category $\mathcal{C}$ and a functor $F : \mathcal{C} \longrightarrow \mathbf{Mod}_R$ whose values FX are Cauchy R-modules. The coend

$$E'_F = \int^X FX \otimes_R (FX)^*$$

becomes an R-coalgebra and we have

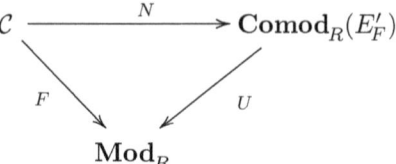

(since we can apply our previous theory to F regarded as going from $\mathcal{C}^{\mathrm{op}}$ to $\mathcal{V} = \mathbf{Mod}_R^{\mathrm{op}}$). Notice that N actually lands in $\mathbf{Comod}_R(E'_F)_c$.

If $\mathcal{C}$ is a tensor category and F is a tensor functor then E'_F becomes an R-bialgebra and N becomes a tensor functor. If $\mathcal{C}$ is left autonomous then E'_F becomes a Hopf algebra with invertible antipode. If $\mathcal{C}$ is a tortile tensor category then E'_F becomes a cotortile R-bialgebra (quantum group!) and N becomes a balanced tensor functor.

An important case of Tannaka duality is the characterization of those $F : \mathcal{C} \longrightarrow \mathbf{Mod}_R$ equivalent to $U_H : \mathbf{Comod}_R(H)_c \longrightarrow \mathbf{Mod}_R$ for some Hopf algebra H. This can be investigated by looking at when the functor $N : \mathcal{C} \longrightarrow \mathbf{Comod}_R(E'_F)_c$ is an equivalence.

The question arises here as to whether $E'_F \cong \mathcal{C}$ when the equality $F = U_C : \mathbf{Comod}(C)_c \longrightarrow \mathbf{Mod}_R$ holds for a coalgebra C. We cannot use the technique of Example 16.1 since, although C is a C-comodule, it is generally not Cauchy as an R-module.

Proposition 16.3 *If C is a coalgebra over a field R and U denotes the forgetful functor $U : \mathbf{Comod}(C)_c \longrightarrow \mathbf{Mod}_R$, then there is a coalgebra isomorphism*

$$E'_U \cong C \,.$$

Proof. We need to show that C has the universal property of E'_U; that is, the assignment $f \longmapsto (f \otimes 1) \circ \delta$ determines a natural bijection between R-module morphisms $F : C \longrightarrow X$ and families of R-module morphisms $\theta_M : U(M) \longrightarrow X \otimes U(M)$ natural in $M \in \mathbf{Comod}(C)_c$.

We need to apply the *fundamental theorem on coalgebras* (see Sweedler [Swe69, p.46]): (when R is a field) "the sub-coalgebra generated by an element of C is Cauchy".

Given a family θ_M, we must define $f(c)$ for each $c \in C$. Let M be any sub-coalgebra of C which contains c and is finite dimensional. Such M exist

by the above *fundamental theorem*, and can be regarded as C-comodules. Put $f(c) = (1 \otimes \varepsilon)\theta_M(c)$. This is independent of the choice of M since θ_M is natural in M. The proof that this gives the inverse to $f \longmapsto (f \otimes 1) \circ \delta$ is now easy.   □

Exercise 16.1 *Give a precise definition of the "obvious pair of arrows" in the defining equalizer for E_F.*

Exercise 16.2 *Complete the proof of Proposition 16.3.*

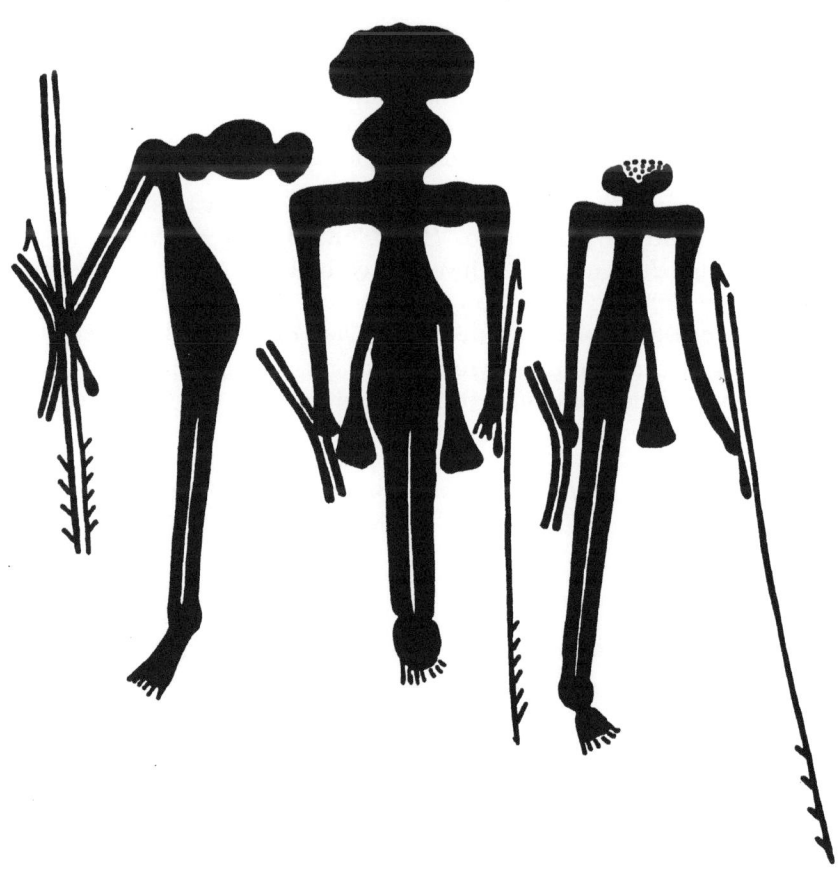

Bradshaw: Clothes Peg Figure Period, [Wal94, Plate 67].

17

Adjoining an antipode to a bialgebra

Tannaka duality allows the possibility of taking an R-bialgebra A, applying some categorical construction to $\mathbf{Comod}_R(A)_c$, and asking whether the result again has the form $\mathbf{Comod}_R(B)_c$ for some R-bialgebra B.

An example of an appropriate categorical construction is adjoining left-dual objects to a tensor category. To each tensor category $\mathcal{C}$, there is a left autonomous tensor category $\mathcal{A}_\ell(\mathcal{C})$ and a tensor functor $\mathcal{C} \longrightarrow \mathcal{A}_\ell(\mathcal{C})$ which induces a natural equivalence between the category of tensor functors $\mathcal{A}_\ell(\mathcal{C}) \longrightarrow \mathcal{D}$ and the category of tensor functors $\mathcal{C} \longrightarrow \mathcal{D}$ for all left autonomous tensor categories $\mathcal{D}$. (See [JS91a], [JSV96] and [KSW02].)

Suppose that $F : \mathcal{C} \longrightarrow \mathbf{Mod}_R$ is a tensor functor whose values FX are Cauchy R-modules. Then we obtain a corresponding tensor functor $\hat{F} : \mathcal{A}_\ell(\mathcal{C}) \longrightarrow \mathbf{Mod}_R$.

Proposition 17.1 $E'_{\hat{F}}$ is the reflection of the R-bialgebra E'_F into the category of Hopf R-algebras.

Proof. Let H be a Hopf R-algebra. Then we have that $\mathbf{Comod}_R(H)_c$ is a left autonomous tensor category. Thus we have an equivalence between tensor functors $\mathcal{A}_\ell(\mathcal{C}) \longrightarrow \mathbf{Comod}_R(H)_c$, over $\mathbf{Mod}_R$ and tensor functors $\mathcal{C} \longrightarrow \mathbf{Comod}_R(H)_c$ over $\mathbf{Mod}_R$. By the left adjoint property of E'_-, it follows that bialgebra morphisms $E'_{\hat{F}} \longrightarrow H$ correspond to bialgebra morphisms $E'_F \longrightarrow H$, as required. $\qquad\square$

This gives a construction for adjoining an antipode to a bialgebra over a field R; that is, a construction for a left adjoint to the inclusion of the category $\mathbf{Hopf}_R$ of Hopf algebras in the category $\mathbf{Big}_R$ of bialgebras. Given a bialgebra A, put $F = U_A : \mathbf{Comod}_R(A)_c \longrightarrow \mathbf{Mod}_R$. By Proposition 16.3, we have $A \cong E'_F$. By Proposition 17.1, the Hopf algebra $H = E'_{\hat{F}}$ is the required reflection.

If we require the adjoined antipode to be invertible, we must replace $\mathcal{A}_\ell(\mathcal{C})$ in the above by $\mathcal{A}(\mathcal{C})$ which is the free autonomous tensor category on the tensor category $\mathcal{C}$. And so on.

118

Bradshaw: Sash Bradshaw Group, [Wal94, Plate 54].

18

The quantum general linear group again

Let V be an n-dimensional vector space over a field $\mathbf{k}$. Given an invertible $q \in \mathbf{k}$, let (y, z) be the tortile Yang–Baxter operator on V defined in Chapter 14. By Example 13.1 and Example 13.2, there are strict tensor functors

$$M : \widetilde{\mathcal{B}} \longrightarrow \mathbf{Mod_k} \quad , \quad G : \widetilde{\mathcal{T}} \longrightarrow \mathbf{Mod_k}$$

taking $(+, c_{+,+}, \theta_+)$ to be (V, y, z) (where we are identifying $\widetilde{\mathcal{B}}$ with the subcategory of $\widetilde{\mathcal{T}}$ whose objects are positively signed sets, with arrows being ribbons which do not bend around).

Applying Tannaka duality ideas (Chapter 16) to M and G, we obtain a cobalanced bialgebra E'_M and a cotortile bialgebra E'_G.

Theorem 18.1 *There are $\mathbf{k}$-bialgebra isomorphisms (see Example 9.8):*

$$E'_M \cong \mathrm{M}_q(n) \quad , \quad E'_G \cong \mathrm{GL}_q(n) .$$

Proof. Let A be the bialgebra E'_M. It comes equipped with a universal linear function $\delta_Z : MZ \longrightarrow A \otimes MZ$ for $Z \in \widetilde{\mathcal{B}}$. In particular, we have $\delta = \delta_+ : V \longrightarrow A \otimes V$, and $\delta_{+,+} : V \otimes V \longrightarrow A \otimes V \otimes V$ is the composite

$$V \otimes V \xrightarrow{\ \delta \otimes \delta\ } A \otimes V \otimes A \otimes V \xrightarrow{\ 1 \otimes \sigma \otimes 1\ } A \otimes A \otimes V \otimes V \xrightarrow{\ \mu \otimes 1 \otimes 1\ } A \otimes V \otimes V$$

while $y : V \otimes V \longrightarrow V \otimes V$ becomes a comodule morphism

$$\begin{array}{ccc}
V \otimes V & \xrightarrow{\ \delta_{+,+}\ } & A \otimes V \otimes V \\
{\scriptstyle y}\Big\downarrow & & \Big\downarrow{\scriptstyle 1 \otimes y} \\
V \otimes V & \xrightarrow{\ \delta_{+,+}\ } & A \otimes V \otimes V
\end{array}$$

Putting $\delta(\varepsilon_i) = \sum_j x_{ij} \otimes \varepsilon_j$, it is a straightforward, but tedious, matter to check that commutativity of the above square is equivalent to the elements

$X = \{x_{ij} \mid i,j = 1, \ldots, n\}$ satisfying the defining relations for the quantum matrix monoid $M_q(n)$ (see Example 9.8). We therefore have a bialgebra morphism $M_q(n) \longrightarrow A$ which can be seen to be invertible.

To introduce an antipode to the bialgebra A and thereby obtain the Hopf algebra H, we must introduce a left dual $\kappa : V^* \longrightarrow A \otimes V^*$ for $\delta :$ $V \longrightarrow A \otimes V$. Since $\widetilde{\mathcal{T}}$ is the free autonomous tensor category on $\widetilde{\mathcal{B}}$, we have $H = E'_G$ (as in Chapter 17). If we put

$$\kappa(\varepsilon_i^*) = \sum_j w_{ij} \otimes \varepsilon_j^*$$

and express what it means for $e : V^* \otimes V \longrightarrow \mathbf{k}$ and $d : \mathbf{k} \longrightarrow V^* \otimes V$ to be H-comodule morphisms, we obtain the conditions

$$\sum_m w_{im}\, x_{jm} = \delta_{ij} \qquad \text{and} \qquad \sum_m x_{mi}\, w_{mj} = \delta_{ij} \,,$$

which mean that the matrix (w_{ij}) is the inverse of the transpose of (x_{ij}). The Hopf algebra H is therefore obtained from A by adjoining elements w_{ij} subject to the above two conditions. By means of a quantum Cramer's Rule (checked in Chapter 3 for the $n = 2$ case), we can take $w_{ij} = t\, \det_q(X_{ij})$ (see Example 9.8) where t is an adjoined inverse for $\det_q(X)$. In this way we see that $H \cong \mathrm{GL}_q(n)$.   □

Corollary 18.2 $M_q(n)$ *is a cobalanced bialgebra and* $\mathrm{GL}_q(n)$ *is a cotortile bialgebra.*

The cobraiding given by $\gamma : \mathrm{GL}_q(n) \otimes \mathrm{GL}_q(n) \longrightarrow \mathbf{k}$, and cotwist given by $\tau : \mathrm{GL}_q(n) \otimes \mathrm{GL}_q(n) \longrightarrow \mathbf{k}$, satisfy the equations

$$y(\varepsilon_i \otimes \varepsilon_j) = \sum_{m,r} \gamma(x_{im}, x_{jr})\, \varepsilon_m \otimes \varepsilon_r$$

$$q^n \varepsilon_i = \sum_m \tau(x_{im})\, \varepsilon_m \,.$$

This means:

$$\gamma(x_{im}, x_{jr}) = \begin{cases} 1 & \text{for } i \neq j,\, m = j,\, r = i \\ (q - q^{-1}) & \text{for } i < j,\, m = i,\, r = j \\ q & \text{for } i = j = m = r \\ 0 & \text{otherwise} \end{cases}$$

$$\tau(x_{ij}) = q^n \delta_{ij} \,.$$

19

Solutions to Exercises

Chapter 4

4.2   For $a \in \mathbb{Z}/(2)$, $b \in \mathbb{Z}/(5)$, put $x = a \otimes b \in \mathbb{Z}/(2) \otimes_{\mathbb{Z}} \mathbb{Z}/(5)$. So we have:

$$5x = 5(a \otimes b) = a \otimes 5b = a \otimes 0 = a \otimes (0 \times 0) = 0a \otimes 0 = 0 \otimes 0 = 0 \,,$$
$$2x = 2(a \otimes b) = 2a \otimes b = 0 \otimes b = 0 \,.$$

Hence $x = (5 - 2 \times 2)x = 0 - 0 = 0$. Elements of the form x generate, so $\mathbb{Z}/(2) \otimes_{\mathbb{Z}} \mathbb{Z}/(5) = \{0\}$.

4.3   (a)   Let $\otimes$ denote $\otimes_{\mathbb{Z}}$. The multiplication and unit are given by

$$(R \otimes S) \otimes (R \otimes S) \xrightarrow[\quad 1 \otimes \sigma \otimes 1 \quad]{\approx} (R \otimes R) \otimes (S \otimes S) \xrightarrow{\mu \otimes \mu} R \otimes S$$
$$\mathbb{Z} \cong \mathbb{Z} \otimes \mathbb{Z} \xrightarrow{\eta \otimes \eta} R \otimes S \,.$$

This makes it clear that μ is an abelian group morphism, so we automatically have distributivity. Now in terms of generating elements, the multiplication is $(r \otimes s)(r' \otimes s') = (rr') \otimes (ss')$ and the unit is $1 = 1 \otimes 1$. Associativity and unit conditions only need to be checked on generators where they clearly follow from these conditions in R and S.

  (b)   Yes, $\varphi(r) = r \otimes 1$ does define a ring morphism $\varphi : R \longrightarrow R \otimes S$.

$$\varphi = \left(R \cong R \otimes \mathbb{Z} \xrightarrow{1_R \otimes \eta} R \otimes S \right)$$

is clearly an abelian group morphism. It remains to check that multiplication and unit are preserved:

- $\varphi(rr') = (rr') \otimes 1 = (r \otimes 1)(r' \otimes 1) = \varphi(r)\,\varphi(r')$;
- $\varphi(1) = 1 \otimes 1 = 1$ using the definition in (a).

  (c)   Let $R \xrightarrow{\varphi} R \otimes_{\mathbb{Z}} S \xleftarrow{\psi} S$, with $\varphi(r) = r \otimes 1$ and $\psi(s) = 1 \otimes s$; these are ring morphisms as in (b). These give our "coprojections". Now given $R \xrightarrow{f} T \xleftarrow{g} S$ with f, g ring morphisms

and T commutative, we must show there is a unique ring morphism $h : R \otimes S \longrightarrow T$ with $h \circ \varphi = f$, $h \circ \psi = g$. These last equations force us to define $h(r \otimes s) = h\big((r \otimes 1)(1 \otimes s)\big) = h(r \otimes 1)h(1 \otimes s) = f(r)\,g(s)$. It is easily checked that $R \times S \longrightarrow T$, $(r, s) \longmapsto f(r)\,g(s)$ is bilinear. So h does give an abelian group morphism. It remains to show h preserves multiplication and unit.

$$
\begin{aligned}
h\big((r \otimes s)(r' \otimes s')\big) &= h\big(rr' \otimes ss'\big) = f(rr')\,g(ss') \\
&= f(r)\,f(r')\,g(s)\,g(s') \\
&= f(r)\,g(s)\,f(r')\,g(s') \\
&= h(r \otimes s)\,h(r' \otimes s')\,.
\end{aligned}
$$

So h is a ring morphism. Since the definition of h was forced, it is unique.

4.4 A module $M : R \longrightarrow S$ is an abelian group with an abelian group morphism $\mu : R \otimes M \otimes S \longrightarrow M$, written $\mu(r \otimes m \otimes s) = rms$, satisfying $r'(rms)s' = (r'r)m(ss')$, $1m1 = m$. One can see that this agrees with the definition given in Chapter 4 (given left R-, right S-scalar multiplications satisfying $(rm)s = r(ms)$ we define $rms = (rm)s$; all the distributive laws precisely summarize to trilinearity (over $\mathbb{Z}$); conversely, given μ, define the two scalar multiplications by $rm = rm1$ and $ms = 1ms$). Now a left $R \otimes S^{\mathrm{op}}$-module is an abelian group M with an abelian group morphism $\overline{\mu} : (R \otimes S) \otimes M \longrightarrow M$, written $\overline{\mu}(r \otimes s \otimes m) = (r \otimes s)m$, satisfying $(1 \otimes 1)m = m$ and $(r'r \otimes ss')m = (r' \otimes s')\big((r \otimes s)m\big)$. Clearly to give the abelian group morphisms μ, $\overline{\mu}$ are the "same thing" via the diagram:

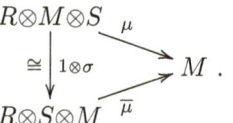

Moreover the conditions on μ directly translate to those on $\overline{\mu}$.

4.5 Suppose $R \longrightarrow^M S \longrightarrow^N T \longrightarrow^L U$. As an abelian group we have:

$$
M \otimes_S N \otimes_T L = B
$$

where B is the subset of the abelian group $\mathcal{F}_{\mathbb{Z}}(M \times N \times L)$ consisting of all elements of the forms:

$$
\begin{aligned}
&(m + m', n, l) - (m, n, l) - (m', n, l)\,, \\
&(m, n + n', l) - (m, n, l) - (m, n', l)\,, \\
&(m, n, l + l') - (m, n, l) - (m, n, l')\,, \\
&(ms, n, l) - (m, sn, l)\,, (m, nt, l) - (m, n, tl)\,.
\end{aligned}
$$

The equivalence class of (m, n, l) is denoted by $m \otimes n \otimes l$. We now define $r(m \otimes n \otimes l)\mu = (rm) \otimes n \otimes (l\mu)$ yielding $M \otimes_S N \otimes_T L : R \rightarrowtail U$. Then

$$\operatorname{Hom}_R^U(M \otimes_S N \otimes_T L, K) \cong \mathbf{Mult}\,(M, N, L; K).$$

Chapter 5

5.2   We use the Fundamental Theorem of Morita Theory. Suppose M is finitely generated and projective. By Theorem 5.2, we have the morphisms $d : R \longrightarrow M^* \otimes_R M$, $e : M \otimes_R M^* \longrightarrow R$ satisfying

$$\left(M^* \xrightarrow{\;d \otimes 1\;} M^* \otimes_R M \otimes_R M^* \xrightarrow{\;1 \otimes e\;} M^* \right) = 1_{M^*}$$
$$\text{and} \quad (e \otimes 1) \circ (1 \otimes d) = 1_M \,.$$

Put $d' = \left(R \xrightarrow{\;d\;} M^* \otimes_R M \xrightarrow{\;\sigma\;} M \otimes_R M^* \right)$, $e' = \left(M^* \otimes_R M \xrightarrow{\;\sigma\;} M \otimes_R M^* \xrightarrow{\;e\;} R \right)$. We can now apply Theorem 5.2(iii) (replacing M, N, e, d with M^*, M, e', d' respectively) and by (iv) M^* is finitely generated and projective.

5.3   Notice that $\rho_L^M : M^* \otimes_R L \longrightarrow \operatorname{Hom}_R(M, L)$ is "natural" in M (and in L too for that matter), meaning that for any module morphism $f : M \longrightarrow N : R \rightarrowtail S$, the following "$f$-square" commutes:

$$
\begin{array}{ccc}
N^* \otimes_R L & \xrightarrow{\;\rho_L^N\;} & \operatorname{Hom}_R(N, L) \\[2pt]
{\scriptstyle f^* \otimes 1}\Big\downarrow & & \Big\downarrow {\scriptstyle \_\, \circ f} \\[2pt]
M^* \otimes_R L & \xrightarrow[\;\rho_L^M\;]{} & \operatorname{Hom}_R(M, L) \,.
\end{array}
$$

Suppose now that M is a retract of a Cauchy module N; so we have $i : M \longrightarrow N$, $r : N \longrightarrow M$, $r \circ i = 1_M$ and ρ_L^N invertible. We can show that the composite

$$\operatorname{Hom}(M, L) \xrightarrow{\;\_\, \circ r\;} \operatorname{Hom}(n, L) \xrightarrow{\;(\rho_L^N)^{-1}\;} N^* \otimes L \xrightarrow{\;i^* \otimes 1\;} M^* \otimes L$$

is an inverse for ρ_L^M. This is seen as follows:

$$\rho_L^M \circ (i^* \otimes 1) \circ (\rho_L^N)^{-1} \circ (\_\, \circ r) = (\_\, \circ i) \circ \rho_L^N \circ (\rho_L^N)^{-1} \circ (\_\, \circ r)$$
$$\text{(by the } i\text{-square)}$$
$$= (\_\, \circ i) \circ 1_{\operatorname{Hom}(N,L)} \circ (\_\, \circ r) = (\_\, \circ i) \circ (\_\, \circ r)$$
$$= \_\, \circ (ri) = \_\, \circ 1_M = 1_{\operatorname{Hom}(M,L)}$$

and

$$(i^* \otimes 1) \circ (\rho_L^N)^{-1} \circ (\_ \circ r) \circ \rho_L^M = (i^* \otimes 1) \circ (\rho_L^N)^{-1} \circ \rho_L^N \circ (r^* \otimes 1)$$

(by the r-square)

$$= (i^* \otimes 1) \circ (r^* \otimes 1) = (ri)^* \otimes 1 = 1_{M^* \otimes 1_L} = 1_{M^* \otimes L}$$

establish the inverses, as required.

Chapter 6

6.1 (a)   We have that $G \longrightarrow \mathrm{End}_R(M \otimes N)$, whereby $g \longmapsto (m \otimes n \longmapsto (gm) \otimes (gn))$, is a monoid morphism since $1 \longmapsto 1_{M \otimes N}$ and $gh \longmapsto (m \otimes n \longmapsto (hm) \otimes (hn) \longmapsto (ghm) \otimes (ghn))$. It extends to a unique R-algebra morphism $R(G) \longrightarrow \mathrm{End}_R(M \otimes_R N)$. So $M \otimes_R N$ is an $R(G)$-module.

(b)   Let $\hat{\mu}(g) : \mathrm{Hom}_R(M, L) \longrightarrow \mathrm{Hom}_R(M, L)$ be the R-module morphisms given by $\hat{\mu}(g)(\mu)(m) = g\mu(g^{-1}m)$. Then we have that $\hat{\mu}(1)(\mu)(m) = \mu(m)$, and $\hat{\mu}(gh)(\mu)(m) = gh\mu(h^{-1}g^{-1}m) = g\hat{\mu}(h)(\mu)(g^{-1}m) = (\hat{\mu}(g) \circ \hat{\mu}(h))(\mu)(m)$.
So $\hat{\mu} : G \longrightarrow \mathrm{End}_R(\mathrm{Hom}_R(M, L))$ is a monoid morphism. Hence $\mathrm{Hom}_R(M, L)$ becomes an $R(G)$-module.

(c) $\mathrm{ev}_M(g \circ (m \otimes \mu)) = \mathrm{ev}_M(gm \otimes g\mu) = (g\mu)(gm) = g\,\mu(g^{-1}gm) = g\,\mu(m) = g\,\mathrm{ev}_M(m \otimes \mu)$, so ev_M preserves the $R(G)$-action.

(d)   We also need $d : N \longrightarrow \mathrm{Hom}_R(M, M \otimes_R N)$, $n \longmapsto (m \longmapsto m \otimes n)$ to be an $R(G)$-module morphism. We have

$$d(gn)(m) = m \otimes gn = g((g^{-1}m) \otimes n)$$
$$= gd(n)(g^{-1}m) = (gd(n))(m) ,$$

so $d(gn) = gd(n)$. The required isomorphism is the restriction of

$$
\begin{array}{ccc}
\mathrm{Hom}_R(N, \mathrm{Hom}_R(M, L)) & \cong & \mathrm{Hom}_R(M \otimes_R N, L) \\
f & \longmapsto & e \circ (1_M \otimes f) \\
(g \circ \_) \circ d & \longmapsfrom & g
\end{array}
$$

to $R(G)$-module morphisms f and g; since e and d are such, so then are $e \circ (1_M \otimes f)$ and $(g \circ \_) \circ d$ when f and g are. (This will be generalized from $R(G)$ to an arbitrary Hopf algebra in Chapter 9.)

Chapter 7

7.1 There is a little abuse of notation here since the four parts to part (a), for example, are elements of $C\otimes_R C$, $R\otimes_R C\otimes_R C$, $C\otimes_R C\otimes_R R$, $C\otimes_R R\otimes_R C$ respectively. But these modules are canonically isomorphic, and so "=" really means "corresponds under the canonical isomorphism to".

(a) To prove the first part:

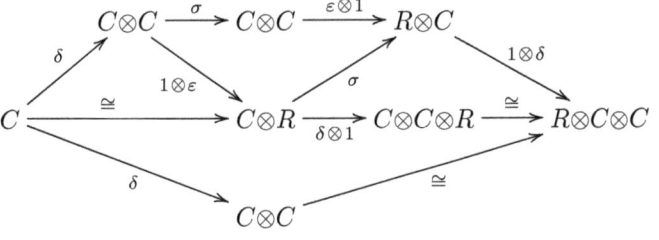

(b) Similarly for the second:

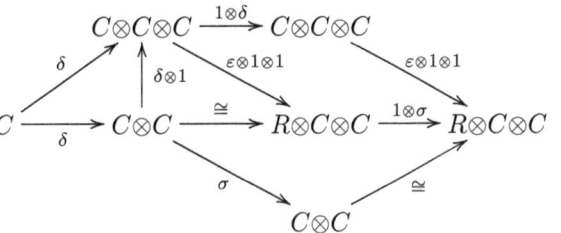

(c) For the third:

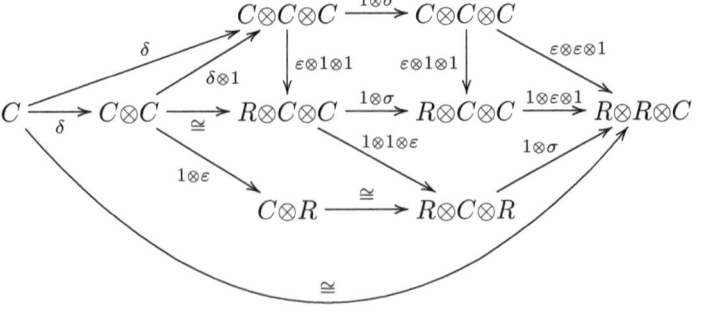

Chapter 9

9.1　(a)　Suppose I is a coideal of C, that is, a submodule satisfying $\delta(I) \subseteq I \otimes C + C \otimes I$ and $\varepsilon(I) = 0$. The composite morphism

$$C \xrightarrow{\ \delta\ } C \otimes C \xrightarrow{\ \rho \otimes \rho\ } C/I \otimes C/I$$

maps I to 0 since $(\rho \otimes \rho)\delta(I) \subseteq (\rho \otimes \rho)(I \otimes C + C \otimes I) = \rho(I) \otimes \rho(C) + \rho(C) \otimes \rho(I) = 0 \otimes C/I + C/I \otimes 0 = 0$. So there exists a unique module morphism $\delta : C/I \longrightarrow C/I \otimes C/I$ such that $\delta \circ \rho = (\rho \otimes \rho) \circ \delta$. Similarly, $\varepsilon(I) = 0$ implies that there exists a unique $\varepsilon : C/I \longrightarrow R$ with $\varepsilon \circ \rho = \varepsilon$. These properties of δ, ε will mean $\rho : C \longrightarrow C/I$ is a coalgebra morphism once we know C/I is a coalgebra. To prove coassociativity of $\delta : C/I \longrightarrow C/I \otimes C/I$, take the coassociativity diagram for C/I and precompose with $\rho : C \longrightarrow C/I$:

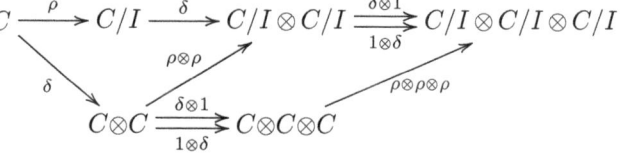

The result commutes by coassociativity of C. But ρ is surjective; so the coassociativity diagram for C/I commutes. Similarly we can prove $\varepsilon : C/I \longrightarrow R$ is a counit. If C is a bialgebra and I is also an ideal, certainly C/I becomes an algebra. All that remains to check are the extra bialgebra conditions (see Proposition 7.5). The main one, showing that δ preserves multiplication, is obtained by precomposing the diagram with $\rho \otimes \rho$ and using the corresponding condition for C. This gives the result since $\rho \otimes \rho$ is surjective.

Since ρ is a bialgebra morphism, the only possible way C/I can become a Hopf algebra is to have $\nu \circ \rho = \rho \circ \nu$. This forces us to ask whether $\nu(I) \subseteq I$ for the antipode of C.

(b)　Put $B = R\langle x, y, z \rangle$. The given equations define algebra morphisms $\delta : B \longrightarrow B \otimes B$, $\varepsilon : B \longrightarrow R$ since B is free as an algebra. By Proposition 7.5, it remains to see that these morphisms make B a coalgebra.

First look at the coassociativity:

$$(\delta \otimes 1)\delta(x) = (\delta \otimes 1)(x \otimes x) = x \otimes x \otimes x$$
$$= (1 \otimes \delta)(x \otimes x) = (1 \otimes \delta)\delta(x) .$$

Similarly for y:

$$(\delta \otimes 1)\,\delta(z) = (\delta \otimes 1)(1 \otimes z + z \otimes x)$$
$$= 1 \otimes 1 \otimes z + (1 \otimes z + z \otimes x) \otimes x$$
$$= 1 \otimes (1 \otimes z + z \otimes x) + z \otimes x \otimes x$$
$$= (1 \otimes \delta)(1 \otimes z + z \otimes x) = (1 \otimes \delta)\delta(z)\ .$$

Then check the counit conditions:

$$(\varepsilon \otimes 1)\,\delta(x) = (\varepsilon \otimes 1)(x \otimes x) = \varepsilon(x)x = x$$
$$= (1 \otimes \varepsilon)(x \otimes x) = (1 \otimes \varepsilon)\delta(x)\ ,$$

and similarly for y:

$$(\varepsilon \otimes 1)\delta(z) = (\varepsilon \otimes 1)(1 \otimes z + z \otimes x) = z + 0x = z$$
$$= 0 + z = (1 \otimes \varepsilon)(1 \otimes z + z \otimes x) = (1 \otimes \varepsilon)\,\delta(z)\ .$$

(c)   We have:

$$\delta(xy - 1)\,(x \otimes x)(y \otimes y) - 1 \otimes 1 = xy \otimes xy - 1 \otimes 1$$
$$= (xy - 1) \otimes xy + 1 \otimes xy - 1 \otimes 1$$
$$= (xy - 1) \otimes xy + 1 \otimes (xy - 1) \subseteq I \otimes B + B \otimes I\ .$$

(d)   We must check:

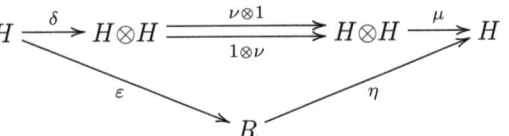

$$\mu(\nu \otimes 1)\,\delta(x) = \mu(\nu \otimes 1)(x \otimes x) = \mu(y \otimes x) = yx \equiv 1 = \nu \varepsilon(x)$$
$$\equiv xy = \mu(x \otimes y) = \mu(1 \otimes \nu)(x \otimes x) = \mu(1 \otimes \nu)\,\delta(x)\ .$$

Similarly for y:

$$\mu(\nu \otimes 1)\,\delta(z) = \mu(\nu \otimes 1)(1 \otimes z + z \otimes x) = \mu(1 \otimes z + (-zy) \otimes x)$$
$$= z - zyx \equiv 0 = \eta \varepsilon(z) = 0$$
$$= -zy + zy = \mu(1 \otimes (-zy) + z \otimes y)$$
$$= \mu(1 \otimes \nu)(1 \otimes z + z \otimes x) = \mu(1 \otimes \nu)\,\delta(z)\ .$$

So H is a Hopf algebra. By Proposition 9.1, ν reverses both multiplication and comultiplication. The formulas for $\nu^r(z)$ are trivial for $r = 0$ or 1. Also $\nu^{2n}(z) = x^n z y^n$ implies:

$$\nu^{2n+1}(z) = \nu(y)^n \nu(z) \nu(x)^n = x^n(-zy)y^n = -x^n z y^{n+1}$$

which gives:

$$\nu^{2n+2}(z) = -\nu(y)^{n+1}\nu(z)\nu(x)^n = -x^{n+1}(-zy)y^n$$
$$= x^{n+1}zy^{n+1} \ .$$

So the formulas follow by induction.

If ν had finite order we would have either $x^n z = zx^n$, or $x^n z = -zx^{n+1}$, which are both false in H.

(e)(i) Evaluate each of ν, δ and ε at $(x^n zy^n - z)$:

$$\nu(x^n zy^n - z) = \nu(y)^n \nu(z)\nu(x)^n - \nu(z) = x^n(-zy)y^n + zy$$
$$= -x^n zy^{n+1} + zy = (x^n zy^n - z)(-y) \in I_n \ ,$$
$$\delta(x^n zy^n - z) = (x^n \otimes x^n)(1\otimes z + z\otimes x)(y^n \otimes y^n) - 1\otimes z - z\otimes x$$
$$= (x^n \otimes x^n z + x^n z\otimes x^{n+1})(y^n \otimes y^n) - 1\otimes z - z\otimes x$$
$$= x^n y^n \otimes x^n zy^n + x^n zy^n \otimes x^{n+1}y^n - 1\otimes z - z\otimes x$$
$$= 1\otimes x^n zy^n + x^n zy^n \otimes x - 1\otimes z - z\otimes x$$
$$= 1\otimes(x^n zy^n - z) + (x^n zy^n - z)\otimes x$$
$$\in I_n \otimes H + H\otimes I_n \ ,$$
$$\varepsilon(x^n zy^n - z) = 0 \ .$$

So I_n is a Hopf ideal in H. Also compute:

$$\nu(x^n - 1) = y^n - 1 \equiv -y^n(x^n - 1) \in J_n \ ,$$
$$\delta(x^n - 1) = x^n \otimes x^n - 1\otimes 1 = (x^n - 1)\otimes x^n + 1\otimes(x^n - 1)$$
$$\in J_n \otimes H + H\otimes J_n \ ,$$
$$\varepsilon(x^n - 1) = 1^n - 1 = 0 \ .$$

So J_n is a Hopf ideal in H.

(e)(ii) $\nu^{2n}(x) = x$ and $\nu^{2n}(y) = y$ since ν just switches x and y.

$$\nu^{2n}(z) = x^n zy^n \equiv z \quad (\text{mod } I_n)$$
$$\nu^{2n}(z) = x^n zy^n \equiv |z| = z \quad (\text{mod } J_n) \ .$$

Chapter 10

10.1 Let M be a Cauchy R-module. The diagrams (coassociativity and counit) showing E to be a coalgebra are:

$$MM^* \xrightarrow{\ 1\otimes d\otimes 1\ } MM^*MM^* \underset{1\otimes d\otimes 1\otimes 1}{\overset{1\otimes 1\otimes 1\otimes d\otimes 1}{\rightrightarrows}} MM^*MM^*MM^*$$

$$MM^* \xrightarrow{1\otimes d\otimes 1} MM^*MM^* \underset{1\otimes1\otimes e}{\overset{e\otimes1\otimes1}{\rightrightarrows}} MM^* \, .$$

These follow from functionality of $\otimes_R$ and Theorem 5.2.

Certainly $\delta \longmapsto \hat{\delta}$ characterizes a bijection between R-linear functions $\delta : M \longrightarrow C\otimes_R M$ and R-linear functions $\omega : M\otimes_R M^* \longrightarrow C$. The inverse assignment $\omega \longmapsto {}^\vee\omega$ is given by:

$$ {}^\vee\omega \;=\; \left(M \xrightarrow{1\otimes d} MM^*M \xrightarrow{\omega\otimes1} CM\right) \, .$$

(That these assignments are mutually inverse follows from the properties of e and d in Theorem 5.2.)

It remains to see that coaction axioms on δ translate precisely to coalgebra morphisms on ω. We shall do the translation for the coaction axiom:

$$ M \xrightarrow{\delta} CM \underset{1\otimes\delta}{\overset{\underline{\delta}\otimes1}{\rightrightarrows}} CCM $$

(where $\underline{\delta}$ is the comultiplication of C). This is equivalent to:

$$ MM^* \xrightarrow{\delta\otimes1} CMM^* \underset{1\otimes\delta\otimes1}{\overset{\underline{\delta}\otimes1\otimes1}{\rightrightarrows}} CCMM^* \xrightarrow{1\otimes1\otimes e} CC \, .$$

Using:

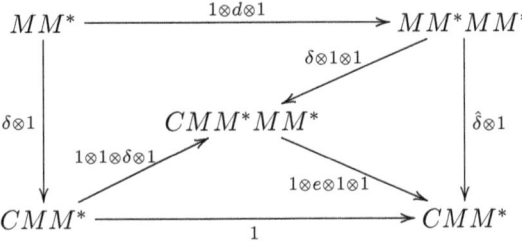

and

we see that the axiom becomes:

$$
\begin{array}{ccc}
MM^* & \xrightarrow{\hat{\delta}} & C \\
{\scriptstyle 1\otimes d\otimes 1}\downarrow & & \downarrow{\scriptstyle \underline{\delta}} \\
MM^*MM^* \xrightarrow{\hat{\delta}\otimes1} CMM^* & \xrightarrow{1\otimes\hat{\delta}} & CC
\end{array}
$$

which is a coalgebra morphism axiom on $\hat{\delta}$.

Chapter 11

11.1 (a) $c_{A,I} = c_{A,I} \circ c_{A,I}$ from the triangle below. Since $c_{A,I}$ is invertible, $c_{A,I} = 1_A$. Similarly $c_{I,A} = 1_A$.

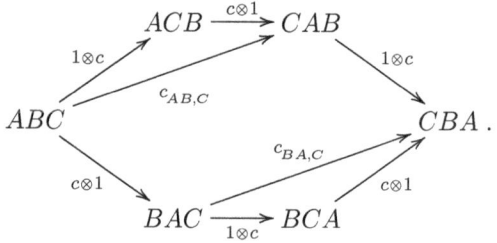

The hexagon can be subdivided into commutative regions thus:

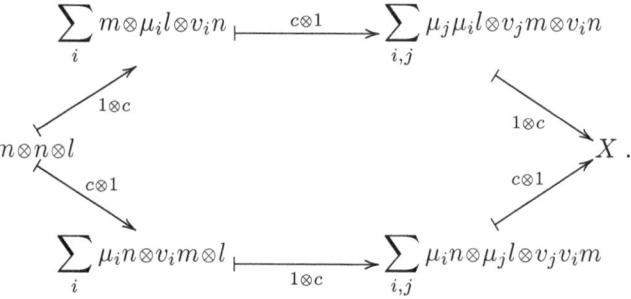

(b)   Put $\gamma = \sum_i \mu_i \otimes v_i \in A \otimes A$ so that $c_{M,N}(m \otimes n) = \sum_i (\mu_i n) \otimes (v_i n)$:

$$
\begin{array}{ccc}
\sum_i m \otimes \mu_i l \otimes v_i n & \xrightarrow{\;c \otimes 1\;} & \sum_{i,j} \mu_j \mu_i l \otimes v_j m \otimes v_i n \\
\nearrow \scriptstyle{1 \otimes c} & & \searrow \scriptstyle{1 \otimes c} \\
m \otimes n \otimes l & & X \\
\searrow \scriptstyle{c \otimes 1} & & \nearrow \scriptstyle{c \otimes 1} \\
\sum_i \mu_i n \otimes v_i m \otimes l & \xrightarrow{\;1 \otimes c\;} & \sum_{i,j} \mu_i n \otimes \mu_j l \otimes v_j v_i m
\end{array}
$$

The hexagon gives us the condition:

$$
X = \sum_{i,j,k} \mu_j \mu_i l \otimes \mu_k v_i n \otimes v_k v_j m = \sum_{i,j,k} \mu_k \mu_j l \otimes v_k \mu_i n \otimes v_j v_i m
$$

in $A \otimes A \otimes A$. Diagrammatically this becomes:

$$
R \xrightarrow{\;\gamma \otimes \gamma \otimes \gamma\;} A^{\otimes 6} \;\underset{\sigma_{536142}}{\overset{\sigma_{315264}}{\rightrightarrows}}\; A^{\otimes 6} \xrightarrow{\;\mu \otimes \mu \otimes \mu\;} A^{\otimes 3} .
$$

(c)

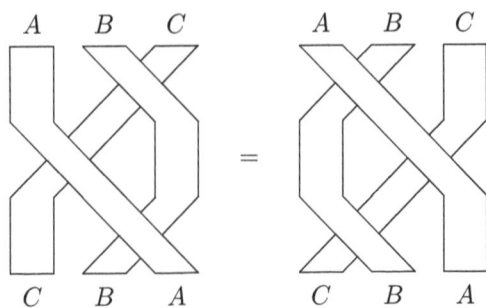

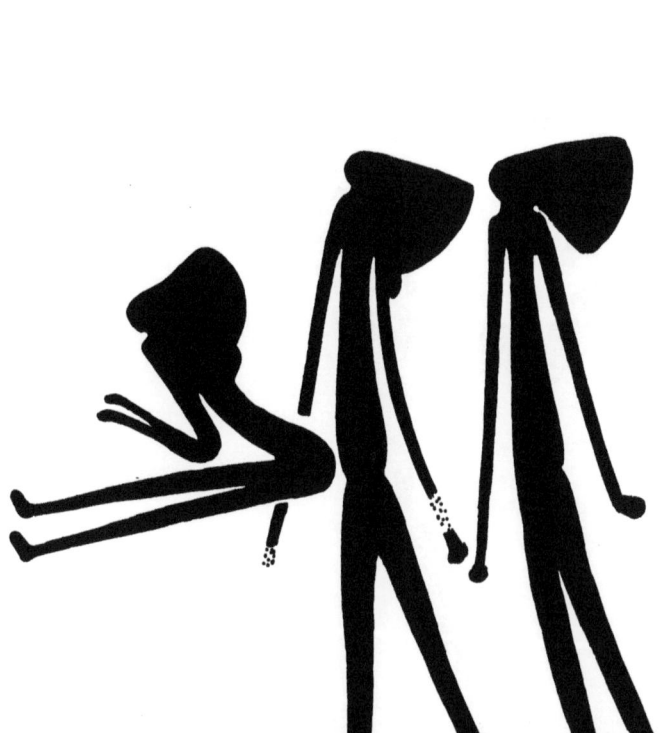

Bradshaw: Elegant Action Figure Group, [Wal94, Plate 79].

Bradshaw: "Broad Clothes Peg Figure Group", Clothes Peg Figure Period, [Wal94, Plate 71].

References

[Abe80]   Eiichi Abe. *Hopf Algebras*, volume 74 of *Cambridge Tracts in Mathematics*. Cambridge University Press, Cambridge, 1980.   **MR594432**

[Dri87]   Vladimir G. Drinfel'd. Quantum groups. In *Proceedings of the International Congress of Mathematicians, Vol. 1, 2 (Berkeley, Calif., 1986)*, pp. 798–820, Amer. Math. Soc., Providence, RI, 1987.
MR934283

[Dri89]   Vladimir G. Drinfel'd. Quasi-Hopf algebras and Knizhnik–Zamolodchikov equations. In *Problems of Modern Quantum Field Theory (Alushta, 1989)*, Res. Rep. Phys., pp. 1–13. Springer, Berlin, 1989. Previously: *Acad. Sci. Ukr.* (Preprint ITP-89-43E, 1989). **MR1091757**

[EK66]   Samuel Eilenberg and G. Max Kelly. Closed categories. In *Proc. Conf. Categorical Algebra (La Jolla, Calif., 1965)*, pp. 421–562. Springer, New York, 1966.   **MR0225841**

[FRT88]   Ludwig D. Fadde'ev, Nikolai Yu. Reshetikhin, and Leon A. Takhtajan. Quantization of Lie groups and Lie algebras. In *Algebraic Analysis*, Vol. I, pp. 129–139. Academic Press, Boston, MA, 1988.
MR992450

[FY89]   Peter J. Freyd and David N. Yetter. Braided compact closed categories with applications to low-dimensional topology. *Adv. Math.*, 77(2):156–182, 1989.   **MR1020583**

[JS91a]   André Joyal and Ross Street. The geometry of tensor calculus. I. *Adv. Math.*, 88(1):55–112, 1991.   **MR1113284**

[JS91b]   André Joyal and Ross Street. An introduction to Tannaka duality and quantum groups. In *Category Theory (Como, 1990)*, volume 1488 of *Lecture Notes in Mathematics*, pp. 413–492. Springer, Berlin, 1991.   **MR1173027**

[JS91c]   André Joyal and Ross Street. Tortile Yang–Baxter operators in tensor categories. *J. Pure Appl. Algebra*, 71(1):43–51, 1991.   **MR1107651**

[JS93]   André Joyal and Ross Street. Braided tensor categories. *Adv. Math.*, 102(1):20–78, 1993.   **MR1250465**

[JS95]   André Joyal and Ross Street. The category of representations of the general linear groups over a finite field. *J. Algebra*, 176(3):908–946, 1995.   **MR1351369**

134

[JSV96]   André Joyal, Ross Street, and Dominic Verity. Traced monoidal categories. *Math. Proc. Cambridge Philos. Soc.*, 119(3):447–468, 1996.
MR1357057

[Kas95]   Christian Kassel. *Quantum Groups*, volume 155 of *Graduate Texts in Mathematics*. Springer, New York, 1995.   MR1321145

[KS74]   G. Max Kelly and Ross Street. Review of the elements of 2-categories. In *Category Seminar (Proc. Sem., Sydney, 1972/1973)*, volume 420 of Lecture Notes in Mathematics, pp. 75–103, Springer, Berlin, 1974.   MR0357542

[KSW02]   Piergiulio Katis, Nicoletta Sabadini, and Robert F. C. Walters. Feedback, trace and fixed-point semantics. *Theor. Inform. Appl.*, 36(2):181–194, 2002.   MR1948768

[Mac71]   Saunders Mac Lane. *Categories for the Working Mathematician*, volume 5 of *Graduate Texts in Mathematics*, Springer, New York, 1971.   MR0354798

[Maj95]   Shahn Majid. *Foundations of Quantum Group Theory*. Cambridge University Press, Cambridge, 1995, (paperback, 2000).   MR1381692

[Man88]   Yuri I. Manin. *Quantum groups and noncommutative geometry*. Université de Montréal Centre de Recherches Mathématiques, Montreal, QC, 1988.   MR1016381

[Shu94]   Shum Mei Chee. Tortile tensor categories. *J. Pure Appl. Algebra*, 93(1):57–110, 1994.   MR1268782

[SR72]   Neantro Saavedra Rivano. *Catégories Tannakiennes*, volume 265 of *Lecture Notes in Mathematics* Springer, Berlin, 1972.   MR0338002

[SS93]   Steven Shnider and Shlomo Sternberg. *Quantum Groups: From CoAlgebras to Drinfel'd Algebras*. Graduate Texts in Mathematical Physics, II. International Press, Cambridge, MA, 1993.   MR1287162

[Swe69]   Moss E. Sweedler. *Hopf Algebras*. Mathematics Lecture Note Series. W. A. Benjamin, Inc., New York, 1969.   MR0252485

[Tur88]   V. G. Turaev. The Yang–Baxter equation and invariants of links. *Invent. Math.*, 92(3):527–553, 1988.   MR939474

[Ulb89]   Karl-Heinz Ulbrich. Tannakian categories for non-commutative Hopf algebras. In *International Conference on Hopf Algebras*, Beer Sheva, January 1989.

[Wal94]   Grahame L. Walsh. *Bradshaws: Ancient Rock Paintings of Northwestern Australia*. Edition Limitée, 1994. Privately commissioned by the Bradshaw Foundation.

[Yet01]   David N. Yetter. *Functorial Knot Theory*, volume 26 of *Series on Knots and Everything*. World Scientific Publishing Co. Inc., River Edge, NJ, 2001.   MR1834675

Index

Bradshaw: Schematized Bradshaw Group, [Wal94, Plate 45].